Advances in Particle Therapy

Series in Medical Physics and Biomedical Engineering

Series Editors: John G. Webster, E. Russell Ritenour, Slavik Tabakov, and Kwan-Hoong Ng

Recent books in the series:

Clinical Radiotherapy Physics with MATLAB: A Problem-Solving Approach
Pavel Dvorak

Advances in Particle Therapy: A Multidisciplinary Approach
Manjit Dosanjh and Jacques Bernier (Eds)

Radiotherapy and Clinical Radiobiology of Head and Neck Cancer
Loredana G. Marcu, Iuliana Toma-Dasu, Alexandru Dasu, and Claes Mercke

Problems and Solutions in Medical Physics: Diagnostic Imaging Physics
Kwan-Hoong Ng, Jeannie Hsiu Ding Wong, and Geoffrey D. Clarke

Advanced and Emerging Technologies in Radiation Oncology Physics
Siyong Kim and John W. Wong (Eds)

A Guide to Outcome Modeling In Radiotherapy and Oncology: Listening to the Data
Issam El Naqa (Ed)

Advanced MR Neuroimaging: From Theory to Clinical Practice
Ioannis Tsougos

Quantitative MRI of the Brain: Principles of Physical Measurement, Second edition
Mara Cercignani, Nicholas G. Dowell, and Paul S. Tofts (Eds)

A Brief Survey of Quantitative EEG
Kaushik Majumdar

Handbook of X-ray Imaging: Physics and Technology
Paolo Russo (Ed)

Graphics Processing Unit-Based High Performance Computing in Radiation Therapy
Xun Jia and Steve B. Jiang (Eds)

Targeted Muscle Reinnervation: A Neural Interface for Artificial Limbs
Todd A. Kuiken, Aimee E. Schultz Feuser, and Ann K. Barlow (Eds)

Emerging Technologies in Brachytherapy
William Y. Song, Kari Tanderup, and Bradley Pieters (Eds)

Environmental Radioactivity and Emergency Preparedness
Mats Isaksson and Christopher L. Rääf

The Practice of Internal Dosimetry in Nuclear Medicine
Michael G. Stabin

Advances in Particle Therapy

A Multidisciplinary Approach

Edited by
Manjit Dosanjh
Jacques Bernier

CRC Press
Taylor & Francis Group
Boca Raton London New York

CRC Press is an imprint of the
Taylor & Francis Group, an **informa** business

CRC Press
Taylor & Francis Group
6000 Broken Sound Parkway NW, Suite 300
Boca Raton, FL 33487-2742

First issued in paperback 2020

© 2018 by Taylor & Francis Group, LLC
CRC Press is an imprint of Taylor & Francis Group, an Informa business

No claim to original U.S. Government works

ISBN-13: 978-0-367-57155-9 (pbk)
ISBN-13: 978-1-138-06441-6 (hbk)

Visit the Taylor & Francis Web site at
http://www.taylorandfrancis.com

and the CRC Press Web site at
http://www.crcpress.com

Contents

Foreword

A T A TIME WHEN radiotherapy is undergoing profound changes considerable efforts have been made to give hadron therapy the place it deserves in the battle against cancer. For too many years and in many countries, numerous particle therapy programmes have remained at the level of discussion only: indeed, uncertainties surrounding their real therapeutic indices and fears about a low cost/benefit ratio have led many national health authorities to remain careful about investing in these technologies. However, recent, significant developments in both the diagnostic and therapeutic spheres, linked to hadrons in a direct or indirect manner, have propelled major institutions and laboratories involved in translational and clinical research to intensify their R&D programmes on particle therapy.

To be effective, this evolution requires an intense and continuous exchange of information and educational programmes among the various scientific communities. The main objective of this textbook is to be part of this effort both by strengthening the knowledge of readers in their own field of expertise (e.g., biology, physics and clinics) and by increasing their familiarity with the know-how developed in the other domains of hadron therapy with which they are called to interact and collaborate.

Towards this aim, the first part of this textbook places hadrons in their historical, biological and technological contexts. In the second part, particular emphasis has been given to the interactions between imaging and particle therapy, as well as to the advantages that can be expected from multi-disciplinary collaborative network in this domain. The most recent developments in clinical practice have been described in depth, with special attention paid to various examples of hadron therapy indications, also in consideration of the expected risks of complications in normal tissues and to issues related to health economics.

The final section is dedicated to various future projections, and revisits a number of fundamental aspects in precision medicine and future technologies. We hope the exhaustive review of the most burning challenges in radiotherapy that concludes the textbook will help increase understanding of the real dimension and place of radiation sciences in this era of globalisation.

The editors express our gratitude to the scientists and clinicians who accepted our invitation to contribute: they are all preeminent in their area of expertise and should be commended for their commitment to promoting modernity and making the dream of curing cancer with particles an option available to many more patients.

Manjit Dosanjh and Jacques Bernier

About the Editors

 **Manjit Dosanjh** went to the Massachusetts Institute of Technology (MIT) in Boston as a postdoctoral fellow after obtaining her degree in biochemistry/chemistry at the University of Leeds and PhD in biochemical engineering from the University of Birmingham, England. She has held positions as a senior scientist at LBNL Berkeley, BEST professor at Jackson State University and visiting professor at University of Padua and University of Cagliari, Italy.

She joined CERN, Geneva, in 2000 and is actively involved in applying particle physics-derived technologies in the field of life sciences and is currently Senior Advisor for Medical Applications. In 2002, she was key in launching the European Network for Light Ion Hadron Therapy (ENLIGHT), a multi-disciplinary platform that strives for a coordinated effort towards particle therapy research in Europe. In 2006, she was appointed Coordinator of ENLIGHT (www.cern.ch/enlight).

Dosanjh also co-chairs with Professor Jacques Bernier the International Conference on Translational Research in Radio-Oncology and Physics for Health – ICTR–PHE, organised every two years in Switzerland.

Dosanjh is actively involved in helping non-profit science, education and gender-related organisations in Geneva and is the UN representative for GWI (International Federation of University Women).

 **Jacques Bernier** completed his training at the MD Anderson Cancer Center, Houston (Texas), and the Curie Institute, Paris, after obtaining his degree in radio-oncology at the University of Liege, Belgium. In 1988, he moved to Switzerland, where he was appointed Chair of the Radio-Oncology Division, Tessin Cantonal Hospital. In 1995, he received a Privat-Docent Chair from the Geneva University. In the early 2000s, he founded the Oncology Institute of Southern Switzerland (IOSI) in Bellinzona. In 2006, he joined the Swiss Genolier Medical Network, where he has since chaired the Radio-Oncology Department. Throughout most of his career, Bernier has been heavily involved with translational and clinical research. In 1993, he received the Yalow–Berson Award, in St Louis, Missouri, for his laboratory work on interferons and interleukins. In 2010, Bernier was awarded the 'Claudius Regaud Medal' by the European Society of Therapeutic Radiology and Oncology (ESTRO).

A course director since 1990, and now member of the Core Faculty at the European School of Oncology, Bernier also co-chairs the International ICTR-PHE Conferences, organised every other year in Geneva. In 2009, he pioneered the application of breast cancer intra-operative electron-therapy (IORT), as innovative approach in Switzerland. In 2014, he created the Genolier Swiss Oncology Network (GSON), and in 2016, he received the accreditation from the Swiss Cancer Network as member of the core team of the GSON. Since 2013, he is member of the Board of the Genolier Breast Unit, accredited by the Swiss Society of Senology. In July 2016, he was elected President of the Euro–Asian Society of Mastology – EURAMA, in Milan, Italy. He is currently President of the Genolier Cancer Centre in Switzerland.

Contributors

Ugo Amaldi
TERA Foundation
Geneva, Switzerland

Jacques Balosso
Department of Radiation Oncology and
    Medical Physics
University Hospital of Grenoble Alpes
    (CHU-GA)
La Tronche, France

Jacques Bernier
Department of Radiation Oncology
Genolier Cancer Centre
Genolier, Switzerland

Eleanor A. Blakely
BioEngineering & BioMedical Sciences
Lawrence Berkeley National
    Laboratory
Berkeley, California

Jeffrey Buchsbaum
Radiation Research Program
National Cancer Institute
Bethesda, Maryland

Valentin Calugaru
Department of Radiation Oncology
Institut Curie—Centre de protonthérapie
    d'Orsay (CPO)
Orsay, France

Abdulhamid Chaikh
Department of Radiation Oncology and
    Medical Physics
University Hospital of Grenoble Alpes
    (CHU-GA)
La Tronche, France

C. Norman Coleman
Radiation Research Program
National Cancer Institute
Bethesda, Maryland

Stephanie E. Combs
Department of Radiation Oncology
Technical University of Munich (TUM)
Munich, Germany

James D. Cox
The University of Texas
MD Anderson Cancer Center
Houston, USA

Anne P.G. Crijns
Department of Radiation Oncology
University Medical Centre Groningen
University of Groningen
Groningen, the Netherlands

J. Debus
Department of Radiation Oncology and
    Radiation Therapy
Heidelberg University Hospital
Heidelberg Ion-Beam Therapy
    Center (HIT)
Heidelberg, Germany

G. Dedes
Department of Experimental
    Physics—Medical Physics
Ludwig-Maximilians-Universität München
    (LMU Munich)
Munich, Germany

Manjit Dosanjh
Directorate Office for Accelerators and
    Technologies (ATS-DO)
CERN
Geneva, Switzerland

Adriano Garonna
TERA Foundation
Geneva, Switzerland

Cai Grau
Department of Oncology and
    Danish Centre for Particle
    Therapy (DCPT)
Aarhus University Hospital
Aarhus, Denmark

David R. Grosshans
Departments of Radiation and
    Experimental Radiation Oncology
The University of Texas
MD Anderson Cancer Center
Houston, Texas

Madelon Johannesma
Department Healthcare Innovation &
    Advice
CZ Health Insurance
Tilburg, the Netherlands

Bleddyn Jones
Gray Laboratory
CRUK-MRC Oncology Centre
University of Oxford
Oxford, United Kingdom

Tadashi Kamada
National Institute of Radiological
    Sciences
National Institute for Quantum and
    Radiological Sciences
Chiba, Japan

Philippe Lambin
Department of Radiation Oncology
    (MAASTRO Clinic, D-Lab)
GROW—School for Oncology and
    Developmental Biology
Maastricht University Medical Centre
Maastricht, the Netherlands

G. Landry
Department of Experimental
    Physics—Medical Physics
Ludwig-Maximilians-Universität München
    (LMU Munich)
Munich, Germany

Johannes A. Langendijk
Department of Radiation Oncology
University Medical Centre Groningen
University of Groningen
Groningen, the Netherlands

Yolande Lievens
Department of Radiation Oncology
Ghent University Hospital
and
Faculty of Medicine and Health Sciences
Ghent University
Ghent, Belgium

John H. Maduro
Department of Radiation Oncology
University Medical Centre Groningen
University of Groningen
Groningen, the Netherlands

Ramona Mayer
Former Medical Director of MedAustron
Wiener Neustadt, Austria

Masashi Mizumoto
Department of Radiation Oncology
University of Tsukuba Hospital
Tsukuba, Japan

Bernd Mößlacher
BBM consulting GmbH
Stockerau, Austria

Radhe Mohan
Department of Radiation Physics
The University of Texas
MD Anderson Cancer Center
Houston, Texas

Christina T. Muijs
Department of Radiation Oncology
University Medical Centre Groningen
University of Groningen
Groningen, the Netherlands

Cary Oberije
Department of Radiation Oncology
    (MAASTRO Clinic, D-Lab)
GROW—School for Oncology and
    Developmental Biology
Maastricht University Medical Centre
Maastricht, the Netherlands

Roberto Orecchia
Medical Director
National Centre for Oncological
    Hadrontherapy
Pavia, Italy

and

Scientific Direction
European Institute of Oncology
Milan, Italy

Yoshiko Oshiro
Department of Radiation Oncology
University of Tsukuba Hospital
and
Department of Radiation Oncology
Tsukuba Medical Centre Hospital
Tsukuba, Japan

K. Parodi
Department of Experimental
    Physics—Medical Physics
Ludwig-Maximilians-Universität München
    (LMU Munich)
Munich, Germany

M. Pinto
Department of Experimental
  Physics—Medical Physics
Ludwig-Maximilians-Universität München
  (LMU Munich)
Munich, Germany

David A. Pistenmaa
International Cancer Expert Corps
Washington, DC

Richard Pötter
Department of Radiation Oncology
Medical University of Vienna
Vienna, Austria

Erik Roelofs
Department of Radiation Oncology
  (MAASTRO Clinic, D-Lab)
GROW—School for Oncology and
  Developmental Biology
Maastricht University Medical Centre
Maastricht, the Netherlands

Hideyuki Sakurai
Department of Radiation Oncology
University of Tsukuba Hospital
Tsukuba, Japan

Marco Schippers
Department Large Research Facilities
  (GFA)
Paul Scherrer Institut (PSI)
Villigen, Switzerland

K. Seidensaal
Abteilung für RadioOnkologie und
  Strahlentherapie
Universitätsklinikum Heidelberg
Heidelberg, Germany

Juliette Thariat
Department of Radiation Oncology
Centre François Baclesse
Caen, France

Beate Timmermann
Clinic for Particle Therapy
University Hospital Essen
West German Cancer Center (WTZ)
Essen, Germany

Hirohiko Tsujii
National Institute of Radiological Sciences
National Institute for Quantum and
  Radiological Sciences
Chiba, Japan

Yvonka van Wijk
Department of Radiation Oncology
  (MAASTRO Clinic, D-Lab)
GROW—School for Oncology and
  Developmental Biology
Maastricht University Medical Centre
Maastricht, the Netherlands

Stanislav Vatnitsky
MedAustron
Wiener Neustadt, Austria

Damien Charles Weber
Medical Director Proton Therapy Centre
Paul Scherrer Institute (PSI)
Villigen, Switzerland

Joachim Widder
Department of Radiation Oncology
Medical University of Vienna
Vienna, Austria

CHAPTER 1

From Röntgen Rays to Carbon Ion Therapy

The Evolution of Modern Radiation Oncology

K. Seidensaal and J. Debus

CONTENTS

DISCOVERY OF X-RAYS AND BIRTH OF RADIATION ONCOLOGY

The association of radiation and medicine began after an experiment by Wilhelm Conrad Röntgen, a professor at Würzburg University, who accidentally discovered in 1895 a new type of rays which have the capability of visualising what is hidden inside the organism. He called them X-rays in tribute to their mystery. Just days later, the very first and most famous X-ray image of his wife's hand wearing a ring was created in a 20-minutes exposure to a Crooks tube, an experimental electrical device. Röntgen published his famous paper 'Über eine neue Art von Strahlung' ('On a New Type of Ray') in 1896 and was invited to present his results to the German Emperor Wilhelm I just two weeks later. Because the potential of his discovery was easily understood by the public, the application of X-rays spread faster than any scientific discovery previously known. Within a short period of time, radiographs were being used for diagnosis of wounded soldiers in military hospitals.

Just months after the discovery, the first attempt to treat a breast cancer patient by X-rays was performed by Emil Grubbe, a student at the University of Chicago. His work was followed by the attempt of Lyon physician Victor Despeignes to treat a patient with stomach cancer. The origins of radiation oncology, however, are commonly assigned to Leopold Freund, who successfully treated a five-year-old patient with *Naevus pigmentosus piliferus* in Vienna just months after the discovery of X-rays by Röntgen. Freund concentrated on dermatology because he was convinced that the penetration depth of X-rays was insignificant beyond the skin; he was the first one to apply scientific methods to the development of treatment protocols.

MADAME CURIE AND FRENCH CONTRIBUTIONS TO RADIOBIOLOGY AND RADIATION ONCOLOGY

The discovery of X-rays was followed by the discovery of radium in 1898 by Marie und Pierre Curie; it yielded them the Nobel Prize in 1903. After the tragic death of her husband and the outbreak of First World War, Marie Curie devoted herself to a new task. Using her celebrity and influence, she initiated the construction of portable X-ray machines to assist surgeons treating the wounded at the battlefield. She often undertook long journeys, driving the X-ray unit herself to teach doctors and nurses the handling of equipment, thereby saving the lives of many French soldiers with the use of approximately 200 stationary and 20 mobile X-ray units.

The radiobiological knowledge that made modern radiation oncology possible was acquired in France at the Institute Curie after the end of the First World War. At this time in Lyon, the law of Bergonié and Tribondeau was discovered. This law describes the correlation of greater reproductive activity and sensitivity to radiation during the M-phase of the cell cycle. Claudius Regaud, a physician and biologist, postulated long before the discovery of DNA that chromatin is the target within the cell. Additionally, Regaud observed in animal experiments that subdividing the dose in subsets that were applied subsequently resulted in different, less harmful effects on healthy tissue than applying the entire dose at once. The continuously proliferating spermatozoa, his model tumour cells, were destroyed nonetheless, leaving the model animal infertile. Inspired by these findings, by 1934, the radiologist Henri Coutard developed a fractionation concept using 30 sessions for treatment of laryngeal carcinoma; this formed the basis of modern radiation oncology. Furthermore, Regaud started treating different types of tumours with radium sources which he placed close to the tumour – for example, by utilising interstitial needles – and thus initiated the development of brachytherapy.

TECHNICAL ADVANCES IN HIGH-ENERGETIC PHOTON THERAPY

Coolidge tubes, the successor of X-ray emitting Crooks tubes, were developed in 1913 and named after the inventor; these were the standard source for cancer treatment until the 1950s. Unfortunately, due to still quite low energy levels, deep-seated tumours could not be treated because most of the dose was applied to the skin. Post World War II, a substantial advance in radiation oncology occurred after the introduction of 'Cobalt bombs'; the radioactive isotope Cobalt-60 was placed in a lead shielding which could be opened by a shutter, thus allowing treatment by photons with an energy of 1.2 megaelectron volts. Although obsolete in relation to current standards, those units are currently still operating in several Third World countries. The Cobalt bomb was succeeded by the 10 megaelectron volts betatron and the electron linear accelerator, which was developed by William Webster Hansen together with the two brothers Sigurd and Russel Varian. Nowadays, Varian is a manufacturer and provider of radiotherapeutic equipment.

Radiation oncologists next started to use shielding blocks to shape the radiation field and reduce the dose delivered to normal tissues and critical organs. This was followed by the invention of the multileaf collimator (MLC) which consists of movable leafs arranged in pairs that allow the shaping of individual conformal fields simply and without time- and cost-consuming preparation of shielding blocks. Subsequently with the invention of computer tomography (CT), more powerful computers and advanced dose-calculation algorithms, two-dimensional (2D) planning was replaced by three-dimensional (3D) conformal treatment planning which allowed dose escalation trials. The first inverse treatment planning algorithm was developed by Steve Webb in 1989. It permitted radiation oncologists to define dose constraints to certain organs at risk and prescribe dose to the treated target volume (tumour), and then a planner would issue importance factors and an optimisation algorithm would calculate a plan which best met all the necessary criteria. A seminal paper published in 1988 by Anders Brahme of Karolinska Institutet is generally considered as the starting point of the field of intensity-modulated radiotherapy (IMRT). The principle of static (step and shoot) IMRT was developed next; modulation of the fields yielded improvement in dose coverage of the tumour and reduced the dose to neighbouring organs at risk. One of the first IMRT programs in Europe originated at the German Cancer Research Center and included a significant contribution by Thomas Bortfeld, currently the Chief of the Physics Division at the Massachusetts General Hospital. Gradually, target volume definition became more accurate thanks to modern imaging modalities such as magnetic resonance (MR) and positron-emission tomography (PET)-CT imaging. The first helical IMRT was developed in Wisconsin in 1993 and became commercial in 2002; it introduced modern image-guided radiotherapy. Volumetric-modulated arc therapy (VMAT) with dynamic MLCs provided an additional technique to perform IMRT using reduced treatment time. The first proposal that optimal delivery of IMRT might be achieved by a short-length linac mounted on a robotic arm arose in 1999 and was accomplished with the Cyberknife of Accuray which was developed at Stanford University and is used for modern radiosurgery. Recently, the combination of (MRI) and a linear accelerator promises to put image-guided radiotherapy at new level, thereby providing real-time imaging as well as target volume and plan adaptation; these machines are already beginning to operate in different institutions around the world. In summary, it took more than 100 years to develop conformal radiotherapy with photons. Currently, scientists are working closely at the physical limits of this technology.

CYCLOTRON, SYNCHROCYCLOTRON AND SYNCHROTRON – THE BEGINNINGS OF PARTICLE THERAPY

The golden age of physics started in the 1920s and continued for the next five decades. Extensive research in general and physics in particular was conducted; never before had scientific developments influenced the history of civilisation to this extent. Ernest Rutherford, one of the greatest experimentalists who managed to prove the existence of the nucleus in 1911 and theorise about the existence of neutrons, requested that physicists provide a 'copious supply' of higher energetic particles than those from natural radioactivity. At this time, Ernest Lawrence, an associate professor at Berkeley University and later a Nobel prize laureate as well as a member of the Manhattan Project, learned about linear acceleration by switching potentials. He was scanning illustrations of a publication by the Norwegian Rolf Widerøe in a German journal on electrical engineering, a language he did not understand. His idea to adapt this type of acceleration by bending the path of charged particles into circular trajectories and circulate them many times laid the groundwork for the construction of the cyclotron in 1930. His new accelerator which he generously shared provided synthetic radionuclides for nuclear medicine and radiation oncology.

The phase stability principle was discovered by the Russian scientist Wladimir Iossifowitsch Weksler and the American scientist Edwin Mattison McMillan, who shared the Atoms for Peace Award in 1963. This principle describes that gradually and continuously increasing the oscillation period of particles with the number of turns in the accelerator leads to formation of a stable and tight bundle of accelerated particles. Subsequently, Lawrence adapted this principle in the development of the synchrocyclotron, which began operating in 1946 and allowed acceleration of protons to 55% of the speed of light (200 megaelectron volts). For the first time, the particle energy was sufficient to provide for protons which could penetrate the patients' relevant depth, and here also was the basis for application to treated tumour target volumes. The spiral form of orbits in the synchrocyclotron makes a large uniform magnetic field necessary in contrast with a synchrotron which requires several smaller magnets that are placed around a hollow doughnut-shaped circular accelerator. The first synchrotrons were constructed by the two discoverers of the phase stability principle; until 1957, energies up to 10 gigaelectron volts were reached. The advantage of these among others is that the weight of the magnets is lower, so beams of much higher energy can be created. Discovery of the model of strong focusing finally lead to the birth of the European Organization for Nuclear Research (CERN) in Europe.

PROTON THERAPY

The idea to use high-energy protons for therapy is older than one might assume; it was proposed first by the physicist Robert R. Wilson, the founder of Fermi National Accelerator Laboratory (Fermilab) and member of the Manhattan Project, in his 1946 publication 'Radiological Use of Fast Protons' in *Radiology* during his work on the design of the Harvard Cyclotron Laboratory. He postulated that maximum irradiation dose could be placed within the tumour, thus sparing healthy tissue based on the 'Bragg Peak' phenomenon which was first described by William Bragg for alpha particles. This was followed by extensive research which managed to confirm Wilson's predictions. The first patients were subsequently treated in 1954 at Lawrence Berkley National Laboratory (LBL) and in 1957 in Uppsala, Sweden. Initially, treatment was performed in and restricted to research facilities using particle accelerators conducted for physics research.

With progressing technical advances in imaging, computers, accelerators and treatment-delivery techniques, proton therapy became more accessible to the routine medical treatment of cancer patients in the 1970s and was approved as a therapy option for certain tumours by the Federal Drug Administration (FDA) in 1988. The first hospital-based oncological particle centre that opened in 1989 was Clatterbridge Centre for Oncology in the United Kingdom, followed in 1990 by the Loma Linda University Medical Center (LLUMC) in the United States. Currently, over 60 centres worldwide offer proton therapy to cancer patients and more than 30 additional proton centres are in the planning phase or under construction. The use of protons enables radiotherapy to be more precise by decreasing the severity of acute and late side effects, thus making it favourable, for example, in the treatment of paediatric cancer patients; however, because the relative biological effectiveness (RBE) of protons is similar to that of photons, scientific attention turned to heavier ions in hopes of increasing biological impact due to higher linear energetic transfer (LET).

CARBON ION-TREATMENT

John H. Lawrence, the brother of Nobel-prize winner and cyclotron inventor Ernest Lawrence, was an American physicist and physician who pioneered the field of nuclear medicine as well as particle therapy. Between 1935 and 1938, he conducted the first biomedical studies and demonstrated the greater biological effect by dense tissue ionisation of heavy particles in normal and cancerous tissue. In the following years, he demonstrated the therapeutic advantages of heavy-charged particles with higher energy together with Cornelius A. Tobias, a nuclear physicist and member of LBL best known for his radiobiological studies and application of the high LET. Since 1952, the very first patients were treated with argon-, neon-, silicon- and helium-particle beams until carbon ion was found to have the ideal radiobiological characteristics. The Bevalac, a 1974-onward combination of the Super Heavy Ion Linear Accelerator (SuperHILAC) linac and the Bevatron, a proton accelerator, enabled clinical trials with heavy ions of more than 1,400 cancer patients at what is now LBL before it was decommissioned in 1993.

The foundation for fast-neutron therapy was also laid at LBL in 1938. After a break during the Second World War when the cyclotron was used for the war effort, research continued in the 1960s. It was the first high LET radiation therapy which was applied clinically with maximum use in the 1970s and 1980s. Effectiveness and favourable local control were shown for different tumour entities, but tremendous late reactions resulted in stopping the use in almost all the centres in Europe, Japan and the United States.

In Japan in 1984, the government began constructing the first heavy-ion facility for routine medical use called the National Institute of Radiological Sciences (NIRS) which employed many of the LBL and Bevalac scientists. The Heavy Ion Medical Accelerator in Chiba (HIMAC) was established in 1994, and provided, similarly as the Bevalac, passive beam irradiation (passive scattering) with protons and carbon ions. It was unique until 1997. Further advances included the installation of a pencil beam raster scanning (PBS) facility and markerless respiration-gated PBS, the carrying out of nearly 70 treatment protocols with treatment of approximately 1,000 patients per year. The Microdosimetric Kinetic Model (MKM) was developed, updated (MKM2010) and implemented within Japan; efforts were made to provide for dose translations to the European local effect model (LEM). The Japan Carbon-ion Radiation Oncology Study Group (J-CROS) is today composed of five carbon-ion therapy centres with horizontal and vertical fixed beams and one gantry offering PBS as well as passive scattering. NIRS is a pioneer which has performed major breakthroughs and paradigm changes in radiation oncology with more than 20 years of experience in carbon-ion irradiation.

The 'Helmholz Gesellschaft für Schwerionenforschung' (GSI) was founded in 1969 in Western Germany; the era of particle treatment started when Gerhard Kraft, a trained nuclear physicist and radiobiologist and a fellow of Cornelius Tobias at LBL, introduced ion therapy in Europe after his return from the United States. During the following years, Germany achieved a pioneering position in proton and carbon-ion therapy. At GSI, the innovative pencil beam, active raster scanning technique was developed using magnetic fields to deflect the beam in horizontal and vertical direction and conform it over the targeted volume, thus performing intensity modulation and more flexible therapy treatment planning compared to the previously implemented passive scattering. Furthermore, optimisation of inverse treatment planning for biological parameters due to the quantitative calculation of the RBE in carbon-ion treatment with the theoretical LEM, which is today widespread throughout Europe, was created there. After promising basic research results, translation to clinical practice allowed treatment of more than 400 patients from 1997 to 2009. The research and experience of GSI were implemented in the Heidelberg Ion-Beam Therapy Center (HIT) which was developed and built by the University Hospital Heidelberg, the German Cancer Research Center (DKFZ), the Helmholtz-Zentrum Dresden-Rossendorf (HZDR), and the company SIEMENS from 2003 to 2009. Currently, more than 4,400 patients have been treated at HIT with excellent results using carbon-ion and proton-beam irradiation by concentrating on rare radioresistant malignancies such as nonsquamous cell tumours of the head and neck – especially adenoidcystic carcinoma; chordoma; chondrosarcomas of the skull base and pelvis as well as more common entities such as glioma; skull base meningioma; and prostate cancer. Several clinical phase I–III studies are investigating the safety and efficacy of particle therapy in combined boost-concepts or alone, such as PROMETHEUS for inoperable hepatocellular carcinoma; MARCIE for anaplastic meningioma; Cinderella for recurrent glioma; ISAC for pelvine chordoma; and OSCAR for inoperable osteosarcoma which is the first carbon-ion treated paediatric study cohort worldwide. Pre-clinical studies investigate the use of additional heavy ions such as helium and oxygen for cancer treatment. Complementing two treatment locations with horizontal beam position, the first worldwide rotating proton- and carbon-ion gantry was created using astronomy telescope technology and implemented in the year 2012; it weighed over 600 tons and allowing a 360-degree rotation of the beam. A similar second proton- and carbon-ion centre with three horizontal treatment places and one 45-degree place started treating patients in 2015 in Marburg and is operated partially by HIT. Further centres treating cancer patients with carbon-ion irradiation are MedAustron in Austria and CNAO in Italy as well as two additional centres located in China.

In summary, together hadron therapy and high precision X-ray therapy costimulated their own development over the last century. The new technologies allow conformal radiotherapy with submillimetre accuracy. The big challenge of radiation oncology remains the biological optimisation and selection of the treatment regarding dose, volume and time.

REFERENCES

Amaldi, U. *Particle Accelerators: From Big Bang Physics to Hadron Therapy*, 2015. doi:10.1007/978-3-319-08870-9.

Bortfeld, T. IMRT: A review and preview. *Phys. Med. Biol.* 51(13) (2006): R363–R379.

Brahme, A. Optimization of stationary and moving beam radiation therapy techniques. *Radiother. Oncol.* 12(2) (1988): 129–140.

Brown, A. and H. Suit. The centenary of the discovery of the Bragg peak. *Radiother. Oncol.* 73(3) (2004): 265–268.

Combs, S. E., O. Jakel, T. Haberer, and J. Debus. Particle therapy at the Heidelberg Ion Therapy Center (HIT)—Integrated research-driven university-hospital-based radiation oncology service in Heidelberg, Germany (in English). *Radiother. Oncol.* 95(1) (2010): 41–44.

Coolidge, W. D. A powerful röntgen ray tube with a pure electron discharge. *Phy. Rev.* 2(6) (1913): 409–430.

Coutard, H. Principles of X ray therapy of malignant diseases. *Lancet* 224(5784) (1934): 1–8.

Del Regato, J. A. Claudius Regaud (in English). *Int. J. Radiat. Oncol. Biol. Phys.* 1(9–10) (1976): 993–1001.

Dieterich, S. and I. C. Gibbs. The cyberknife in clinical use: Current roles, future expectations (in English). *Front Radiat. Ther. Oncol.* 43 (2011): 181–194.

Ebner, D. K. and T. Kamada. The emerging role of carbon-ion radiotherapy (in English). *Front. Oncol.* 6(140) (2016).

Elements of general radiotherapy for practitioners. *J. Am. Med. Assoc.* XLIII(13) (1904): 905–1005.

Elsasser, T., M. Kramer, and M. Scholz. Accuracy of the local effect model for the prediction of biologic effects of carbon ion beams in vitro and in vivo (in English). *Int. J. Radiat. Oncol. Biol. Phys.* 71(3) (2008): 866–872.

The First Cyclotron. American Institute of Physics. https://history.aip.org/exhibits/lawrence/first_text.htm.

Ginzton, E. L. *Varian Associates: An Early History*. Palo Alto, CA: Varian Associates, 1998.

Glasser, O. W. C. Roentgen and the discovery of the roentgen rays (in English). *AJR Am. J. Roentgenol.* 165(5) (1995): 1033–1040.

Grubbé, E. H. Priority in the therapeutic use of X-rays. *Radiology* 21(2) (1933): 156–162.

Haber, A. H. and B. E. Rothstein. Radiosensitivity and rate of cell division: Law of Bergonie and Tribondeau. *Science* 163(3873) (1969): 1338–1339.

Haberer, T. H., W. Becher, D. Schardt, and G. Kraft. Magnetic scanning system for heavy ion therapy. *Nucl. Instrum. Meth. Phys. Res. Sect. A Accel. Spectrom. Detect Assoc. Equip.* 330(1) (1993): 296–305.

Jeraj, M. and V. Robar. Multileaf collimator in radiotherapy. *Radiol. Oncol.* 38(3) (2004): 235–240.

Jones, D. T. L. and A. Wambersie. Radiation therapy with fast neutrons: A review. *Nucl. Instrum. Meth. Phys. Res. Sect. A Accel. Spectrom. Detect Assoc. Equip.* 580(1) (2007): 522–525.

Kacperek, A. Protontherapy of eye tumours in the UK: A review of treatment at Clatterbridge. *Appl. Radiat. Isot.* 67(3) (2009): 378–386.

Kamada, T., H. Tsujii, E. A. Blakely, J. Debus, W. De Neve, M. Durante, O. Jakel et al. Carbon ion radiotherapy in Japan: An assessment of 20 years of clinical experience (in English). *Lancet Oncol.* 16(2) (2015): e93–e100.

Lambert, B. John H. Lawrence, 87; Led in radiation research. *New York Times*, September 9, 1991.

Laugier, A. The first century of radiotherapy in France (in French). *Bull. Acad. Natl. Med.* 180(1) (1996): 143–160.

Lawrence, E. O. and M. Stanley Livingston. The production of high speed light ions without the use of high voltages. *Phys. Rev.* 40(1) (1932): 19–35.

Lentle, B. and J. Aldrich. Radiological sciences, past and present. *Lancet* 350(9073) (1997): 280–285.

Lischalk, J. W., L. Konig, M. C. Repka, M. Uhl, A. Dritschilo, K. Herfarth, and J. Debus. From rontgen rays to carbon ion therapy: The evolution of modern radiation oncology in Germany. *Int. J. Radiat. Oncol. Biol. Phys.* 96(4) (2016): 729–735.

Page, B. R., A. D. Hudson, D. W. Brown, A. C. Shulman, M. Abdel-Wahab, B. J. Fisher, and S. Patel. Cobalt, linac, or other: What is the best solution for radiation therapy in developing countries? *Int. J. Radiat. Oncol. Biol. Phys.* 89(3) (2014): 476–480.

Particle therapy co-operative group; a non-profit organisation for those interested in proton, light ion and heavy charged particle radiotherapy. https://www.ptcog.ch/index.php.

Röntgen, W. C. Ueber eine neue art von strahlen. *Annalen der Physik.* 300(1) (1898): 12–17.

Scholz, M., A. M. Kellerer, W. Kraft-Weyrather, and G. Kraft. Computation of cell survival in heavy ion beams for therapy. The model and its approximation (in English). *Radiat. Environ. Biophys.* 36(1) (1997): 59–66.

Slater, J. D. Development and operation of the Loma Linda University Medical Center proton facility (in English). *Technol. Cancer Res. Treat.* 6(4) (2007): 67–72.

Smith, A. R. Proton therapy. *Phys. Med. Biol.* 51(13) (2006): R491–R504.

Trombetta, M. Madame Maria Sklodowska-Curie—Brilliant scientist, humanitarian, humble hero: Poland's gift to the world. *J. Contemp. Brachyther.* 6(3) (2014): 297–299.

Webb, S. Conformal intensity-modulated radiotherapy (IMRT) delivered by robotic linac—Testing IMRT to the limit? (in English). *Phys. Med. Biol.* 44(7) (1999): 1639–1654.

Webb, S. Optimisation of conformal radiotherapy dose distributions by simulated annealing (in English). *Phys. Med. Biol.* 34(10) (1989): 1349–1370.

Wilson, R. R. Radiological use of fast protons (in English). *Radiology* 47(5) (1946): 487–491.

Yarris, L. 1930s: The rad lab—From a small wooden building to a national laboratory. http://history.lbl.gov/1930s/.

CHAPTER 2

Radiobiology and Hadron Therapy

What Do We Know and What Do We Need to Know?

Eleanor A. Blakely and Manjit Dosanjh

CONTENTS

INTRODUCTION

The use of ionising radiation to treat cancer has a long history since the first treatment of cancer with X-rays in the late 1800s. Many types of radiation have been employed to achieve control of tumour viability. The depth–dose profile of four current types of external beam radiations – a 'low' linear energy transfer (LET) beam of photons, two 'high' LET

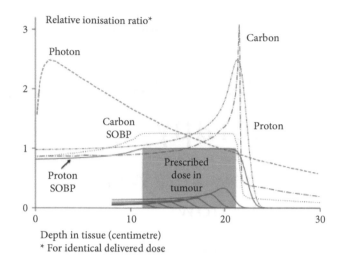

FIGURE 2.1   Depth-dose profiles of different radiation types.

unmodified and spread-out Bragg peak (SOBP) beams of protons, and carbon ions – are compared in Figure 2.1. LET is a physical parameter of ionising radiation that describes the ionisation density along the tracks emanating from a radiation source. The figure illustrates the dramatic differences in dose distributions arising from each of the radiation sources. Photons demonstrate a high initial dose that declines with penetration into the depth of the absorbing material of the body. In contrast, the proton and carbon particle beams demonstrate superficial low initial energy deposition that is maximally absorbed at depth in a Bragg peak of energy absorption where the primary ions dump their energies as they stop with significantly reduced energy deposited beyond the stopping peaks. These depth-dose profiles are dependent on the initial beam energies, can be broadened to cover tumour volumes, and have proven to provide a significant advantage to the physical targeting of deep-seated tumours while sparing radiation dose to surrounding normal tissues. Figure 2.2 from the work of Dr. Harry Heckman illustrates heavy ion particle tracks coming in from the left and stopping in photographic emulsion as the particles slow down to a stopping point in the extreme example of a very high atomic number uranium beam where dark delta rays pre-dominate in the image.

Radiobiology is a scientific field that measures a large and diverse number of biological effects from exposures to different radiations from the electron magnetic spectrum at several levels of biological scale and organisation from submolecular to cells, tissues, whole organisms and even populations. Radiobiology can provide essential information regarding preclinical responses to radiation-based treatments for disease that can guide a physician's selection of ionising radiation dose and treatment regime time-course. The medical field of radiotherapy for the treatment of cancer has significantly benefited from radiobiological evidence underlying the mechanisms involved in the clinical outcome. Radiobiology has been essential for the implementation of new treatment modalities involving radiation alone or in combination with chemotherapy and has uncovered

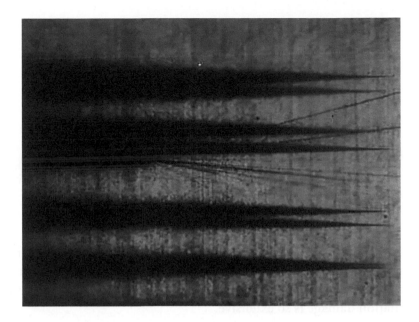

FIGURE 2.2 Charged-particle tracks in photographic emulsion. (From Tobias, C.A., *Radiat. Res.*, 103, 1–33, 1985. With permission.)

inherent heterogeneities within tumour and normal tissues responses that have led to the realisation that personalised medicine for individual patients is likely critical to future medical care.

Radiobiology provides support for clinical therapy in order to (1) understand basic mechanisms of radiation action for designing the rationale for new treatment strategies and protocols for various cancer types, sites and stages; (2) answer very specific applied technical questions arising in the implementation of a new modality including dose per fraction, intensity of dose delivery and overall treatment time; and (3) provide information on what are safe and effective applications of radiation treatment in diagnostic imaging and radiotherapy to avoid acute and chronic toxicities since radiation can cure cancer but can also cause cancer and other late-appearing complications.

The quantitative determination of radiation dose-effects were especially useful once Puck and Marcus (Puck et al., 1956) generated the first *in vitro* radiation survival curve describing the relationship between the radiation dose and the proportion of mammalian cells growing in Petri dishes that survive to form discreet circular colonies that could be fixed, stained and counted. Dose is the dependent variable and is usually plotted on a linear scale on the abscissa while the fractional survival is plotted on the ordinate on a logarithmic scale. The requirement to calibrate the dose of a new test radiation modality that results in the same radiation effect as that from a dose of a reference photon radiation modality led to the development of the concepts of a relative biological effectiveness (RBE) ratio such as $RBE_{50\%survival} = D_{50\%reference}/D_{50\%test}$ either in air or under hypoxia and the oxygen enhancement ratio (OER) for hypoxic and aerobic dose responses that yield the same biological effect (in this case, cell survival) as illustrated in Figure 2.3. Reduced oxygen levels which frequently occur in rapidly growing

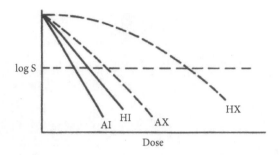

FIGURE 2.3 RBE and OER calculation illustrated with cell survival curves. S – survival, X – photon, I – ion, A – aerobic, and H – hypoxic.

tumours result in much greater biological radioresistance to photons, but much less so for radiations with increased ionisation qualities. This means particle beams with LET values near 100 KeV/μm can eradicate tumour cells regardless of the oxygen status since the radiation damage is so extensive.

$$\text{oerobic RBE}_{10\%S} = \left[\frac{\text{ADX}_{10\%S}}{\text{ADI}_{10\%S}} \right] \qquad \text{OER} - \text{X}_{10\%S} = \left[\frac{\text{HDX}_{10\%S}}{\text{ADX}_{10\%S}} \right]$$

$$\text{hypoxic RBE}_{10\%S} = \left[\frac{\text{HDX}_{10\%S}}{\text{HDI}_{10\%S}} \right] \qquad \text{OER} - \text{I}_{10\%S} = \left[\frac{\text{HDI}_{10\%S}}{\text{ADI}_{10\%S}} \right]$$

$$\text{OER}_{10\%S} = \left[\frac{\text{OER} - \text{X}_{10\%S}}{\text{OER} - \text{I}_{10\%S}} \right]$$

RBE values are highly dependent on many parameters. The RBE can vary with the type of radiation; the type of cell or tissue exposed; the biological effect under investigation; the dose, dose rate and fractionation; ambient oxygen level; and the presence of other chemicals. For particle beams, the RBE also depends on the ion atomic mass, energy and mode of beam delivery and modification to shape the beam to the tumour target. An increased RBE in itself does not offer therapeutic advantage unless there are differential effects between normal and tumour tissue in the treatment plan.

We know at the molecular level that high LET particle beams cause clustered damage from densely ionising damage to DNA along the particle track. DNA double strand breaks increase with increasing LET, and this damage is more difficult to repair, thereby resulting in higher levels of residual damage, mutation, chromosome aberrations and cell death. This damage can be a consequence of both direct and indirect ionisations and is critically dependent on chromatin structure and repair capability.

Radiation treatment-planning physics is a highly technical and precise field based primarily on radiation measurements of dose; however, the radiation physicists actually measure physical dose and cannot directly measure biological effective dose for radiations that may have a variable radiation quality of dense ionisations. The physicists require biological measurements to calibrate the resulting biological consequences of physical doses or a theoretical model that can bridge this gap. The number of theoretical models that have collated information and best-fit parameters for data analysis has grown significantly, and these models hold a significant promise of being able to predict outcomes and to be integrated into treatment-planning software in the clinic.

A major decision for this research is which biological model should be studied. Both two-dimensional (2D) and three-dimensional (3D) culture models of normal or tumour human cells are pertinent to the treatment of site-specific diseases with radiation. This is because gene expression evident in 2D *in vitro* may significantly change in 3D-growth *in situ* or *in vivo*. Limitations for each exist because often this work requires years of basic laboratory studies during which both normal or tumour human cells can significantly change spontaneously during extended culturing. Many investigators use immortalised human or rodent cell lines or tumours due to their ease of culturing and long stability in radiation response, but the relevance and limitations of these biological models with regard to their gene expression responses to radiation exposures *in vivo* and whether one can generalise from the responses of biological models derived from non-human species still requires further confirmation.

Differences between human and rodent cell lines that resulted in reported differences in RBE between the German and Japanese charged particle radiotherapy programs which are currently leading the field have been reported (Fossati et al., 2012). As additional new ion beam therapy facilities come on line, comparative studies have been completed between model systems at each facility to document differences in the validation of biologically effective doses of individual ion beams and promote integration of biological parameters and theoretical models into the treatment planning process. This demonstrates an international convergence for the need of a standardised pre-clinical protocol in the commissioning of new particle facilities.

Current ion therapy treatment planning for cancer is still evolving to keep pace with rapidly emerging technological advances in several disciplines (Kamada et al., 2015). There is a huge potential range of additional clinical benefits possible from simultaneous improvements in patient-specific imaging; in the radiobiological estimation of individual responses of tumour and normal tissue responses of patients or of relevant experimental models; and in further defining the tumour-driven selection of ion beam species, modes of beam delivery, dose fractionation and overall time-course. In addition, an increasing number of novel mechanisms of particle beam radiobiology have been reported that are distinctly different than those observed with conventional radiations. In many cases, these differences appear to hold potential benefits to patient outcome.

Particle radiobiology has matured as a scientific specialty since the late 1800s in parallel with the development of an extraordinary number of technical advances that have increased our capabilities to investigate mechanisms of action. Early health effects from alpha particle-emitting radioisotopes were recognised first, and soon after the nucleus of a cell was identified as more radiosensitive than the cytoplasm (Jacob et al., 1970). Disrupting the genetic information in the DNA was clearly a key lesion leading to loss of viability. After the cyclotron was invented and accelerators became available, detailed systematic studies revealed acute and late effects of particle radiation exposures.

What has become quite clear is that the damage to cells and tissues are distinctly different for particle radiations compared to conventional photon radiations even at the same level of effect due to submicroscopic differences in the energy deposition pattern. Early research measuring DNA strand breaks by various types of ionising radiations using alkaline unwinding and alkaline and neutral filter elution techniques always yielded RBE values for the total number of DNA-induced strand breaks that were less than one despite significant differences in cellular dose-dependent survival. Light microscopy allowed the visualisation of chromosomes within the nucleus of dividing cells; and confocal microscopy, fluorescence microscopy and transmission and scanning electron microscopy have each provided additional information about the structure of DNA as well as information regarding the sequence of a complex series of proteins that bind to DNA after radiation exposure to facilitate the DNA repair process.

Enhanced double-strand breaks in DNA molecules were identified as a prominent feature of decreased repair capacities after exposure to densely-ionizing radiations (Roots et al., 1989, 1990). The number of particle-radiation-induced breaks in chromatin fibers were visualized by premature chromosome condensation (Goodwin et al., 1989, 1992, 1994, 1996), and chromosomal damage was scored with traditional Giemsa-stained techniques and analyses as well as by visualizing specific chromosomes with re-arrangements identifications made possible with fluorescently-labeled immune probes. In addition, pulse field gel electrophoresis studies revealed that high LET particle damage increased the production of small DNA fragments (Rydberg, 1996); however, until very recently, technical limitations in imaging resolution have prevented the visualisation of megabase 3D domains of chromatin fibers in intact cells.

RECENT NEW INFORMATION ON THE BASIC STRUCTURE OF DNA

The classic 'hierarchical chromatin folding' model of DNA indicated that the primary 11-nanometre DNA core nucleosome polymers assemble into 30-nanometre fibers that fold into 120-nanometre chromonema, 300- to 700-nanometre chromatids and ultimately mitotic chromosomes (Kuznetsova et al., 2016). Recently a novel technique called 'chromEMT' has been reported (Ou et al., 2017) which combines electron microscopy tomography (EMT) with a labeling method (chromeEM) that selectively enhances the contrast of DNA. Using chromEMT with advances in multitilt EMT allowed imaging of chromatin ultrastructure and 3D packing of DNA in both human interphase and mitotic chromosomes. The chromatin was discovered to be a flexible and disordered 5–24 nanometre diameter granular chain that is packed together at different concentration densities (rather than with higher-order folding as was previously understood) within interphase nuclei and in mitotic chromosomes. This feature appears to determine global accessibility and activity of DNA.

Independently, superresolution light microscopy – combining structured illumination microcopy (SIM) and spectral precision distance microscopy (SPDM) – has been developed to explore the 3D arrangement of phospho-H2AX (histone-2AX)-labeled chromatin in the nuclear volume of human HeLa cells (with its well-annotated genome) and its dynamic evolution during the DNA-damage response (DDR) from X-rays or clustered regularly interspaced short palindromic repeat (CRISPR), double strand break (DSB). Cas9-mediated DSBs in human cells with genome-wide sequencing analysis at accuracies of 10–20 nanometres (Natale et al., 2017). With this approach, heterochromatin exhibited DNA decondensation while retaining heterochromatin histone marks, indicating that chromatin structural and molecular determinants were uncoupled during repair.

The key structural factor CTCF (CCCTC-binding factor) was found to be flanking the phospho-H2AX nanodomains that arrange into higher-order clustered structures of discontinuously phosphorylated chromatin. CTCF knockdown impaired the spreading of the phosphorylation throughout the 3D-looped nanodomains. Co-staining of phospho-H2AX with phosphor-Ku70 and TUNEL (Terminal deoxynucleotidyl transferase dUTP Nick End Labeling) revealed that clusters rather than nanofoci represent single DSBs. This work provided evidence that each chromatin loop is a nanofocus whose clusters corresponded to previously known phospho-H2AX foci. The subfoci structure of the local chromatin at the DSB sites likely denote elementary DNA repair units along the carbon-ion trajectories where there are multiple DSBs in proximity and bear a striking similarity to the disordered 5–24 nanometre diameter chromatin structure reported by Ou et al. (2017). This work also challenges the belief that each γ-H2AX focus represents one DSB since isolated subfoci were found outside of the ion track that may represent delta ray-induced damage from individual particle tracks. The authors point out that counting the number of γ-H2AX foci from densely ionising radiations using conventional microscopy may underestimate the actual number of DSBs in the DNA due to the number of large clustered DSB-damaged foci.

CONSEQUENCES OF CLUSTERED DAMAGE – ACUTE MOLECULAR SIGNALING VIA THE RADIATION-INDUCED PHOSPHOPROTEOME

It is not surprising that the novel clustered characteristics of particle-induced DNA damage triggers a unique set of DDR post-lesion formation-signaling cascades and cell cycle arrest and recruitment of DNA repair factors compared to X-ray damage. A systematic study to decipher acute signaling events induced by different radiation qualities using high-resolution mass spectrometry-based proteomics has recently been published by Winter et al. (2017). Two hours after exposure of stable isotope labeling by amino acids in cell culture (SILAC)-labeled human lung adenocarcinoma A549 cells to X-rays, protons or carbon ion showed extensive alterations of the phosphorylation status despite protein expression remaining largely unchanged. Phosphorylation events were similar for proton and carbon irradiation; however, a distinctly different number of sites responded differentially for X-rays. The results were also validated with targeted spike-in experiments. This information will provide unique insight into the differential regulation of phosphorylation sites for radiations of different quality and how they may be further optimised for cancer radiotherapy.

PROTEOGENOMICS

The U.S. National Cancer Institute's Clinical Proteomic Tumor Analysis Consortium (CPTAC) is collaborating with the U.S. Department of Defense (DoD) and the Veteran Affairs (VA) Veterans Health System to incorporate proteogenomics as part of cancer patient treatment regimes. These three organisations have announced the formation of the Applied Proteogenomics Organizational Learning and Outcomes (APOLLO) Network which aims to build a system in which VA and DoD cancer patients routinely undergo genomic and proteomic profiling with the goal of matching their tumour types to targeted therapies. Applied proteogenomics is a developing new weapon in the war against cancer and is considered the keystone to everything the VA and DoD will be doing in precision oncology. APOLLO will characterise and compare tumours made available through the APOLLO network to develop a deeper understanding of cancer biology, identify potential therapeutic targets and identify pathways important for cancer detection and intervention.

It has become increasingly obvious that genomics has traditionally dominated clinical "omics" work, but to understand the features present in the genome, epigenome, and transcriptome, data is required on the proteome including posttranslational modifications. Past work has indicated that potentially meaningful proteomic changes are not present at the genomic level which suggests that information that could enable better-targeted treatments might be missed without proteomic analysis.

HUMAN CANCER PATHOLOGY ATLAS

A future endpoint that may be useful to consider is the human cancer transcriptome (Uhlen et al., 2017). The open-access Human Pathology Atlas database (www.proteinatlas.org/pathology) allows for genome-wide exploration of the impact of individual proteins on clinical outcome in major human cancers. If existing patient biopsy materials are available from patients, specific tumor protein profiles could be correlated with clinical outcome and could be used to prognostically identify patients for whom heavy charged-particle radiotherapy is appropriate.

EMPHASIS ON THE RBE-RATIO OF GAMMA-TO-ION DOSES TO YIELD THE SAME EFFECT

Currently, most particle treatment plans have incorporated improved Monte Carlo beam transport codes describing how primary ion depth–dose distributions are modified by secondary and tertiary particle interactions with absorbing materials and the increasingly broader spectrum of more diverse radiation types and energies of stopping ion beams. The biological input is virtually exclusively based on experimentally measured linear–quadratic parameters associated with best fits of data from laboratory biological models exposed to individual or mixed components of the stopping beams to theoretical biophysical algorithms or amorphous track models predicting risk of cell death, chromosomal changes or cancer induction. Acute and late-appearing aspects of these endpoints are rolled into a single RBE number that is highly diverse depending on the many biological and physical variables important for both cancer patients treated with particles or for astronauts traveling in space to distant planets (Durante et al., 2008). There are numerous previous reviews of cellular and tissue particle radiobiology (Amaldi et al., 1994; Blakely, 2001; Blakely et al., 1984, 1998, 2009; Brahme, 1998, 2014; Leith et al., 1983; Linz, 1995, 2012; Raju, 1980, 1995; Skarsgard, 1983, 1998; Tobias et al., 1997; Tsujii, 2014), and valuable today are also the recently available digital radiobiology data libraries of Friedrich et al. (2013).

EMPHASIS ON THE OER

Tumour hypoxia, especially among its cancer stem cells (CSCs), is long associated with radioresistance, and heavy charged-particle therapy is recognised as one of the key therapeutic approaches to treat such radioresistance regardless of the oxygen levels inherent in the heterogeneous tumour micro-environment. The mechanisms underlying hypoxic radioresistance are complex, but several unique molecular mechanisms of action have been uncovered for carbon-ion irradiations compared to photons as noted by Wozny et al. (2017).

Figure 2.4 presents aerobic and hypoxic survival curves obtained in track segment experiments at varying depths of penetration along the full range of beams of carbon or argon ions with 4-centimetre extended Bragg peaks. These curves can be used to calculate the OER which is the ratio of the dose of hypoxic survival to the dose of aerobic survival at the same level of effect. At high levels of LET near 150 KeV/μm, cell survival has a maximal RBE and a reduced dependence on the presence of oxygen (Figure 2.5). This means that ion beams at those LET values are more effective at eradicating radioresistant hypoxic cells than conventional radiations.

Under hypoxia, the protein hypoxia-inducible factor-1 (HIF-1) is the key transcriptional regulator of the cellular response controlling oxygen homeostasis. Ogata et al. (2011) studying HIF-1α expression in normoxic human lung adenocarcinoma cells showed that photon irradiation enhances the phosphorylation of AKT (protein kinase B) (previously reported by Harada et al., 2013) whereas carbon-ion irradiation decreases it, leading to reduction of HIF-1α resistance in normoxia. Subtil et al. (2014) used adenocarcinoma DNA microarrays to demonstrate that photons but not carbon-ion irradiation significantly altered the mechanistic or mammalian target of rapamycin (mTOR) pathway. Photons increase the phosphorylation of the mTOR protein, but carbon ions significantly decrease its phosphorylation, thereby inhibiting HIF-1α expression.

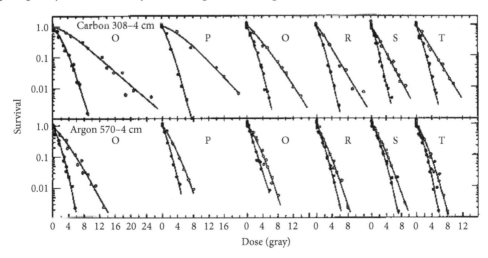

FIGURE 2.4 Aerobic and hypoxic cell survival curves for 308 MeV/u carbon or 570 MeV/u argon ions at different depths along the full range of a beam with a 4-cm SOBP. (From Tobias, C.A., *Radiat. Res.*, 103, 1–33, 1985. With permission.)

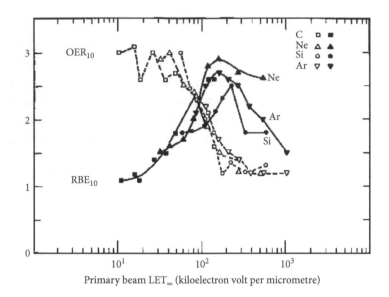

FIGURE 2.5   Plot of RBE and OER versus LET.

Recent investigations of head and neck squamous cell carcinoma (HNSCC) cell lines *in vitro* have uncovered one molecular explanation of the oxygen effect. Wozny et al. (2017) have demonstrated that there is a major role for HIF-1α through reactive oxygen species (ROS) production in the radioresistance of HNSCC and their related CSCs in response to photons under normoxia and hypoxia as well as carbon ions under hypoxia. Under hypoxia, HIF-1α is expressed earlier in CSCs compared to non-CSCs. The combined effect of hypoxia and photons enhances a synergistic and earlier HIF-1α expression in both CSCs and non-CSCs whereas the combined effect of hypoxia and carbon ions is not as great. Wozny et al. (2017) report no oxygen effect for carbon-ion exposures, and the ROS formed in the track are likely insufficient to stabilise HIF-1α but if HIF-1α is inhibited in hypoxic cells, they become radiosensitive to either gamma or carbon ions.

RADIOBIOLOGICAL *IN VIVO* DATA FOR CLINICAL TRIALS – IMMUNE AND LATE EFFECTS

Clinical radiotherapy protocols for cancer treatment by design are developed slowly and carefully to avoid adverse outcomes to the patient. An example would be the development of Phase I/II hypo-fractionated treatments for peripheral Stage I non-small cell lung carcinoma with carbon-ion radiotherapy (CIRT) by the Japanese National Institute of Radiological Sciences (NIRS) in which over the course of 20 years from 1994 to 2003 doses of 59.4–95.4 gray equivalents CIRT in 18 fractions over 6 weeks was systematically reduced to single doses of 28–50 gray equivalents with improved outcome in local control and overall survival (Kamada et al., 2015).

In the Western world, clinical trials must finally progress through Phase III trials fully randomised against the currently accepted gold standard treatment before being accepted, but in fact, several specific improvements in particle therapy have not been fully vetted against conventional radiotherapy. Phase III trials of proton radiotherapy for various tissue-specific sites have only recently, and in some instances, have never been completely randomised against intensity-modulated radiotherapy (IMRT), image-guided radiation therapy (IGRT), or stereotactic body radiation therapy (SBRT) due primarily to the concept that such comparisons so obviously favour the particle physics that they would be unethical (Suit et al., 2008).

THEORETICAL MODELLING OF PARTICLE EFFECTS

Recently, the quantity and quality of DNA damage upon a wide range of ion energies relevant to radiation therapy have been assessed systematically on a detailed mechanistic basis with simulations by Friedland et al. (2017) using the biophysical Monte Carlo code PARTRAC (PARticles TRACks), and detailed event-by-event track structure simulations that have been benchmarked against experimental data on interaction physics, radiation chemistry, biophysics and biochemistry of DNA-damage inductions. The complexity of DNA damage on the nanometre scale, and its clustering on the micrometre scale vary with increasing LET. Friedland et al. (2017) conclude that both the nanometre and micrometre scales are likely related to the induction of biological effects such as chromosome aberrations or cell killing. It would be informative to have this work extended to the evaluation of ion-induced tumorigenesis (Chang et al., 2016).

EMPHASIS ON ANATOMICAL AND FUNCTIONAL
TISSUE IMAGING IN TREATMENT PLANNING

Modern radiotherapy has improved significantly with advances in the resolution of four-dimensional (4D) computed tomography (CT) (De Ruysscher et al., 2015) in the capacity of dynamic and functional positron emission tomography (PET) (Grimes et al. 2017; Peet et al., 2012), and magnetic resonance imaging (MRI) (Kumar et al., 2016 and Parodi et al., 'Imaging and Particle Therapy: Current Status and Future Perspective' in this book).

Enhanced imaging technologies offer potential prevention of radiotherapy-induced dysfunction in survivors of ion-beam radiotherapies for paediatric brain tumours (Ajithkumar et al., 2017). With this, patient-specific tumour characteristics including metabolic states and regional oxygen tension can be incorporated into MRI and CTs to manage malignancies in radiotherapy in order to improve treatment planning by differentiating tumour volume from surrounding normal tissues that enables dose escalation and by sparing normal tissues as the tumour responds to initial treatment.

The availability of a commercial X-ray image-guided system for pre-clinical *in vivo* studies has proven the feasibility to investigate relevant radiobiological effects in the laboratory (Du et al., 2016; Ford et al., 2017; Wu et al., 2017). It is important to tailor the scale of radiation treatments to animal models for appropriate conclusions to be drawn.

RISK OF SECOND CANCERS

Of all cancers diagnosed each year in the United States, 6%–10% are second malignant neoplasms (SMNs). Currently, the risks associated with charged particle radiation-induced SMN in comparison to photon-induced risks are unknown. This is an important factor in defining the future of particle radiotherapy. Epidemiologic studies by themselves will not provide timely assessments of risks. There is a decades-long latency for radiation-induced solid tumours to develop whereas changes to treatment planning and therapy equipment are continuously evolving. Prior exposure to other carcinogens such as smoking complicates the estimation of radiation-induced SMN. Recent work indicates that genetic factors play a role in susceptibility to SMN, and other than prostate and cervical cancers where surgery alone can be an alternative treatment, there are no good control groups that can be used as a reference. Patients with childhood cancers in particular may be at increased risk due to their small body size, age at treatment and greater exposure to scattered radiation due to the small distance between treatment volume and nearby organs (Schneider et al., 2008). Several investigators have predicted that the risks of SMN from dose distributions from the primary fields for IMRT and proton therapy are comparable or lower after particle radiation therapy (Schneider et al., 2000; Taddei et al., 2010). Protons actually are reported to have a slightly reduced risk of secondary malignancies compared to photons in a seven-year follow-up study tracking more than 500 matched proton vs. photon patients (Chung et al., 2013).

Modeling SMN risks from animal studies is essential for future progress. The cumulative dose to the normal tissue can still be substantial despite the fact that the dose-distribution characteristics of particle radiation offers the advantage of more precise treatment planning protocols, minimising the dose to nontargeted tissues. High-charge, high-energy (HZE) ions have been reported to have high RBE values of up to 70 at low doses for some radiation-induced solid tumors in rodent models (e.g., Chang et al., 2016), but RBE values for hematologic malignancies appear to be low (e.g., Weil et al, 2009). The limitations to the available animal studies must however, be acknowledged (Barcellos-Hoff et al, 2015, Bielefeldt-Ohmann et al., 2012, Chang et al., 2016, Dicello et al., 2004; Weil et al., 2009). Most of these studies were designed to answer questions about cancer risks to spaceflight crewmembers from galactic cosmic radiation exposures. Relatively low doses were used compared to those used in radiotherapy, and the ions and energies examined reflect those in the galactic cosmic rays (GCR) spectrum with the exposures usually to the whole body. In fact, we have relatively sparse data on carcinogenesis dose-responses for radiation qualities and doses relevant to particle therapy as well as data on the effects of fractionation and partial body exposures relevant to particle therapy (Ando et al., 2005; Mohan et al., in press).

IMMUNE RESPONSES TO PARTICLE RADIATION

The demonstrated ability of radiation therapy to drive immunogenic modulation and promote immune-mediated killing of tumour cells in a variety of human carcinomas of distinct origin and genotype gives it broad clinical applicability for cancer therapy. There is a surge of interest in immunotherapy for cancer which offers an opportunity for the field of radiation oncology because mounting evidence suggests that radiation-induced cell death simultaneously contributes to an immunologically active process known as immunogenic cell death wherein apoptotic and necrotic dying tumour cells release a variety of tumour-associated antigens (TAAs) that can potentially be exploited to stimulate robust tumour-specific immune responses for effective disease control. Ionising radiation (RT) causes changes in the tumour microenvironment that can lead to intratumoural as well as distal immune modulation – the so-called abscopal phenomenon.

Several factors can influence the ability of radiation to enhance immunotherapy including the dose of radiation per fraction and the number of fractions as well as the volume of the irradiated tumour tissue and target location; however, the impact of these variables is not well understood and more research is needed to add to what little is known about the combined effects of immunotherapy and ion-beam therapy (Crittenden et al., 2015).

CONCLUSIONS AND THE NEED FOR MORE INTERNATIONAL COLLABORATIVE RESEARCH TO INTEGRATE RADIOBIOLOGICAL ADVANCES IN ION-BASED RADIOTHERAPIES

There have been several efforts to unite worldwide research on hadron therapy. One such example is the establishment of the European Network for Light Ion Hadron Therapy (ENLIGHT) that was launched in 2002 to catalyse efforts and collaboration in order to steer European research efforts in using ion beams for radiation therapy. ENLIGHT was envisaged not only as a common multi-disciplinary platform where participants could share knowledge and best practice, but also as a provider of training and education and as an instrument to lobby for funding in critical research and innovation areas. Over the years, the network has evolved, adapting its structure and goals to emerging scientific needs. It has been instrumental in catalysing collaborations between European centers to promote particle therapy particularly with carbon ions (Dosanjh et al., 2016). Due to the length of time needed to accumulate epidemiological data, a concerted multi-centre international effort should be established for long-term follow-up of charged particle radiotherapy patients. Animal studies to determine the carcinogenic efficacies of charged particles for radiation qualities and fractionated partial body exposures relevant to charged particle radiotherapy are needed. Animal studies and genetic epidemiological studies designed to address the open questions should be undertaken.

Ion-beam therapy with protons or carbon ions is recognised globally as offering a transformative new modality in cancer treatment for numerous specific tumour types and sites such as head and neck, pancreas, liver, lung, breast and prostate. For many complex reasons, the potential of this emerging cancer therapy has not been fully realised worldwide. Clinical evidence provided by treatment of over 150,000 patients by 2015 persuasively indicates overall optimism for particle therapy and in several cases even curative treatments with acceptable toxicities. Many of these patients were treated under Phase I/II trials or not on protocol at all during a period when the field was slowly changing during optimisation studies. The Western gold standard of level-one evidence provided by Phase III trials with protons or carbon ions randomised against conventional, state-of-the-art radiotherapy is still lacking.

Basic radiobiology provided an essential initial impetus for the implementation of ion therapy from 1970 to 1990, quantitating relative dose ratios for the determination of how

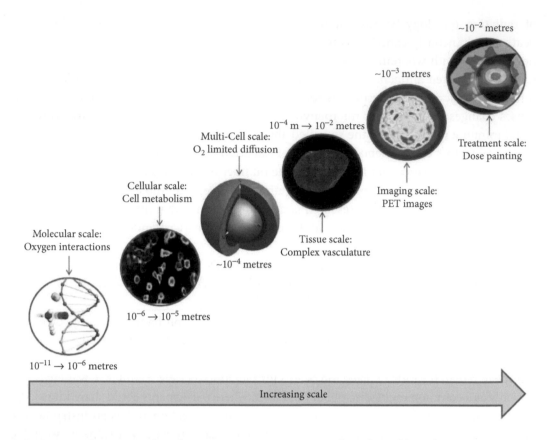

FIGURE 2.6   Oxygen-mediated treatment resistance as a multiscale problem. (From Grimes, D.R. et al., *Br. J. Radiol.*, 90, 20160939, 2017. With permission conveyed through Copyright Clearance Center.)

much more effective the ions were compared to photons to yield the same biological outcome for a large number of endpoints. Novel mechanisms of action were discovered that contributed to protocol designs. The literature in the field of carbon-ion radiotherapy significantly increased from only a few publications each year beginning in the early 1970s up to 100–173 papers per year since 2012. There is a multi-dimensional issue involved here in order to take quantitative information on radiation mechanisms at the nanometric or micrometric level and then scale the information to anatomical imaging of patients that is resolved at the millimetre level for treatment planning (Figure 2.6). There are, however, many aspects of ion-beam radiotherapy still to be explored in order to advance the field by fully integrating the unique physical and biological advances of hadrons into ion-therapy treatment planning and clinical trials.

ACKNOWLEDGEMENTS

Supported by NIH/NCI P20CA183640, the Lawrence Berkeley National Laboratory under Contract No. DE-AC02-05CH11231 with the U.S. Department of Energy, and Medical Applications, CERN. (EB) would like to acknowledge helpful discussions with Dr. Polly Chang.

REFERENCES

Ajithkumar, T., Price, S., Horan, G., Burke, A., and Jefferies, S. Prevention of radiotherapy-induced neurocognitive dysfunction in survivors of paediatric brain tumours: The potential role of modern imaging and radiotherapy techniques. *Lancet Oncol.* 18, no. 2 (2017): e91–e100.

Alpen, E.L., Powers-Risius, P., Curtis, S.B., and DeGuzman, R. Tumorigenic potential of high-Z, high-LET charged-particle radiations. *Radiat. Res.* 136, no. 3 (1993): 382–391.

Amaldi, U. and Larsson, B. Hadrontherapy in oncology. *Proceedings of the First International Symposium on Hadron Therapy*, Como, Italy, October 18–21, 1993. New York: Elsevier, 1994.

Ando, K., Koike, S., Oohira, C., Ogiu, T., and Yatagai, F. Tumor induction in mice locally irradiated with carbon ions: A retrospective analysis. J. *Radiat. Res.* 46, no. 2 (2005): 185–190.

Barcellos-Hoff, M.H, Blakely, E.A., Burma, S., Fornace, A.J.Jr., Gerson, S., Hlatky, L., Kirsch, D.G., Luderer, U., Shay, J., Wang, Y., and Weil, M.M., Concepts and challenges in cancer risk prediction for the space radiation environment, *Life Sciences in Space Res* 6, 92–103 (2015).

Bielefeldt-Ohmann, H., Genik, P.C., Fallgren, C.M., Ullrich, R.L., and Weil, M.M. Animal studies of charged particle-induced carcinogenesis. *Health Phys.* 103, no. 5 (2012): 568–576.

Blakely, E.A. New measurements for hadrontherapy and space radiation: Biology. *Phys. Med.* 17, no. Suppl 1 (2001): 50–58.

Blakely, E.A. and Chang, P.Y. Biology of charged particles. *Can. J.* 15, no. 4 (2009): 271–284.

Blakely, E.A. and Kronenberg, A. Heavy-ion radiobiology: New approaches to delineate mechanisms underlying enhanced biological effectiveness. *Radiat. Res.* 150, no. 5 Suppl (1998): S126–S145.

Blakely, E.A., Ngo, H.Q.H., Curtis, S.B., and Tobias, C.A. Heavy-ion radiobiology: Cellular studies. *Adv. Radiat. Biol.* 11 (1984): 295–389.

Brahme, A. Aspects on the development of radiation therapy and radiation biology since the early work of Rolf Wideroe. *Acta Oncol.* 37, no. 6 (1998): 593–602.

Brahme, A. *Biologically Optimized Radiation Therapy*. London, UK: World Scientific Publishing, 2014.

Chang, P.Y., Cucinotta, F.A., Bjornstad, K.A., Bakke, J., Rosen, C.J., Du, N., Fairchild, D.G., Cacao, E., and Blakely, E.A. Harderian gland tumorigenesis: Low-dose and LET response. *Radiat. Res.* 185, no. 5 (2016): 449–460.

Chung, C.S., Yock, T.I., Nelson, K., Xu, Y., Keating, N.L., and Tarbell, N.J. Incidence of second malignancies among patients treated with proton versus photon radiation. *Int. J. Radiat. Oncol. Biol. Phys.* 87, no. 1 (2013): 46 52.

Crittenden, M., Kohrt, H., Levy, R., Jones, J., Camphausen, K., Dicker, A., Demaria, S., and Formenti, S. Current clinical trials testing combinations of immunotherapy and radiation. *Semin. Radiat. Oncol.* 25, no. 1 (2015): 54–64.

De Ruysscher, D., Sterpin, E., Haustermans, K., and Depuydt, T. Tumour movement in proton therapy: Solutions and remaining questions: A review. *Cancers (Basel)* 7, no. 3 (2015): 1143–1153.

Dicello, J.F., Christian, A., Cucinotta, F.A., Gridley, D.S., Kathirithamby, R., Mann, J., Markham, A.R. et al. In vivo mammary tumourigenesis in the Sprague-Dawley rat and microdosimetric correlates. *Phys. Med. Biol.* 49, no.16 (2004): 3817–3830.

Dosanjh, M., Cirilli, M., Myers, S., and Navin, S. Medical applications at CERN and the ENLIGHT network. *Front Oncol.* 6 (2016): 9.

Du, S., Lockamy, V., Zhou, L., Xue, C., LeBlanc, J., Glenn, S., Shukla, G. et al. Stereotactic body radiation therapy delivery in a genetically engineered mouse model of lung cancer. *Int. J. Radiat.* 96, no. 3 (2016): 529–537.

Durante, M. and Cucinotta, F.A. Heavy ion carcinogenesis and human space exploration. *Nat. Rev. Cancer* 8, no. 6 (2008): 465–472.

Ford, E., Emery, R., Huff, D., Narayanan, M., Schwartz, J., Cao, N., Meyer, J. et al. An image-guided precision proton radiation platform for preclinical in vivo research. *Phys. Med. Biol.* 62, no. 1 (2017): 43–58.

Fossati, P., Molinelli, S., Matsufuji, N., Ciocca, M., Mirandola, A., Mairani, A., Mizoe, J. et al. Dose prescription in carbon ion radiotherapy: A planning study to compare NIRS and LEM approaches with a clinically-oriented strategy. *Phys. Med. Biol.* 57, no. 22 (2012): 7543–7554.

Friedland, W., Schmitt, E., Kundrat, P., Dingfelder, M., Baiocco, G., Barbieri, S., and Ottolenghi, A. Comprehensive track-structure based evaluation of DNA damage by light ions from radiotherapy-relevant energies down to stopping. *Sci. Rep.* 7 (2017): 45161.

Friedrich, T., Scholz, U., Elsaesser, T., Durante, M., and Scholz, M., Systematic analysis of RBE and related quantities using a database of cell survival experiments with ion beam irradiation, *J. Radiat. Res.* 54, 494–514 (2013).

Goodwin, E., Blakely, E., Ivery, G., and Tobias, C. Repair and misrepair of heavy-ion-induced chromosomal damage. *Adv. Space Res.* 9, no. 10 (1989): 83–89.

Goodwin, E.H. and Blakely, E.A. Heavy ion-induced chromosomal damage and repair. *Adv. Space Res.* 12, no. 2–3 (1992): 81–89.

Goodwin, E.H., Bailey, S.M., Chen, D.J., and Cornforth, M.N. The effect of track structure on cell inactivation and chromosome damage at a constant LET of 120 keV/micrometer. *Adv. Space Res.* 18, no. 1–2 (1996): 93–98.

Goodwin, E.H., Blakely, E.A. and Tobias, C.A. Chromosomal damage and repair in G1-phase Chinese hamster ovary cells exposed to charged-particle beams. *Radiat. Res.* 138, no. 3 (1994): 343–351.

Grimes, D.R., Warren, D.R., and Warren, S. Hypoxia imaging and radiotherapy: Bridging the resolution gap. *Br. J. Radiol.* 90 (1076) (2017): 20160939.

Harada, R., Kawamoto, T., Ueha, T., Minoda, M., Toda, M., Onishi, Y., Fukase, N. et al. Reoxygenation using a novel CO_2 therapy decreases the metastatic potential of osteosarcoma cells. *Exp. Cell Res.* 319, no. 13 (2013): 1988–1997.

Imaoka, T., Nishimura, M., Kakinuma, S., Hatano, Y., Ohmachi, Y., Yoshinaga, S., Kawano, A., Maekawa, A., and Shimada, Y. High relative biologic effectiveness of carbon ion radiation on induction of rat mammary carcinoma and its lack of H-ras and Tp53 mutations. *Int. J. Radiat. Oncol. Biol. Phys* 69, no. 1 (2007): 194–203.

Jacob, S.T., Muecke, W., Sajdel, E.M., and Munro, H.N. Evidence for extranucleolar control of RNA synthesis in the nucleolus. *Biochem. Biophys. Res. Commun.* 40, no. 2 (1970): 334–342.

Kamada, T., Tsujii, H., Blakely, E.A., Debus, J., De Neve, W., Durante, M., Jakel, O. et al. Carbon ion radiotherapy in Japan: An assessment of 20 years of clinical experience. *Lancet Oncol.* 16, no. 2 (2015): e93–e100.

Kumar, S., Liney, G., Rai, R., Holloway, L., Moses, D. and Vinod, S.K. Magnetic resonance imaging in lung: A review of its potential for radiotherapy. *Br. J. Radiol.* 89, no. 1060 (2016): 20150431.

Kuznetsova, M.A. and Sheval, E.V. Chromatin fibers: From classical descriptions to modern interpretation. *Cell Biol. Int.* 40, no. 11 (2016): 1140–1151.

Leith, J.T., Ainsworth, E.J., and Alpen, E.L. Heavy-ion radiobiology: Normal tissue studies. *Adv. Radiat. Biol.* 10 (1983): 191–236.

Linz, U. *Ion Beams in Tumor Therapy*. New York: Chap& Hall, 1995.

Linz, U. *Ion Beam Therapy: Fundamentals, Technology, Clinical Applications*. Heidelberg, Germany: Springer, 2012.

Mohan, R., Held, K., Story, M., Grosshans, D., Capala, J. "Critical Review: Proceedings of the National Cancer Institute Workshop on Charged Particle Radiobiology" in *International Journal of Radiation Oncology, Biology and Physics* (accepted for publication Dec 11, 2017—In Press).

Natale, F., Rapp, A., Yu, W., Maiser, A., Harz, H., Scholl, A., Grulich, S. et al. Identification of the elementary structural units of the DNA damage response. *Nat. Commun.* 8 (2017): 15760.

Ogata, T., Teshima, T., Inaoka, M., Minami, K., Tsuchiya, T., Isono, M., Furusawa, Y., and Matsuura, N. Carbon ion irradiation suppresses metastatic potential of human non-small cell lung cancer A549 cells through the phosphatidylinositol-3-kinase/Akt signaling pathway. *J. Radiat. Res.* 52, no. 3 (2011): 374–379.

Ou, H.D., Phan, S., Deerinck, T.J., Thor, A., Ellisman, M.H., and O'Shea, C.C. ChromEMT: Visualizing 3D chromatin structure and compaction in interphase and mitotic cells. *Science* 357, no. 6349 (2017): eaag0025.

Peet, A.C., Arvanitis, T.N., Leach, M.O., and Waldman, A.D. Functional imaging in adult and paediatric brain tumours. *Nat. Rev. Clin. Oncol.* 9, no. 12 (2012): 700–711.

Puck, T.T. and Marcus, P.I. Action of x-rays on mammalian cells. *J. Exp. Med.* 103, no. 5 (1956): 653–666.

Raju, M.R. *Heavy Particle Radiotherapy*. San Francisco, CA: Academic Press, 1980.

Raju, M.R. Proton radiobiology, radiosurgery and radiotherapy. *Int. J. Radiat. Biol.* 67, no. 3 (1995): 237–259.

Roots, R., Holley, W., Chatterjee, A., Irizarry, M. and Kraft, G. The formation of strand breaks in DNA after high-LET irradiation: A comparison of data from in vitro and cellular systems. *Int. J. Radiat. Biol.* 58, no. 1 (1990): 55–69.

Roots, R., Holley, W., Chatterjee, A., Rachal, E. and Kraft, G. The influence of radiation quality on the formation of DNA breaks. *Adv. Space Res.* 9, no. 10 (1989): 45–55.

Rydberg, B. Clusters of DNA damage induced by ionizing radiation: Formation of short DNA fragments. II. Experimental detection. *Radiat. Res.* 145, no. 2 (1996): 200–209.

Schneider, U., Lomax, A., and Lombriser, N. Comparative risk assessment of secondary cancer incidence after treatment of Hodgkin's disease with photon and proton radiation. *Radiat. Res.* 154, no. 4 (2000): 382–388.

Schneider, U., Lomax, A., and Timmermann, B. Second cancers in children treated with modern radiotherapy techniques. *Radiother. Oncol.* 89, no. 2 (2008): 135–140.

Skarsgard, L.D. *Pion and Heavy Ion Radiotherapy: Pre-clinical and Clinical Studies*. New York: Elsevier Biomedical, 1983.

Skarsgard, L.D. Radiobiology with heavy charged particles: A historical review. *Phys. Med.* 14, no. Suppl 1 (1998): 1–19.

Subtil, F.S., Wilhelm, J., Bill, V., Westholt, N., Rudolph, S., Fischer, J., Scheel, S. et al. Carbon ion radiotherapy of human lung cancer attenuates HIF-1 signaling and acts with considerably enhanced therapeutic efficiency. *FASEB J.* 28, no. 3 (2014): 1412–1421.

Suit, H., Kooy, H., Trofimov, A., Farr, J., Munzenrider, J., DeLaney, T., Loeffler, J., Clasie, B., Safai, S., and Paganetti, H. Should positive phase III clinical trial data be required before proton beam therapy is more widely adopted? No. *Radiother. Oncol.* 86, no. 2 (2008): 148–153.

Taddei, P.J., Howell, R.M., Krishnan, S., Scarboro, S.B., Mirkovic, D., and Newhauser, W.D. Risk of second malignant neoplasm following proton versus intensity-modulated photon radiotherapies for hepatocellular carcinoma. *Phys. Med. Biol.* 55, no. 23 (2010): 7055–7065.

Tobias, C. and Tobias, I. *People and Particles*. San Francisco, CA: San Francisco Press, 1997.

Tsujii, H. *Carbon-Ion Radiotherapy*. Tokyo, Japan: Springer, 2014.

Uhlen, M., Zhang, C., Lee, S., Sjostedt, E., Fagerberg, L., Bidkhori, G., Benfeitas, R. et al. A pathology atlas of the human cancer transcriptome. *Science* 357, no. 6352 (2017): eaan2507.

Weil, M.M., Bedford, J.S., Bielefeldt-Ohmann, H., Ray, F.A., Genik, P.C., Ehrhart, E.J., Fallgren, C.M. et al. Incidence of acute myeloid leukemia and hepatocellular carcinoma in mice irradiated with 1 GeV/nucleon (56)Fe ions. *Radiat. Res.* 172, no. 2 (2009): 213–219.

Winter, M., Dokic, I., Schlegel, J., Warnken, U., Debus, J., Abdollahi, A., and Schnolzer, M. Deciphering the acute cellular phosphoproteome response to irradiation with X-rays, protons and carbon ions. *Mol. Cell Proteomics* 16, no. 5 (2017): 855–872.

Wozny, A.S., Lauret, A., Battiston-Montagne, P., Guy, J.B., Beuve, M., Cunha, M., Saintigny, Y. et al. Differential pattern of HIF-1alpha expression in HNSCC cancer stem cells after carbon ion or photon irradiation: One molecular explanation of the oxygen effect. *Br. J. Cancer* 116, no. 10 (2017): 1340–1349.

Wu, C.C., Chaudhary, K.R., Na, Y.H., Welch, D., Black, P.J., Sonabend, A.M., Canoll, P., Saenger, Y.M. et al. Quality assessment of stereotactic radiosurgery of a melanoma brain metastases model using a mouselike phantom and the small animal radiation research platform. *Int. J. Radiat. Oncol. Biol. Phys.* 99, no. 1 (2017): 191–201.

CHAPTER 3

Challenges Surrounding Relative Biological Effectiveness for Particle Therapy

Where Do We Go from Here?

Bleddyn Jones, David R. Grosshans and Radhe Mohan

CONTENTS

INTRODUCTION

Particle therapy delivered either with protons, carbon, or other ions has unique biological effects when compared to photon therapy. To optimise the clinical outcomes of particle therapy, it is necessary to understand the biological characteristics of particle beams (PB) including how these relate to physical factors and introduce this knowledge into the treatment planning process. Currently, protons are believed to have biological effects relatively like photons while carbon-ion treatments incorporate more detailed biological information within the medical dose prescription.

Because the patients treated with photons far outnumber those treated with particle therapy, attempts are made to relate the biologic effectiveness of PB to standard photon references. The enhanced biological effects produced by PB are quantified by their relative biological effectiveness (RBE) which defines the dose modification necessary to achieve the same biological endpoint. RBE is defined as the dose of the reference radiation divided by the particle dose required to achieve the same bioeffect. Many factors influence RBE such as the dose used; the local ionisation density or clustering which is quantified by the linear energy transfer (LET); the quality of the reference radiation; the mixture of Bragg curves; the position and depth within an arrangement of beams; and also the radiobiological characteristics of the cell, tissue, or tumour. This multitude of factors introduces uncertainties into the appropriate values to be used for clinical treatment planning.

What do we know about RBE? Many experiments in different laboratories over the past 70 years have provided some useful general conclusions. RBE is a complex function of LET and dose. RBE increases with LET until a maximum or ultimate value (RBE_U) is reached at higher LETs after which it decreases possibly due to dose wasting where ionisation clustering is excessive. The maximum value of RBE does not appear to vary much with the ion species (defined by its atomic number ['Z']), but there is a progressive decrease in RBE with increasing dose. Each ion species appears to have its own turnover point at a certain LET (LET_U). The LET_U value increases with Z, so protons with a $Z = 1$ show increments in RBE at lower LET values than each further ion, although the displacements of LET_U become less with increasing Z. For example, protons reach their maximum RBE when LET_U is around 30 keV.μm^{-1} whereas helium ($Z = 2$) and carbon ions ($Z = 6$) have LET_U values of around 110 and 200 keV.μm^{-1}, respectively.

Another important feature is that the magnitude of the RBE at any dose is dependent on the radiobiological characteristics of the biosystem. In general, biosystems which are very radiosensitive and have high α/β ratios show small RBE values close to unity compared with the high RBE values of 3–8 demonstrated by radioresistant systems with small α/β ratios [1–3]. The low LET α/β ratios obtained using the reference radiation reflect the repair capacity and intrinsic radiosensitivity of a biosystem with marked inverse influences on its dose-fraction sensitivity; these will influence the RBE by providing the numerator dose within the RBE definition.

As the PB energy falls with the depth of beam penetration, the interactions with matter increase and are more localised, thereby resulting in an increase in LET. In this way, LET and RBE values are highest around the distal ends of a beam and especially at the distal fall-off of the Bragg curves.

ACUTE- AND LATE-REACTING TISSUES

In radiobiology, a considerable distinction is made between acute- and late-reacting tissues since their α/β ratios differ markedly. Late-reacting tissues with the lowest α/β ratios show greater dose per fraction sensitivity with standard megavoltage treatments; the dependence on dose per fraction is reduced at higher LET values. Such tissues will show a greater change in the numerator of the RBE definition when dose is varied. Acute-reacting tissues show little dose per fraction sensitivity and therefore have a lower RBE at low dose. It is the damage to the late-reacting tissues that dominate the quality of life in surviving patients

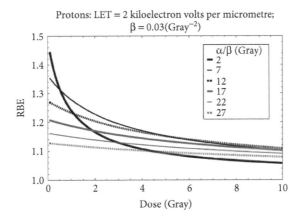

Protons: LET = 2 kiloelectron volts per micrometre;
$\beta = 0.03(\text{Gray}^{-2})$

FIGURE 3.1   Modelled relationship between RBE and dose per fraction for protons at LET values of 2 keV.μm^{-1} which reflects the upper range limit normally found in mid-SOBP used in the clinic. Much higher LET values will be found in distal ends of SOBP. (From Jones, B., *Acta Oncol.*, 56, 1374–1378, 2017.)

following radiotherapy of any type. The distinction between late- and early-reacting tissues with different α/β ratios is critical within particle therapy especially since at low dose, $\text{RBE}_{\text{late}} > \text{RBE}_{\text{acute}}$, but at high dose, it is predicted that $\text{RBE}_{\text{late}} < \text{RBE}_{\text{acute}}$ as can be seen in Figure 3.1 where the typical acute α/β (>7 Gray) and late-reacting α/β of 2 Gray – typical of the central nervous system tissues – curves cross over. It must be appreciated that some radiosensitive tumors especially in children have very high α/β values with lower RBEs observed. For lower LET values of 1–1.5 keV.μm^{-1} within spread-out Bragg peaks (SOBPs), the RBE values are probably below 1.1 [4].

ION-BEAM RELATIVE BIOLOGICAL EFFECTIVENESS MODELS

In Japan, considerable progress was made in developing RBE predictive models based on fast neutron RBE data which closely approximated those obtained for carbon ions at the low end of carbon LET values. In the case of carbon ions, an 'inverted' dose profile was used for each single field (composed of multiple Bragg peaks). It was designed to compensate for the inevitable increase in LET along its path [5]. Deeper regions where the LET is highest receive a reduced dose, but shallower regions are allowed a higher dose because of the lower LET; then a constant RBE multiplier is used to generate an assumed plateau of equal bioeffectiveness over the target region. This RBE was generated from relatively low-dose experiments (using mostly fast neutrons) and could not be easily changed for higher doses per fraction without incorporating a flexible RBE model. More recently, the microdosimetric kinetic model (MKM) based on original insights by Hawkins in the United States has been used in Japan [6]. This compensates for dose per fraction but assumes the RBE is a complex inverse function of the low LET α/β. In contrast, the German carbon-ion beam projects at *Gesellschaft für Schwerionenforschung* (GSI). GSI is situated in Darmstadt, Germany, and subsequently at Heidelberg, have gathered experience in the use of the local effect model (LEM) which continues to be developed in stages [7]. It incorporates microdosimetric theory extrapolated to cell survival curves (CSCs). Again, the dose profile is controlled by the

LET profile, and there has been far less variation in dose per fraction compared with Japan. Comparisons of treatment outcomes will be the most effective way of deciding currently which is the best approach.

Such refinements have not been used for proton beams where a fixed RBE of 1.1 is assumed and a uniform physical dose rather than bioeffect profile is delivered. It remains to be seen if more detailed LET mapping with appropriate dose weighting based on predictive RBE models will be used. This may be indicated where there is sensitive normal tissue in the close vicinity of a tumour target and becomes an urgent consideration (especially when single or few field directions are used) if important functional normal tissue is exposed to distal beam regions. At the very least, before such a sophisticated approach is developed and employed, it may be expedient to assume simpler generic changes in tissue RBE values which are protective. It is also noteworthy that the inclusion of variable RBE models (or LET-based models) in the criteria for optimisation of intensity-modulated proton therapy has been shown to lead to proton plans with lower LET in normal tissues and higher LET in the target for the same physical dose distribution as shown by Mohan et al. [8].

PROTON MODEL COMPARISONS

Numerous models to predict the RBE of proton beams currently exist. Detailed work on the comparisons between nine different 'phenomenological' models of proton RBE prediction is being pursued by Rorvik for a PhD thesis at the University of Bergen, Norway. The various published models are based on different assumptions and utilise different amounts of biological data, so some will only be appropriate for a limited range of α/β biosystems. Whereas some models link the RBE increments to only the α radiosensitivity parameter, others use an inverse relationship between α/β and RBE while independent changes in α and β may offer the most realistic option. In a dose-planning study, the range of predicted optic nerve dose assessments by the various models covered a range of 56–72 Gray while the standard RBE = 1.1 dose was 48 Gray. Full publication of this important work is awaited with interest. Additionally, like carbon-ion therapy, it might be advantageous in clinical practice if more than one model could be used, provided these are sufficiently comprehensive and incorporate full parameter input ranges rather than a limited set. Clinical decisions regarding RBE tissue allocations could perhaps be allowed to proceed if at least two or more such models are in reasonable agreement.

USE OF THE PRODUCT OF DOSE AND LET AS A SURROGATE FOR PROTON RELATIVE BIOLOGICAL EFFECTIVENESS

Some investigators are enthusiastic about applying the product of dose and LET as an indicator of higher risk of radiation damage. This appears to be superficially attractive but since dose and LET are inversely and directly related to RBE, respectively, then it is inevitable that the dose × LET product could be misleading. Evidence for this is presented in Jones (2017) [4]. Alternatively, since measurements and most current proton models indicate that the RBE is essentially a linear function of LET up to the Bragg peak for monoenergetic protons, Mohan et al. [8] have suggested one could assume RBE to be represented by (A + B × LET).

CAN A SINGLE MODEL BE USED FOR ALL ION BEAMS INCLUDING PROTONS?

The insights provided by Katz [9] showed how each ion beam would have its own radius of effect, being dependent on Z/(v/c) where 'v' is particle velocity and 'c' the speed of light, which could provide a basis for comparative radiobiological studies using different ion sources. The Katz, MKM and LEM models can also be used to predict RBE, but are complex and at some stages involve formidable mathematics. A simpler approach has recently been proposed by Jones [10,11] where the unique LET_U position for each ion type is obtained from its Z value (Figure 3.2); furthermore, simple saturation relationships between low LET α and β and their ultimate values α_U and β_U at the LET_U are then scaled with respect to energy efficiency, and also with dose by use of the LQ model to provide the RBE at any value of LET. Such a simple approach appears to give reasonable fits to various ion-beam data sets including protons and is easy to apply with minimal computer coding.

The finding that LET-RBE turnover points seem to exist at unique LET values for different ions needs further rigorous work, since such phenomena are contrary to traditional approaches where it was thought that only LET determined RBE and that all ions could be 'lumped together' as suggested in modern radiobiology textbooks such as Hall and Giaccia [12], where the common turnover point (for all ions) is suggested to be around 100–120 keV.μm^{-1}. Close inspection of multiple ion-beam data sets such as in the work of Sorensen et al. (Figures 3.2 and 3.3) show that the positions of the turnover points shift to higher values of LET as Z increases [13]. A separate statistical study by the first of the present authors using carbon [1] and helium ion data [14] has shown that these LET_U values are significantly different (Table 3.1).

It remains to be seen if an international laboratory such as CERN [16], Brookhaven, Groningen, or elsewhere will generate more useful data on these important relationships

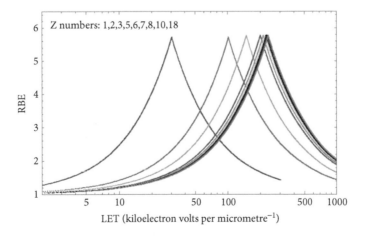

FIGURE 3.2 Graphical display of RBE and LET each with unique turnover point positions for multi-ion data of Todd, assuming $\alpha = 0.14$ Gy^{-1} and $\beta = 0.05$ Gy^{-2} for a dose of 1.5 Gray. From left to right, the ionic elements are shown as follows with colour code and LET_U (rounded to nearest integer for values over 100) in parentheses: Deuterium (red, 30.5); Helium (brown, 103); Lithium (pink, 150); Boron (blue, 200); Carbon (orange, 213); Nitrogen (green, 221); Oxygen (Black, 227); Neon (purple, 232); and Argon (grey, 237). The respective Z numbers are indicated on the graphic.

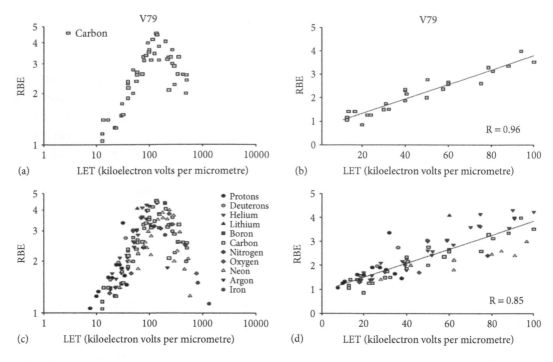

FIGURE 3.3 Graphs—(a–d) as shown in *Acta Oncol.* by Sorensen, Overgaard, and Bassler (2011). For these V–79 cells, there is a clear linear relationship between LET and RBE. Graphic (c) includes multiple ions and superficially gives the impression of an overall relationship, although close inspection shows turnover points for each ion species occur at a higher value of LET with increasing Z.

TABLE 3.1 The Locations of the Combined C Ion and Helium Data are Significantly Different (Mann–Whitney $p = 0.028$, t-test $p < 0.0001$)

Ion and Data Source	Cell Type	Estimated LET$_U$ (keV·µm^{-1}) (mean, standard error)
C ions (Weyrather et al., GSI, Darmstadt, Germany)	CHO	145.81 ± 9.88
	V–79	159.05 ± 3.95
	Combined CHO + V–79 data	152.43 ± 4.29
Helium (Barendsen, the Netherlands)	Human T cells	124.24 ± 0.56

Source: Chapter 12 (page 12–12), Additional Considerations and Conclusions in: *Practical Radiobiology in Proton Therapy Planning*. Institute of Physics Publishing (Bristol and Philadelphia). ISBN:978-0-7503-1338-4 (and 1339-1).

since our present data sets are fragmented in time and place and contain different cell systems. Standardisation is necessary not only in the choice of a wide range of representative biosystems *in vitro* but also with highly selective *in vivo* testing.

CLINICAL ASPECTS

Of the patients treated with particle therapy, the majority have been treated with protons and hence there are more clinical data available. Even so, the publication of clinical results has been limited compared with technical improvement reports in the literature, but there appears to be some evidence that some patients may suffer unexpected adverse outcomes following particle therapy. This may be for a variety of reasons not only including inaccuracy of dose placement (according to the precise Bragg peak positions) and dosimetry, but also due to an incorrect RBE. At the present time, knowledge of particle physics is certainly more comprehensive than particle radiobiology. Even so, the physical influences of the electron density heterogeneities within the human body, the vulnerability of particle dose distributions to inter- and intra-fractional anatomical changes, and the behaviour of low energy particles at the end of range need to be better understood. Then, in terms of the biophysical interactions, it is likely that RBE ranges are much higher than previously anticipated in the case of protons, and should in some way be used to provide better normal tissue constraints. The standard practice of a constant 1.1 proton RBE in all tissue types rather than a variable LET-dependent RBE value as in Figure 3.1 has been criticised [17] and appears to be gaining acceptance with time by an increasing number of researchers and practitioners. Another important argument is that if there is significant dose displacement beyond the planned target volumes, then clinicians could observe both reduced tumour control and increased toxicity to a greater extent than either of these clinical outcomes alone. This does not appear to be the case which suggests that RBE factors may be the main cause of 'unexpected' toxicity.

A greater use of randomised studies to test new ideas against conventional practice is also indicated. For example, the 1.1 RBE is probably incorrect for application in all tissues, tumours and dose per fraction. This will need to be tested against approaches where tissue dose tolerances are changed because of local LET and dose (higher and lower RBEs). If clinical policies are to be changed in the expectation of an improved safety and efficacy, then a randomised trial would be the best way of providing the most convincing evidence.

With existing data sets, observational data is extremely important. Studies of rare complications – brainstem, optic chiasm, brain and spinal necrosis – need to be fully investigated for dose and LET anomalies by estimations of the risk of the complication using existing RBE models [18,19].

Tumour control monitoring is also mandatory especially when the tumour types are known to be highly radiosensitive to standard megavoltage radiotherapy where the RBE may be very close to unity. Even the use of a proton RBE of 1.1 could therefore lead to under-dosage in this category which includes lymphomas and many childhood tumors such as Wilm's tumour, medulloblastoma, rhabdomyosarcoma and ependymoma. It has been suggested that a tumour RBE should not be used because of this risk, although in such patients where life expectancy can be very good, a higher RBE than 1.1 might be applied to protect critical normal tissues [20]. Still to date, published studies suggest comparable rates of tumour control for protons when compared to historic photon data.

CHANGES IN FRACTIONATION

The economic cost of radiotherapy is an increasing issue despite a considerable decrease in the number of treatment fractions made with stereotactic-guided forms of precision megavoltage photon therapy. Although progress has been made with hypofractionation with carbon ions in Japan, with many reports of reduced toxicity commensurate with the reduction in normal tissue RBE at high doses including in the lung and prostate [21], no such progress has been made with proton therapy. Further research to investigate the degree to which the RBE is reduced in late-reacting tissues with hypofractionation may lead to further improvements of the therapeutic ratio as long as an adequate tumour dose can be delivered, often with the advantage of a shorter overall treatment time in the case of more rapidly growing tumours.

SUMMARY AND CONCLUSIONS

Large gaps in our knowledge of RBE remain although many key principles have been identified. The existing models could be improved in terms of their precision by further experimental work and clinical analysis.

Biological effects, and therefore clinical consequences for the same physical dose, depend on the particle species (photons, protons or heavier ions), LET, dose rate, dose per fraction, the duration over which the course of radiotherapy is delivered and other factors. These dependencies occur because of the differences in ionisation densities produced, or the track structure, of each type of particle and its energy. The clinical response also depends on the dose distribution patterns within a patient, that is, volumes of tissues irradiated to different dose levels.

The rationale for defining RBE as the ratio of dose with a reference radiation (typically photons of energy used in clinical practice) to the dose to achieve the same biological effect with particles enables extrapolation from our vast clinical experience with photon therapy; however, biological effects of particles are highly complex functions of many parameters which include the ones mentioned above and the biological endpoint of interest. RBE thus cannot be adequately represented by a simple ratio of photon and particle dose so often expressed as a constant or by very simplistic models. It may be argued that the very concept of RBE is questionable, since biologically, protons and heavier ions are very different from photons. It may be worth considering dropping the 'R' of RBE to study and apply the biological effects of ions independently of photons, but even then there will be relative differences between each Z-specific ion as discussed above.

This may be desirable in the long run, but for now for reasons of simplicity and practicality as well as the existing evidence base, it is necessary to retain the RBE concept by obtaining its values empirically, modelling it and applying it for clinical applications. Progress requires not only new data but also analysis of past high LET – RBE data sets using fast neutrons, protons and a range of ions. Both clinical and laboratory data are fragmented. For biologic studies – not discussed in extensive detail here – many are uncoordinated and contain uncontrolled variables such as the definition of LET used and detailed data on the experimental conditions used. For clinical studies, particularly those using protons, assessments of LET effect which require Monte Carlo calculations are uncommon. This in addition to other variables such as patient heterogeneity, lack of follow-up and so on makes drawing reliable conclusions challenging.

While the challenges are daunting, there still exists a great potential to improve the therapeutic index of particle therapy through a greater understanding of RBE. For protons, if it is believed that RBE increases near the distal regions of the beam, this knowledge can be translated into the clinic. For example, by cautiously incorporating the most realistic biologic effect models into intensity-modulated proton therapy and deliberately placing the most elevated LET regions preferentially within target volumes and away from normal critical tissues, improvements in the therapeutic outcomes are likely to be obtained. This could contribute to improved tumour control and reduce the risk of normal tissue toxicity. It should be noted that such treatments would be best offered as part of prospective clinical trials in which careful assessments of tumour control, tumour and normal tissue response and so on are undertaken. For heavy ions including carbon, the existing biologic effect models may vary greatly in their predictions. This complicates the interpretation of clinical outcome data between institutions in Europe and Asia. If improved laboratory studies can potentially be informative as to the appropriateness of each model, these should be pursued. Similarly with protons, careful evaluation of outcome data in the setting of prospective trials – perhaps incorporating different RBE allocations than 1.1 – should inform future treatment protocols.

REFERENCES

1. Weyrather, W.K., S. Ritter, M. Scholz, and G. Kraft. RBE for carbon track-segment irradiation in cell lines of differing repair capacity. *Int. J. Radiat. Biol.* 75(11) (1999): 1357–1364.
2. Carabe-Fernandez, A., R.G. Dale, and B. Jones. The incorporation of the concept of minimum RBE (RBE_{min}) into the linear-quadratic model and the potential for improved radiobiological analysis of high-LET treatments. *Int. J. Radiat. Biol.* 83 (2007): 27–39.
3. Carabe-Fernandez, A., R.G. Dale, J.W. Hopewell, B. Jones, and H. Paganetti. Fractionation effects in particle radiotherapy: Implications for hypo-fractionation regimes. *Phys. Med. Biol.* 55(19) (2010): 5685–5700.
4. Jones, B. Clinical radiobiology of proton therapy: Modeling of RBE. *Acta Oncologica* 56 (2017): 1374–1378. doi:10.1080/0284186X.2017.1343496.
5. Kanai, T., Y. Furusawa, K. Fukutsu, H. Itsukaichi, K. Eguchi-Kasai, and H. Ohara. Irradiation of mixed beam and design of spread-out Bragg peak for heavy-ion radiotherapy. *Radiat. Res.* 147 (1997): 78–85.
6. Inaniwa, T., M. Suzuki, T. Furukawa, Y. Kase, N. Kanematsu, T. Shirai, and R.B. Hawkins. Effects of dose-delivery time structure on biological effectiveness for therapeutic carbon-ion beams evaluated with microdosimetric kinetic model. *Radiat. Res.* 180(1) (2013): 44–59.
7. Friedrich, T., U. Scholz, T. Elsässer, M. Durante, and M. Scholz. Systematic analysis of RBE and related quantities using a database of cell survival experiments with ion beam irradiation. *J. Radiat. Res.* 54(3) (2013): 494–514.
8. Mohan, R., C.R. Peeler, F. Guan, L. Bronk, W. Cao, and D.R. Grosshans. Radiobiological issues in proton therapy. *Acta Oncol.* 56(11) (2017): 1367–1373.
9. Katz, R., B. Ackerson, M. Homoyooufar, and S.C. Sharma. Inactivation of cells by heavy ion bombardment. *Radiat. Res.* 47 (1971): 402–425.
10. Jones, B. A simpler energy transfer efficiency model to predict relative biological effect (RBE) for protons and heavier ions. *Front. Oncol.* 5 (2016): 184. doi:10.3389/fonc.2015.00184. Erratum in: *Front. Oncol.* 6: 32.
11. Jones, B. Towards achieving the full clinical potential of proton therapy by inclusion of LET and RBE models. *Cancers* (Basel) 7(1) (2015): 460–480.
12. Hall, E.J. and A. Giaccia. *Radiobiology for the Radiologist*. 7th Ed. Philadelphia, PA: Lippincott, Williams, and Wilkins, 2012, p. 106.
13. Sørensen, B.S., J. Overgaard, and N. Bassler. In vitro RBE-LET dependence for multiple particle types. *Acta Oncol.* 50(6) (2011): 757–762.

14. Barendsen, G.W. Responses of cultured cells, tumours and normal tissues to radiations of different linear energy transfer. *Curr. Topics Radiat. Res. Q.* 4 (1968): 293–356.

15. Jones, B. *Practical Radiobiology for Proton Therapy Planning.* Bristol: IOP Publishing, 2017.

16. Dosanjh, M., B. Jones, and S. Myers. A possible biomedical facility at CERN. *Brit. J. Radiol.* 86 (2013): 1025–1029.

17. Jones, B. Why RBE must be a variable and not a constant in proton therapy. *Brit. J. Radiol.* 89(1063) (2016): 20160116. doi:10.1259/bjr.20160116.

18. Peeler, C.R., D. Mirkovic, U. Titt, P. Blanchard, J.R. Gunther, A. Mahajan, R. Mohan, and D.R. Grosshans. Clinical evidence of variable proton biological effectiveness in pediatric patients treated for ependymoma. *Radiother. Oncol.* 121 (2016): 395–401.

19. Gunther, J.R, M. Sato, M. Chintagumpala, L. Ketonen, J.Y. Jones, P.K. Allen, A.C. Paulino, et al. Imaging changes in pediatric intracranial ependymoma patients treated with proton beam radiation therapy compared to intensity modulated radiation therapy. *Int. J. Radiat. Oncol. Biol. Phys.* 93 (2015): 54–63.

20. Jones, B. Editorial: Patterns of failure after proton therapy in medulloblastoma. *Int. J. Radiat. Oncol. Biol. Phys.* 90(1) (2014): 25–26.

21. Tsujii, H., T. Kamada, T. Shira, K. Noda, H. Tsuji, and K. Karasawa (Eds.). *Carbon-Ion Radiotherapy: Principles, Practices, and Treatment Planning.* Tokyo, Japan: Springer, 2014.

CHAPTER 4

Advances in Beam Delivery Techniques and Accelerators in Particle Therapy

Marco Schippers

CONTENTS

INTRODUCTION

Since cyclotrons and synchrotrons have been the first type of machines to accelerate particles to several megaelectron volts, these were also the first to treat patients with cancer by means of a beam of energetically charged ions. The first particle therapy facility in a hospital [1] used a specially developed synchrotron to accelerate protons. Since the late 1990s, commercial companies slowly became interested in producing the equipment. This has increased the number of facilities to more than 50 in 2015 [2].

Today all particle treatments are done with beams of protons or carbon ions. Use of other ions is only performed in dedicated research programs at a few places in the world. The goal of particle therapy is to exploit the possibility of depositing a high dose in the *target* volume (usually the tumour) with a minimum dose in the surrounding healthy tissue. To achieve this goal, use is made of the characteristic shape of the dose profile deposited by the particles as a function of depth: a plateau followed by a peak (the so-called *Bragg peak*) and a drop to zero just before the end of the range. The irradiation is performed by 'filling' the target volume with Bragg peak doses. The particle energy determines the depth at which the Bragg peak dose will be deposited in the patient's body. The clinically useful maximum beam energy for proton therapy is 200–250 megaelectron volts. Protons of these energies have a range of 25–35 centimetres in water which is the typical reference material in radiation therapy. For carbon ions, the maximum energy used is 400–450 megaelectron volts per nucleon. The lowest particle energies which are down to a few megaelectron volts are needed for treatment of very superficially located tissues. In a patient treatment, this energy is usually reached by putting some material into the beam just before the patient.

The choice of the accelerator depends on many parameters. Several techniques are applied in the use of particle beams for treatments; some implications of the beam characteristics are discussed here. First, the major characteristics and recent developments in the fields of cyclotrons and synchrotrons will be discussed, followed by a description of the most important beam handling operations in the beam transport system – to prepare the beam extracted from the accelerator for therapy. It will be shown how the different types of beam operation depend on the type of accelerator. This will be followed by an overview of the beam characteristics that are relevant for the application of therapy.

ACCELERATOR TYPES CURRENTLY IN USE FOR PARTICLE THERAPY

The cyclotron and synchrotron have remained the two types of accelerators currently utilised in particle therapy. Several types of these machines are in use [2]. The eight facilities employing carbon or other ions are using a synchrotron. For protons, either synchrotrons or cyclotrons are employed; a few synchrocyclotrons are also utilised. Although other types of accelerators do exist or are being developed, these are not yet suitable or used for hadron therapy [3–6]; however, recently there is serious interest in a linac developed at CERN [7]. It is very promising that this interesting linac technology has left the laboratory phase and is entering the clinical world.

Cyclotron and Synchrocyclotron

Cyclotrons for proton therapy are single-magnet machines with a typical diameter of five metres and a weight of 200 tons. In between the magnet poles, an oscillating (RF) electric field is constructed between D-shaped electrodes. In the RF field between the Dees, the protons are accelerated in a plane between the magnet poles. Due to the magnetic field and their increasing energy from the RF field, they follow a spirally shaped orbit of a few hundred turns to the outer radius of the magnet from where they are extracted and sent into the beam line. The fundamental property of a cyclotron is that although particles with higher energy are circulating at a larger radius in the cyclotron, all particles have the same revolution frequency; therefore, these original cyclotrons are called *cyclotrons*.

In cyclotrons, the beam extraction efficiency is an important parameter. Apart from radiation damage due to lost beam, it is also important to prevent unnecessary activation of the cyclotron. This will cause a high dose exposure to personnel during servicing of the cyclotron. Beam extraction efficiencies of 70%–80% have been achieved by the ACCEL/ Varian cyclotron [8]. In this 250 megaelectron volts cyclotron, the beam orbit at 0.8 metre extraction radius is very small but due to dedicated local distortions in the 3T (Tesla) magnetic field, the orbit separation is increased by exciting a radial betatron resonance.

It can be advantageous to reduce the size of a cyclotron, but to do this, a stronger magnetic field is needed. This is only possible by using a SC magnet. The first SC (superconducting) proton cyclotron for therapy [9,10] has a 3.5 metres diameter and a weight of 100 tons. Several of these cyclotrons are in operation or under construction; many facilities have indicated serious interest in this SC cyclotron as well. Most developments in this type of cyclotron have been driven by commercial companies. A design and serious plans have been proposed at IBA for a SC cyclotron for carbon ions and protons. At Mevion and IBA, much development has been done to reduce the size of proton cyclotrons. By increasing the magnetic field by a factor of 2–4, cyclotrons of only 30–50 tons tons and a 1–2 metres diameter have been created [11–13]. In the Mevion system, the cyclotron is mounted directly on a rotating gantry [11]. It should be noted, however, that these strong magnetic fields tend to have a decreased strength near the outer cyclotron radius. This requires a varying frequency for the accelerating RF field to keep the RF in phase with the particle revolution frequency. This frequency sweep is repeated at a rate of 500–1,000 Hertz. At each frequency sweep, one group of protons is accelerated from source to extraction and creates a pulse of protons being extracted. Since such small cyclotrons operate in this 500–1,000 Hertz pulsed beam intensity mode, they are called *synchrocyclotrons*.

In the field of SC cyclotrons, also studies have been undertaken to design a cyclotron with a magnet that has no iron yoke [14], but these developments are still in a very early phase.

Currently, the recent significant cyclotron-related advancements such as facility-size reduction, SC cyclotrons and single-room layouts, are being utilised in clinical facilities.

Synchrotron

Proton synchrotrons are composed of a preaccelerator which injects the protons into a ring of 4–8 bending magnets with quadrupole and sextupole magnets in between them. Also in the ring, an RF cavity is mounted in which an oscillating electric field is used for further acceleration of the protons. The RF frequency is varied with the increasing revolution frequency of the protons and the magnets are increasing their strengths synchronous to the acceleration process. For protons, the ring has a typical diameter of 6–8 metres and for ions a ring of ~25 metres diameter is required. The preaccelerator is a separate 6–10 metres-long chain of an ion source followed by accelerators – usually a radio frequency quadrupole (RFQ) accelerator followed by a linac. In the case of ions, one can connect different ion sources in parallel to the preaccelerator so that one can quickly change beam type. The maximum number of particles that can be injected into the ring is limited, but can increase with the injection energy. For the application of one field at the patient, one typically needs 1–3 filling and acceleration sequences ('spills') [15].

The typical extraction efficiencies of synchrotrons using slow extraction is between 80% and 95%. Due to the relatively open and modular character of a synchrotron, these losses can be dealt with rather easily by means of local shielding, and since these are the only relevant beam losses in most synchrotron facilities, the activation is typically much less than in cyclotron-driven facilities which need local shielding at the degrader and ESS (Energy Selection System).

Slow-cycle synchrotrons have been widely applied with high reliability and availability in particle therapy systems since their introduction. Recently, new technologies for beam injection and extraction have been applied to new irradiation procedures such as schemes to enable gating to the patient's respiration and apply spot scanning.

The first synchrotron especially made for therapy with ions heavier than protons was developed at the National Institute of Radiological Sciences (NIRS) [16]. The ion-synchrotrons at Centro Nazionale di Adroterapia Oncologica (CNAO) [17] in Pavia, Italy and at MedAustron [18] in Wiener Neustadt, Austria have been built based on a European Organization for Nuclear Research (CERN) design.

Recently, smaller proton synchrotrons have been developed with diameters down to five metres [19]. Ring diameter reductions have primarily been achieved by using improved magnets, modification of the beam optics and optimisation of interfaces to the injector. Improvements in the injection chain have also reduced the power consumption of synchrotron systems. A recent prototype used a new RFQ [20] which directly injected into the synchrotron, thus reducing the length of the preaccelerator and injection line and its power consumption. This was the strategy adopted in the compact – diameter of 5.5 metres – synchrotron design 'Radiance 330' [21] which has been installed at McLaren, Flint, MI and MGH, Boston, MA. Recently, also Hitachi has developed a compact synchrotron, the first of which has started operation in Hokkaido in 2014 [22].

Additional improvements in the power consumption and treatment time can be achieved via special field-regulation of synchrotron magnets. First, implementations of this technique at the Heidelberg Ion-Beam Therapy Center (HIT) demonstrated a reduction of 30% in cycle time [23]. Although this has the most significant impact in carbon-ion systems, it will also be relevant for proton machines. Another approach to decrease the 'dead time' between the spills is to increase the ramping speed of the magnets, as was studied by Best Medical and Brookhaven National Laboratories [24].

Despite much progress in SC magnet development for synchrotrons in high-energy physics, the application of SC magnets for medical synchrotrons has only recently started to be investigated. This is due to their typically smaller scale and necessarily faster ramping speeds. NIRS researchers are working on a new carbon-ion synchrotron dubbed 'SuperMinimac' with strong SC magnets and a diameter of only seven metres [25] instead of the typical 25-metres diameter. It is expected, however, that for proton–synchrotrons, such a significant reduction in size will still require more research.

At NIRS, Chiba, Japan, one has recently introduced the possibility of changing the energy of the extracted carbon beam within 0.1 second during the extraction, so this process would be within a spill [26]. Although it would reduce treatment times considerably, changing energy during extraction has not yet been achieved in a proton synchrotron. At HIT, other improvements have focused on the stability of the extracted beam intensity [27].

This study is quite important, since the extracted beam intensity from currently used synchrotrons is rather noisy. Especially in proton synchrotrons, it varies typically about 10%–20% whereas this can typically be 1% 5% for an isochronous cyclotron. Especially for continuous scanning, random beam intensity fluctuations should be as low as possible.

It is therefore very promising so see the many ongoing interesting developments in magnets, RF systems, injection and extraction schemes, a higher filling of the ring and a faster and more accurate control of the beam parameters.

USE OF THE BEAM

Gantries

When irradiating the tumour from different directions, one can obtain an optimal precision of the dose distribution in the tumour region while avoiding critical healthy tissue. This can be done by means of a *gantry*. This is a mechanical structure supporting a beam-transport system which can rotate around the patient. Since the inception of the proton facility at Loma Linda, gantries became available (Figures 4.1 and 4.2) [28].

Gantries for particle therapy are large and heavy devices since strong and large magnets are needed for the beam transport. For protons, the gantries have a diameter of up to 12 metres and a mass of 100–200 tons. The gantry used for carbon therapy at HIT [30] has a diameter of 13 metres and a mass of 600 tons. Gantries require special attention with respect to mechanical accuracy – submillimetre precision of the dose deposition – safety, accessibility of the patient as well as for service; beam and monitoring; and presence of medical equipment for imaging and beam optics.

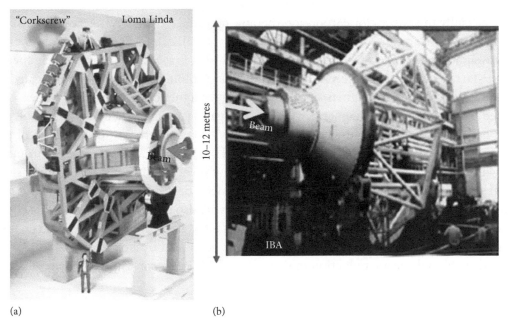

(a) (b)

FIGURE 4.1 (a) A model of the first gantry for proton therapy in Loma Linda [28], as seen from the beamline side; and (b) a gantry from IBA during its construction phase.

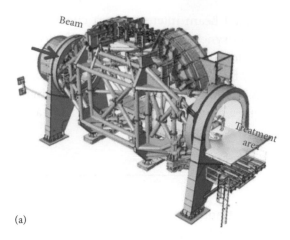

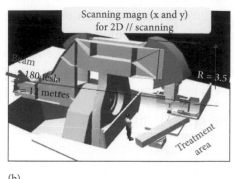

(a) (b)

FIGURE 4.2   (a) Gantries for carbon PBS at HIT. (From Fuchs, R. et al., The heavy ion gantry of the HICAT facility, *Proceedings of the 9th European Particle Accelerator Conference*, EPS-AG, Lucerne, Switzerland, 2550–2552, 2004.); and (b) proton PBS at PSI. (From Pedroni, E. et al., *Z. Med. Phys.*, 14, 25–34, 2004.) The figures do not have the same scale.

Although many developments have been implemented to improve the beam optical characteristic and apply different irradiation techniques, the size and weight of the gantries did not reduce dramatically in recent years except for the first worldwide scanning gantries at the Paul Scherrer Institute (PSI); therefore, motivated by the need for a reduction in the price of equipment, technological research on this topic has gained the interest of several groups and companies [34,35]. The use of SC magnets is one of the developments in which several interesting milestones have recently been achieved.

Especially in gantries for carbon-ion therapy, both the reductions in size and mass can be quite significant [31,34]. The carbon-ion gantry at NIRS uses SC magnets of relatively low weight with a large aperture. Treatments with this first SC gantry in particle therapy were started in May 2017. Recent other studies on proton gantries have investigated the possibility of using SC magnets in which different magnetic multipoles have been combined within a single SC magnet. Several of such gantry designs are being proposed; they can be designed such that they can accept a very broad spectrum (up to ±25%) of proton energies without changing magnet settings [32,33,36,37]. This extreme achromaticity will help to shorten the treatment time and would enable ultra-fast three-dimensional (3D) scanning without an adjustment of the magnetic field in the SC gantry magnet. The more relaxed ramping schedule of such a magnet will also require less cooling capacity. In addition, when mounting a degrader on such a gantry, one could refrain from the usual degrader and energy selection system (ESS) at the beam exit of the cyclotron [36,37]. This would enable a dramatic reduction in size of the whole cyclotron-gantry system.

Distributing the Dose

The beam from an accelerator is not directly applicable for accurate and appropriate dose delivery. In general, the tumour diameter is much larger (10–100 times) than the (lateral) width of the beam (typically <1 centimetre) and much thicker (5–30 times) than the width of the Bragg peak; therefore, special techniques have been developed for spreading the beam in the lateral direction as well as in depth. After aiming the beam from the desired direction by setting the gantry at the correct angle, the beam must be spread in the lateral direction. Currently, most existing facilities are using the passive scattering technique to give the beam the necessary lateral width [46].

In this technique, the beam is aimed at a thin foil made of high-Z material such as a heavy metal like tungsten or lead. Due to the interaction with the atoms in the foil, particles in the beam will undergo the so-called multiple scattering process. This causes an increase in the angular spread of the beam beyond the foil. The angular distribution can be well described as a normal (Gaussian) distribution. At some distance behind this first scatter foil, also the beam profile has the shape of a normal distribution. Some distance behind the scatter foil, a second device is placed in the beam. The goal of this often called 'second scatterer' is to make the beam profile flat at isocentre. This system has to remove particles from the central maximum of the intensity profile peak. This can be done by stopping the central part of the beam in a metal cylinder of sufficient thickness. Due to the angular spread of the beam particles, part of the particles in the non-stopped outer region of the beam – that is in the 'tail' of the Gaussian – will have a direction towards the now 'empty' central part of the beam. When using a proper cylinder diameter in relation to the Gaussian width of the beam just before the cylinder and the proper distance to the isocentre, the hole in the central part of the beam is filled and over a beam diameter of several cm a flat profile can be obtained.

Another device to create a flat profile is a second scattering foil of high-Z material which has a thickness varying with the radial distance to the beam axis. At the beam axis, the foil is usually the thickest; at larger radius the foil can be thinner. In this way, the central part of the beam will get another divergence increase. After some distance, that is, at isocentre, the particles scattered out of the centre will add to the dose at the outer radius of the beam, and as in the cylinder system, part of the particles from the outer part will also contribute to the central part of the beam. In some cases, just some rings of increased thickness will be sufficient to accomplish this. This redistribution of intensity can lead to flat dose distributions of considerable size (~20 centimetres) at isocentre. Some versions of this second scatter foil are covered with a low-Z disk – such as Lucite – with variable thickness (thinnest in centre part) to compensate the differences in energy loss in the high-Z foil with minimal additional scattering.

For situations where a very large field size is needed, beam-wobbling magnets are used at some places. These magnets let the beam continuously sweep over a circle or line pattern in the lateral plane. This continuous motion (so not passive) will effectively increase the field size at the isocentre.

In both scattering methods, the shape of the field at the patient is limited laterally by a dedicated metal collimator which has an aperture of the same lateral shape as the target volume as seen from the beam direction. This collimator is placed just before the patient.

In the passive scattering method, spreading in depth is performed by a rotating wheel in the beam before the scattering foils. The often Lucite wheel has an azimuthally varying thickness. The beam is crossing the wheel at approximately 50% of its radius. The wheel rotation will bring different Lucite thicknesses in the beam, resulting in corresponding shifts of energy, thereby causing the necessary shifts of the Bragg peak depth at the patient. By using the appropriate distribution of the Lucite thicknesses and a decent rotational speed of this range-modulation wheel, a very flat spread-out Bragg peak (SOBP) is obtained with appropriate thickness. Often a separate Lucite plate with a thickness profile is used in front of the collimator before the patient to align the deepest ranges of the beam according the distal limit of the target volume. The passive scattering method is relatively simple (passive!) and rather robust in case of organ motion since the only time-dependent process is the range modulation making the SOBP; however, using a sufficiently high rotational speed of the modulation wheel, the corresponding movement in the dose-application process can be so fast that it does not interfere with organ motions.

It has also been recognised, however, that the scattering method has several disadvantages. A drawback of the scattering is an increased sensitivity to beam steering which may easily cause a tilt in the lateral dose distribution. Also, the dose distribution has some disadvantages since the width of the (SOBP) is the same for all lateral locations. At lateral positions where the target volume is not as thick as the maximum thickness of the volume, also a SOBP-dose will be deposited proximal to the target volume. Last but not least, the dedicated devices needed for this technique have a considerable contribution to the costs of a treatment.

Since its introduction at PSI and GSI the Physics Research Institute GSI in Darmstadt (D) in 1995 [38,39], the pencil beam scanning (PBS) technique is regarded as the most optimal method in most cases. This technique is offered by most suppliers nowadays. Figure 4.3 shows that the number of facilities equipped with PBS has increased dramatically since 2009.

In PBS, the beam is 'actively' scanned in the transverse plane over the tumour cross section. Until now, this is performed in a discrete way by sequentially aiming the beam

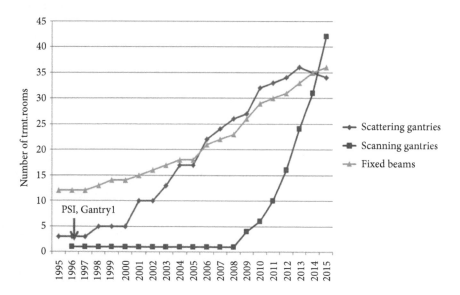

FIGURE 4.3 The number of gantries with beam delivery by scattering and scanning. (Courtesy of Martin Jermann, PTCOG, 2014; From Particle Therapy Cooperative Group, www.ptcog.ch.)

at sequential volume elements in the tumour into which a dose is applied (*spot scanning*). This spot-scan technique is the method that is being used in clinical practice. The other method, *continuous scanning*, is still in development. In this method, the beam is swept over the target volume and the dose is controlled by varying the beam intensity and/or the speed of the sweep as a function of time.

In the spot-scan technique, beam position and the dose per spot are the critical parameters. The beam is aimed at a spot to deposit the spot dose. It is shifted to the next spot when the spot dose has been delivered, so within certain limits the beam intensity may fluctuate a bit; however, in the case of the future continuous scanning technique, the beam intensity must be set very accurately. When applying the spot-scan technique from an accelerator system operating in a pulsed mode at a rate in the order of a kilohertz, it makes sense to use each individual pulse to fill a spot. In that case, the dose per spot must be set for each pulse prior to delivery of the spot dose. Usually there is a limited accuracy on the actually delivered dose per pulse; therefore, in a pulsed beam system, at least two pulses need to be applied at each spot to obtain the required accuracy of the dose per spot. In that case, the dose to be delivered in the second pulse can be adjusted to correct for the error in the first pulse. Of course, the dose delivered by each pulse must be measured to do this correction, but it must be measured accurately anyway. Another method to be applied in pulsed-beam systems is to apply a sequence of low-dose pulses until the desired spot dose has been reached. This method needs more time, however. In both methods, the total irradiation time will be larger than when using a continuous beam, and as is clear from above-described process, a pulsed beam at <1 kilohertz rate cannot be used for continuous scanning.

In general, the PBS technique is considered the one that gives the highest quality dose distributions in the patient; however, to accomplish this advantage, it is clear that a very well-known and accurately adjustable beam intensity and accurate knowledge of the pencil beam position are the essential specifications for the accelerator and the control of dose deposition and beam line. The time dependence of the dose-application process is considered the big disadvantage of the PBS technique since it may interfere with organ motion; however, all kind of mitigation methods are being investigated at this moment. One can use (a combination of) gating or a very fast (continuous) scanning or an active correction of the pencil beam position.

Dose Spreading in Depth

The variation in depth of the Bragg peak location is performed by changing the beam energy. This process is called *range modulation* or *energy modulation*. To obtain a good dose distribution in depth, the Bragg peaks are typically shifted over 10–20 layers in steps of 2–5 millimetres, depending on the width of the Bragg peak and the required variation in depth, that is, the thickness of the tumour. To obtain the desired depth-dose distribution, the relative weight of the different Bragg peaks varies over an order of magnitude for the different energies used. This implies that per energy, the beam intensity and beam-on time are strongly correlated in this process. To limit the total treatment time, each energy change must be done within a second or faster.

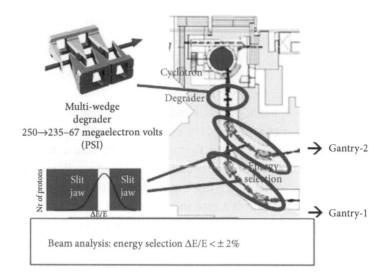

FIGURE 4.4 The degrader and energy selection system (ESS) at PSI which is used for the setting of maximal beam energy per field at Gantry-1 with an energy modulation system in the nozzle and for energy modulation at Gantry-2.

In synchrotrons, the beam energy is set per spill which is not fast enough for modulation except for the carbon synchrotron at NIRS; therefore, at each spill the energy is set to the maximum energy (corresponding to the maximum depth) needed at a certain gantry angle in a treatment. The necessary lower energies are produced by a local range modulator in the beam line to the treatment area. Usually this is a Plexiglas wheel with an azimuthally varying thickness. By rotating the wheel in the beam, the beam is crossed by the different thicknesses and degraded according to these thicknesses.

When use is made of a cyclotron, the energy of the extracted beam cannot be changed within a reasonable time; consequently, the beam is slowed down in a degrader (Figure 4.4) in the beam line following extraction. Such a degrader consists of an adjustable amount of low-Z material such as graphite in the beam. To obtain a homogeneous amount of degrading over the beam cross section, mostly a wedge-shaped system is used. The exact beam energy and the energy spread – which should not be larger than the acceptance of the following beam transport system – of the beam that is sent to the patient are selected in an energy selection system. This system consists of bending magnets of approximately 45 degrees and a slit system at a location where a monochromatic focus coincides with large momentum dispersion. Effectively, the energy selection takes place at this slit. All magnets in the beam transport system following the degrader as well as those in the gantry need to change their strengths according to the degrader setting.

These energy variations must be made quite rapidly. Since the step size in proton range is typically ~5 millimetres, this requires a magnetic field change of approximately 1% in all beam-transport magnets downstream of the degrader. Of course, this speed is limited by power supplies and magnet design. Since this affects the total treatment time, this speed is an important parameter. At PSI, such a step is made within 0.1 second; several companies are working on their system to reach such a speed as well.

ACCELERATORS AND THEIR BEAMS

In multi-room facilities, the beam extracted from the accelerator is sent to one treatment room at a time. Apart from the technical complications linked to split a beam, the main reason for this nonparallel operation is the requirement that each treatment room must have direct control over some accelerator parameters such as beam intensity and beam energy. These parameters are specifically linked to the treatment being performed. In multi-room systems, the switching of the beam to another treatment room can usually be performed within a few seconds. Eventual needed changes in accelerator setting usually hardly contribute to the room-switching time. Usually if the energy change can be made quickly, also the beam transport system setting can be switched rapidly to another room; however, in many cases the procedures to be followed when switching the beam to another room require many checks of the security status. In some systems, operator actions or manual confirmations are needed and sometimes a ramping sequence of magnets in the beam line and gantry that is being switched to, is required. Altogether this may need a fraction of a minute and varies between the different facilities. It is also important to realise that the actual irradiation time (that is, 'beam-on') in a treatment room is usually only a few minutes. In addition to that, no beam can be sent to the treatment room for ~15 minutes due to time needed for patient setup and alignment. One can schedule the program such that the beam is used for treatment in another room during that 'set-up' time; therefore, if possible, parallel operation would only be of advantage in the case of a facility with more than approximately four treatment rooms.

Beam Intensity Issues

The requirements of the beam extracted from the accelerator are determined by the necessary beam intensity, the beam energy and whether the beam is pulsed (i.e., <kilohertz) or continuous. The beam intensity determines the dose rate at the patient and thus the time a treatment requires. The typical time of the dose delivery is between one and several minutes. This time depends on the size of the tumour, treatment technique and beam intensity. A short treatment time is, of course, more comfortable for the patient. It also reduces the chance that the patient moves during the treatment; however, even if the patient positioning can be guaranteed by proper techniques, a too-short delivery time – that is, a very high dose rate – could decrease the accuracy of the dose-delivery process. Due to limitations to measure dose at a high-dose rate because of, for example, space-charge or saturation effects in detectors, the application can become quite inaccurate. This is especially critical for devices that must generate a signal to switch the beam off when the dose to be delivered has been achieved. The reaction time must be such that a too-high dose is prevented. This might be a problem at high intensities since then the reaction time might be too long or the alarm thresholds are not reached due to saturation. Also, the consequence of positioning errors can be very serious: a too high dose at the wrong location. Consequently, there is much reluctance to apply very high dose rates in particle therapy.

The beam intensity at the patient is determined by the extracted beam intensity but also by the transmission through the beam transport system. Typically at the patient, the required beam intensity is in the order of a few tenths to a few nanoampères (nA). In the case of a cyclotron, usually extracted intensities of a few hundred nA are used. The transmission in the beam line varies with energy since there are big energy-dependent transmission losses in the degrader and ESS. This is mainly due to the also occurring multiple scattering in the degrader and increases with the amount of energy loss in the degrader. The beam emittance must be collimated to match the acceptance of the following beam transport system and gantries after leaving the degrader. This can be done by collimators mounted behind the degrader. In addition to the emittance increase due to multiple scattering, one should also consider that the degrading process also causes energy spread in the beam due to the energy-straggling process of the protons in the degrader. This spread can be several times larger than the energy acceptance of the following beam transport system. It is reduced and limited in the ESS. In such a system, usually there are almost no beam losses between ESS and patient. The loss of transmission in the degrader, the following collimators and in the ESS increases when selecting a lower energy. The transmission can be less than 1% when degrading from 230–250 to 70 megaelectron volts [40]. At the patient, this change would imply an intensity variation of a factor 50–100 over the typical energy ranges used; therefore in many facilities, measures are taken to obtain an almost energy-independent beam intensity in the order of 0.1–1 nanoampere at the patient so that there is an energy- independent reaction time of dosimetry and scanning verification systems which are typically located just before the patient.

The most straightforward measure is to adjust the beam intensity in the cyclotron, however, it is an important security issue to prevent unwanted high intensities: one must be sure that the intensity adjustment is done in correct synchronisation with the energy adjustment.

At PSI, the beam intensity at the patient is regulated by adapting the transmission in the beam transport system. The beam intensity extracted from the cyclotron is kept constant at the highest value needed so the one required for the low energies needed, which is the one required for the lowest. As an integrated part of the energy-dependent beam line setting following the degrader setting, the beam is defocused as a function of energy at two dedicated collimators in the beam line. At high-beam energies, the intensity at the patient is reduced by this defocusing, and at low energies, the beam is focused more so that the transmission reduction is less. Using this technique at low energies (80–100 megaelectron volts), reasonable beam intensities are obtained at the patient which do not differ much from the intensity at high energies (180–230 megaelectron volts). To limit the risk, this focusing correction is an integrated part of the energy-dependent setting of the beam transport systems between degrader and patient.

Typically, the beam intensity from a cyclotron is determined by several parameters. Of course, the most important parameter is the intensity obtained from the ion source, but at the PSI cyclotron, the ion source is operated at a constant setting during the day and required variations of the extracted beam intensity are performed by means of a set of deflection plates installed at the first few turns in the central region (Figure 4.5).

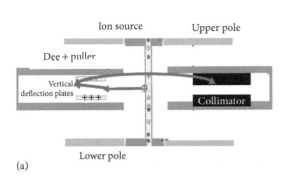

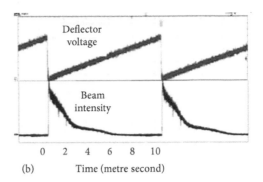

(a)

(b)

FIGURE 4.5 (a) Schematic overview of the vertical deflector in the central region of the cyclotron at PSI. It can bend the beam in the vertical direction so that it is partially intercepted by a vertical collimator; and (b) the deflector voltage (upper curve) and corresponding beam intensity after extraction from the cyclotron (lower curve) as a function of time.

By employing a vertical electric field between these plates, the beam is deflected in the vertical direction. A vertically limiting collimation system can intercept part of the deflected beam and the nonintercepted part of the beam is accelerated and extracted in the normal way. This system can be used for very fast (<100 microseconds) beam-intensity adjustment as needed for continuous scanning and/or fast suppression of the beam intensity.

The beam intensity from a synchrotron using the slow extraction method is determined by the fraction of the beam, that is, in an unstable region of the radial phase space in the ring. Particle in this unstable region will be intercepted by a septum and extracted. The amount of particles in this unstable region is determined by the total filling of the ring and by the amplitude and duration of an action to shift particles into this phase space region. Several types of this action are possible. The classical method is to excite the betatron resonances by slight changes in the focusing which cause a shift of the tune into the direction of a betatron resonance. Particles excited in the resonance will get a large betatron oscillation amplitude and will be split off by the septum, often followed by a septum dipole magnet, and then extracted. This can also be achieved by a betatron core [41] making slight energy changes. Also, here one brings the beam tune close to a radial betatron resonance to extract the particles that received a large oscillation amplitude. Another method to increase the amplitude of the betatron oscillation is the RF Knockout. In this method [42,43], the tune of the synchrotron is held fixed. An electric RF field between a pair of plates deflects particles in the radial direction and excites radial betatron oscillations which increases the radial emittance. The amplitude of this RF voltage, its duration and the actual filling of the ring determine the particle intensity within this emittance increase and thus the intensity of the extracted beam. The advantage of the RF-Knockout method is that the position, emittance, and beam shape of the extracted beam remain constant during a spill and that intensity can be controlled by tuning the Knockout-RF field.

Since the extracted-beam intensity is mostly determined by the processes affecting betatron oscillations, the intensity of the ion source does not play a role in the extracted beam intensity or emittance in case of a synchrotron; however, it is important to get enough intensity from the source and injection system to limit the time of the injection process to fill the ring. To prevent beam losses at the low injection energies, it is useful to have a short injection time and start acceleration to the higher energies as soon as possible. The prevention of such beam losses helps to obtain sufficient accelerated beam intensity.

Beam Shape and Time Structure

The shape of the beam extracted from a cyclotron is not so critical in general. Although the emittances in the horizontal and vertical direction will never be the same, this asymmetry vanishes due to multiple scattering in the degrader.

In older synchrotrons not using the RF-knockout extraction, there is a change in emittance and intensity during the slow-extraction process within a spill, and for all synchrotrons, the emittance of the beam from a synchrotron as obtained with slow extraction is very asymmetric. Typically, the emittance in the direction perpendicular to the bending plane could be 10 times larger than in the bending plane. This is usually in a horizontal plane. Currently, all synchrotrons are oriented such that they are bending and extracting in a horizontal plane. For a system with passive scattering, this can be dealt with, but one should be careful when using a gantry with spot scanning. In that case, the beam shape at the patient as well as the transmission through the gantry could be gantry-angle dependent. This can be resolved in the treatment planning, but it is much more convenient to have a gantry angle-independent dose delivery. This can be achieved by dedicated manipulations before the beam enters the gantry. The emittance can be made more symmetrical by scattering and collimation. This may cause high beam losses, however, but it can also be achieved by rotating the beam emittance to match the gantry angle by using dedicated beam optics and rotating quadrupoles [44].

If the time structure of the beam consists of pulses at a frequency of less than a few kHz, usually such a beam is denoted as pulsed. In this regard, beams from an isochronous cyclotron are considered as continuous ('DC') beams in the normally applied therapy despite the RF frequency-related pulses in the range of 10–100 megahertz. During the spill of a synchrotron, the extracted beams are continuous beams since the beam bunches are diluted over the synchrotron circumference during the storage and the slow extraction process. Several remaining time structures in Hertz and kilohertz ranges are present but can be kept well under control [16].

In general, a pulsed time structure could play a role when the intensity during the pulses is very high. For example in a very specific dose measurement, equipment saturation processes might occur due to the high intensities during the pulses. One should also take care that the pulses from a synchrocyclotron (typical pulse rates are up to kilohertz) may interfere with other time structures at an approximately kilohertz rate in the dose-delivery process such as in scanning or with rotating range modulation wheels. In order to achieve high enough average dose rate at the patient, the beamless period between the pulses of synchrocyclotrons is compensated by having high-beam intensity within the pulses; therefore, the beam intensity during a pulse of a synchrocyclotron can be in the order of about 100 times larger than the intensity of an isochronous cyclotron to obtain a more or less similar treatment time. One should be aware of the response of detectors, beam diagnostics and dosimetry devices when having such a pulse structure, but at the currently used intensities during these pulses, no dose rate-related biological effects have been observed.

With regard to dealing with organ motion, no special devices have been developed in the accelerators or beam delivery systems. It is more a matter of integrating position measurement devices with the existing control system of the accelerator. For example, a beam 'on–off' possibility will be necessary and this will need a dedicated input connection. Apart from beam on–off (that is, the 'gating technique'), other techniques are also being investigated. All these techniques such as on-line pencil beam position corrections or very fast continuous scanning are using existing technology. Currently, most research focuses on the adaption of control systems and development of fast and accurate position measurement systems.

SUMMARY AND OUTLOOK

The layout of a particle therapy facility usually clearly shows separated modules: typically, accelerator; beam-energy adjustment (for cyclotrons); beam-transport system; gantries and/or fixed beam lines; however, the interplay between these modules is extremely important. The PBS method used at the treatment has direct consequences on the accelerator design and several operating characteristics.

Most of the technical developments in recent years have sought to reduce costs as a primary goal. This has put the focus on a reduction in size, but also developments aimed at a simplification of the accelerator and beam handling devices play an important role in this respect. Especially related to simplification, one should be very careful to keep the high quality of the dose distribution at the patient as is currently achieved in most facilities. A dose distribution of high quality has been and will remain the only motivation to apply particle therapy.

Due to its success and breakthrough in the clinical world, particle therapy is entering a phase in which difficult choices must be made in the balance between the goal of widespread clinical application preferably for a low price and further technological developments. In a pure clinical hospital-based environment, there is quite some pressure – and of course the goal – to treat as many patients as possible. Although the financial reasons for this are clear, one should realise that this often tempers the implementation of major improvements. In addition to that, regulations and patient safety require essential procedures on quality verification and treatment control [45]; therefore, facilities in a laboratory environment will play an even more important role in performing the typical large-scale technological research and development.

ACKNOWLEDGEMENTS

The author would like to thank the numerous colleagues at PSI and other particle therapy institutes as well as those at several companies for the countless discussions at conferences and workshops, the detailed information given at site visits and for sharing their experiences and expertise in proton therapy.

REFERENCES

1. Slater, J.M. et al. The proton treatment center at Loma Linda University Medical Center: Rationale for and description of its development. *Int. J. Radiat. Oncol. Biol. Phys.* 22 (1992): 383–389.
2. Particle Therapy Cooperative Group. www.ptcog.ch, January 1, 2016.
3. Schippers, J.M. Beam delivery systems for particle radiation therapy: Current status and recent development reviews of accelerator science and technology. *Rev. Accl. Sci. Tech.* 2 (2009): 179–200.
4. Ma, C.M. and T. Lomax (Eds.) *Proton and Carbon Ion Therapy.* Boca Raton, FL: CRC Press, 2013.
5. Paganetti, H. (Ed.) *Proton Therapy Physics.* Boca Raton, FL: CRC Press, 2012.
6. Schippers, J.M. and A. Lomax. Emerging technologies in protontherapy. *Acta Oncol.* 50 (2011): 838–850.
7. Degiovanni, A. and U. Amaldi. Proton and carbon linacs for hadron therapy. *Proceedings of LINAC2014*, Geneva, Switzerland, FRIOB02, 2014, pp. 1207–1212.
8. Schippers, J.M. et al. The SC 250 MeV cyclotron and beam lines of PSI's new proton therapy facility PROSCAN. *Nucl. Instr. Meth.* B 261 (2007): 773–776.
9. Schillo, M. et al. Compact superconducting 250 MeV proton cyclotron for the PSI PROSCAN proton therapy project. *Cyclotrons and Their Applications 2001, 16th International Conference,* East Lansing, MI, 2001, pp. 37–39.
10. Marti, F. (Ed.). *Cyclotrons and Their Applications 2001, 16th International Conference,* East Lansing, MI, 2001, p. 22.
11. Mevion. www.mevion.com, January 1, 2016.
12. IBA. https://iba-worldwide.com/proton-therapy/proton-therapy-solutions/proteus-one, January 1, 2016.
13. Kleeven, W. et al. The IBA superconducting synchro-cyclotron project S2C2. *Proceedings of the Cyclotrons 2013*, Vancouver, Canada, MO4PB02, 2013, p. 119.
14. Radovinsky, A. et al. Superconducting magnets for ultra light and magnetically shielded, compact cyclotrons for medical, scientific, and security applications. *IEEE Trans. Appl. Supercond.* 24(3) (2014): 4402505.
15. Hiramoto, K. et al. The synchrotron and its related technology for ion beam therapy. *Nucl. Instr. Meth.* B 261 (2007): 786–790.
16. Furukawa, T. et al. Design of synchrotron and transport line for carbon therapy facility and related machine study at HIMAC. *Nucl. Instr. Meth.* A 562 (2006): 1050–1053.
17. Amaldi, U. The Italian hadrontherapy project CNAO. *Phys. Med.* 17(1) (2001): 33–37.
18. Auberger, T. and E. Griesmayer (Eds.). *Das Projekt MedAustron.* Wiener Neustadt, Austria, 2004. ISBN 3-200-00141-0.
19. Umezawa, M. et al. Development of compact proton beam therapy system for moving organs. *Hitachi Rev.* 64(8) (2015): 507.
20. Vretenar, M. et al. A compact high-frequency RFQ for medical applications. *Proceedings of the LINAC2014*, Geneva, Switzerland, THPP04, 2014, pp. 935–938.
21. Wang, F., J. Flanz, and R.W. Hamm. Injection study of the proton-radiance 330 synchrotron with a 1.6 MeV RFQ Linac. *The 19th Particles and Nuclei International Conference*, July 24–29, Cambridge, MA, 2011.
22. Hitachi. www.oia.hokudai.ac.jp/blog/2014/10/24/62/, January 1, 2016.
23. Feldmeier, E. et al. The first magnetic field control (B-train) to optimize the duty cycle of a synchrotron in clinical operation. *Proceedings of IPAC2012*, New Orleans, LA, THPPD002, pp. 3503–3505, 2012.
24. Trbojevic, D. et al. Lattice design of a rapid cycling medical synchrotron for carbon/proton therapy. *Proceedings of IPAC2011*, San Sebastián, Spain, WEPS028, 2011.

25. Iwata, Y. et al. *Workshop at CIEMAT*, SC Gantry at NIRS and new developments in SC magnets for gantries & synchrotrons, November, Madrid, Spain, 2016.
26. Iwata, Y. et al. Multiple-energy operation with quasi-dc extension of flattops at HIMA. *Proceedings of IPAC2010*, Kyoto, Japan, MOPEA008, 2010, pp. 79–81.
27. Schoemers, C. et al. The intensity feedback system at Heidelberg ion-beam therapy centre. *Nucl. Instr. Meth. A*. 795 (2015): 92–99.
28. Koehler, A.M. 1987. *Proceedings of the 5th PTCOG Meeting: International Workshop on Biomedical Accelerators*, Lawrence Berkeley Laboratory, Berkeley, CA, 1987, pp. 147–158.
29. Pedroni, E. et al. The PSI Gantry 2: A second generation proton scanning gantry. *Z. Med. Phys.* 14 (2004): 25–34.
30. Fuchs, R., U. Weinrich, and P. Emde. The heavy ion gantry of the HICAT facility. *Proceedings of the 9th European Particle Accelerator Conference*, Lucerne, Switzerland, EPS, 2004, pp. 2550–2552.
31. Iwata, Y. et al. Design of superconducting rotating-gantry for heavy-ion therapy. *Proceedings of IPAC2012*, New Orleans, LA, THPPR047, 2012.
32. Gerbershagen, A., C. Calzolaio, D. Meer, S. Sanfilippo, and M. Schippers. Advantages and challenges of superconducting magnets in particle therapy particle therapy. *Supercond. Sci. Technol.* 29 (2016): 083001, 15pp. doi:10.1088/0953-2048/29/8/083001.
33. Wan, W. et al. Alternating-gradient canted cosine theta superconducting magnets for future compact proton gantries. *Phys. Rev. ST Accel. Beams* 18 (2015): 103501.
34. Derenchuk, V. Particle beam technology and delivery. *55th Annual AAPM Meeting, Proton Symposium*, Indiana, 2013.
35. Holder, D.J., A.F. Green, and H.L. Owen. A compact superconducting 330 MeV proton gantry for radiotherapy and computed tomography. *Proceedings of IPAC2014*, Dresden, Germany, WEPRO101, 2014, pp. 2201–2204.
36. Gerbershagen, A., J.M. Schippers, D. Meer, and M. Seidel. Novel beam optics concepts in particle therapy gantries utilizing the advantages of superconducting magnets. *Z. Med. Phys.* 26 (2016): 224–237. doi:10.1016/j.zemedi.2016.03.006.
37. Sanfilippo, S., C. Calzolaio, A. Anghel, A. Gerbershagen, and J.M. Schippers. Conceptual design of superconducting combined function magnets for the next generation of beam cancer therapy gantry. *Proceedings of RuPAC2016*, St. Petersburg, Russia, THCDMH0, 2016, pp. 138–140.
38. Pedroni, E. et al. The 200 MeV proton therapy project at the Paul Scherrer Institute: Conceptual design and practical realization. *Med. Phys.* 22 (1995): 37–53.
39. Haberer, T. et al. Magnetic scanning system for heavy ion therapy. *Nucl. Instr. Meth. A* 330 (1993): 296–305.
40. van Goethem, M.J. et al. Geant4 simulations of proton beam transport through a carbon or beryllium degrader and following a beamline. *Phys. Med. Biol.* 54 (2009): 5831–5846.
41. Knaus, P. et al. Betatron core driven slow extraction at CNAO and MEDAUSTRON, *Proceedings of IPAC2016*, Busan, Korea, TUPMR037.
42. Tomizawa, M. et al. Slow beam extraction at TARN II. *Nucl. Instr. Meth. A* 326 (1993): 399.
43. Noda, K. et al. Slow beam extraction by a transverse RF field with AM and FM. *Nucl. Instr. Meth. A* 374 (1996): 269.
44. Benedikt, M., P. Bryant, and M. Pullia. A new concept for the control of a slow-extracted beam in a line with rotational optics. *Nucl. Instrum. Meth A* 430 (1999): 523–533.
45. Schippers, J.M. and M. Seidel. Operational and design aspects of accelerators for medical applications. *Phys. Rev. ST Accel. Beams* 18 (2015): 034801. doi:10.1103/PhysRevSTAB.18.034801.
46. Chu, W.T., B.A. Ludewigt, and T.R. Renner. Instrumentation for treatment of cancer using proton and light ion beams. *Rev. Sci. Instrum.* 64 (1993): 2055–2122.

CHAPTER 5

Requirements for Setting Up a Particle Therapy Centre

Ramona Mayer, Stanislav Vatnitsky and Bernd Mößlacher

CONTENTS

H ADRON THERAPY CENTRES ARE still rare, but as treatment is promising there is a global demand for more facilities. This chapter gives a short overview in planning, staffing and commissioning such a dedicated centre. Issues related to the extension of a radiotherapy department/centre with a single-room proton facility installed next to an existing radiotherapy unit are not covered in this overview.

PRECONDITIONS TO ESTABLISH A HADRON THERAPY CENTRE

Decision: Proton or Carbon Ion Facility

The accelerator system needs to be selected carefully by evaluating all medical and technical requirements specifications in detail during the engineering project phase. Two solutions are conceivable:

1. The line of modern hadron therapy synchrotron-accelerators, like at Centro Nazionale di Adroterapia Oncologica (CNAO), MedAustron, or Heidelberg, designed for providing proton and carbon ion beams with a penetration depth of about 30 centimetres in water-equivalent tissue. The penetration depth is adapted from spill to spill by changing the extraction energy of the synchrotron, covering a range from about 3 to 30 centimetres in human tissue.

2. An industrially available cyclotron-based facility concept designed for continuous 230–250 megaelectron volts proton beams. To adapt to the tumour-specific penetration depth, energy degraders would be applied.

From the medical point of view, both accelerator concepts shall use the state-of-the-art, active pencil beam scanning (PBS) method for particle delivery to the tumour volume. In PBS, the beam is steered across the target volume, one layer at a time, to precisely match the shape of the target. This active scanning technique allows physicians to modulate or change the intensity of the beam at any specific location in the target. PBS is easier to operate and does not require the use of patient-specific devices. PBS enables intensity-modulated proton therapy (IMPT), giving the ability to precisely target the tumour by controlling the intensity and spatial distribution of the dose to the millimetre. The beam with selectable diameter in the range of 4–10 millimetres (size in vacuum) is positioned with a precision of ±0.5 millimetre.

In order to ensure that the dose is applied as prescribed, beam parameters are monitored online and controlled during the entire treatment process. Beam intensity, position and size are supervised by redundant means in front of the patient. In case of deviations from the nominal values, the beam is switched off within 1 milli second, ensuring the safety of the patient.

The main parameters (depending on accelerator concept) are shown in Table 5.1.

TABLE 5.1   Main Parameters of Particle Therapy Accelerators

Parameter	Value
Proton energy range	$\leq$250 megaelectron volts per mass (cyclotron: degraded)
Carbon energy range	120–400 megaelectron volts per mass (synchrotron only)
Beam size (FWHM at isocentre in vacuum)	4–10 millimetres, active PBS
Extraction time	1–10 seconds (synchrotron) or continuous (cyclotron)
Maximum number of p/C extracted	$1 \cdot 10^{10}/4 \cdot 10^8$ (synchrotron) or continuous (p, cyclotron)
Intensity variation	0.01–1
Ion species	p, C^{6+} (synchrotron only)
Transverse field for scanning (PBS)	200 × 200 square millimetres (approx.)

Patient Numbers and Patient Throughput

Due to the complexity of a particle therapy facility, precise data are required to most accurately define the number of potential patients. Therefore, several epidemiologic studies have been performed in Italy, France and Austria (Orecchia and Krengli 1998; Baron et al. 2004; Krengli and Orecchia 2004; Mayer et al. 2004). The studies showed that 13.5%–16% of the patients, who actually are treated by conventional radiotherapy could profit from the therapeutic advantages of hadron therapy.

Another way to select patients for proton therapy is the use of a prediction model for evidence-based selection of patients who will benefit most from this kind of radiotherapy (Langendijk et al. 2013). The model was developed in the Netherlands and is extensively discussed in Chapter 12.

When planning realistic patient numbers for a new particle centre, patient numbers at existing operational facilities (https://www.ptcog.ch) should be considered as reference. From the numbers on the Particle Therapy Co-Operative Group (PTCOG) website, it can be assumed that a patient number of significantly more than 1,000 patients per year cannot be reached easily.

To estimate patient throughput realistically, the following aspects should be taken into consideration. Facility capacity explains the number of treatment sessions which can be delivered per day, and for this estimation the time needed in the treatment room has to be analysed. Several factors, for example, complexity of the treatment (short and simple treatments versus complex and long treatments), number of fields per patient per treatment session, procedures like respiratory gating or general anaesthesia in small children or the number of treatment rooms, have to be taken into account.

Another factor is the total number of fractions per patient, which will be lower in carbon ion settings compared to proton facilities or the modality of patient recruitment. Interconnected hospitals should already be capable of attracting a significant number of patients who are candidates for hadron therapy. Creating a referring network is essential from the very start of the project. Regular contacts to the conventional radiotherapy community, but also to colleagues from oncological centres as well as relevant medical societies, have to be established.

Location and Site Requirements

Before starting a new particle facility it is recommended to conduct a feasibility study which should include an epidemiological study (see Patient numbers and patients throughout), but also access potentialities, existing local medical infrastructure and connection to oncologic hospitals and universities.

Patients in many clinical situations may benefit from both conventional radiotherapy and hadron therapy (e.g., as boost treatment), or conventional radiotherapy can serve as backup to prevent treatment delays in case of an accelerator shutdown. Optimally, both options should be situated under the same roof, or the hadron therapy facility should be situated within a short distance to the next conventional RT unit. The particle facility should also be located near an oncological hospital – cancer management includes not only radiotherapy but also surgery, chemotherapy and critical support services.

If it is intended to also treat children, then the proximity of a paediatric oncology care unit and an anaesthesia department is also of great importance. Another important factor is good connectivity to road networks and public transport.

From the perspective of the municipal and communal infrastructure, there are many different demands on the location. A redundant medium voltage supply with the possibility of also providing pulsed power is a prerequisite. A communal heat supply is an advantage but not a must. Due to the need to deliver heavy loads during the construction phase, a correspondingly powerful access road is necessary. All other requirements on the site correspond to those of a modern hospital location, with a particular emphasis on data communication.

BUILDING OF A CORE TEAM

Generally, the construction, commissioning and – above all – operation of such a highly sophisticated particle therapy centre requires a considerable number of specially trained and thoroughly experienced employees. The special fields that have to be covered are:

- Radiation oncology with a focus on particle therapy

- Medical physicist experts with knowledge in highly advanced conventional and particle therapy

- Radiation technologists with training in highly advanced conventional and particle therapy

- Particle accelerator know-how with a focus on physics, electronics, machinery, software and controls, as well as on operation and maintenance

- Radiation protection experts with a focus on neutron shielding

- Facility technicians for operating the particle accelerator

Due to the limited number of comparable centres in the world, most of those people cannot be found in the labour market but have to be educated and trained. For radiation oncologists and medical physicists, pre-existing expertise in advanced photon therapy is desirable and should be supplemented by specific particle training. This includes courses, for example, from PTCOG or the European Society for Therapeutic Radiology and Oncology (ESTRO), the Paul Scherrer Institute (PSI) winter school or training courses on carbon ion therapy offered, for example, by the National Institute of Radiological Sciences (NIRS). In recent years, the number of dedicated courses offered by select medical centres and universities are also increasing. Additionally, hands-on trainings of at least 6–12 months in active particle facilities are important. Training activities must be continued by regular advanced in-house training of all staff members.

The group of physicists and engineers for the construction, commissioning and operation of the particle accelerator need years of hands-on experience at comparable facilities in order to understand this complex machine. Again, programs at other centres need to be established for their training.

Besides patient treatment, translational research is also essential. Research directions in this field can be illustrated with a very successful example: In the second decade of the twenty-first century a Marie Curie Training project called Particle Training Network for European Radiotherapy (PARTNER) was funded by the European Commission. The project, coordinated by Manjit Dosanjh, CERN (European Organization for Nuclear Research) aimed at the creation of the next generation of experts. The project offered research and training opportunities to young (medical) physicist, biologists and engineers in close collaboration with leading European institutes and research centres and two leading companies in the field of particle therapy. A similar project would be very worthwhile to cover the increasing requirement for specialists.

In conclusion, it must be emphasised, that in order to have a competent group of employees at the time of commissioning and first patient treatment, selection of candidates and educational programmes must start immediately upon the fundamental decision of realizing the project. A corresponding hiring and education plan has to be evaluated and the accruing costs must be considered in the financial planning.

CIVIL ENGINEERING AND TECHNICAL INFRASTRUCTURE

The technical equipment and infrastructure of a cutting-edge facility lie hidden from the general public and are usually inaccessible to patients and visitors. In spite of that, or even exactly because of that, a sound and highly efficient backbone infrastructure is the key factor for running such a centre successfully from both a medical and an economical point of view. Reliability and cost-efficient operation are material to the building infrastructure.

Architectural Concept

Apart from the aesthetic considerations, the radiation protective housing for the accelerator components and the patient irradiation rooms dominates the layout and the construction. Secondary neutrons are the main particles to be shielded. Thus, massive material must be set against them. Patients should perceive that they are in contact with medical equipment only, and for good reason, from a rather emotional point of view. Thus, a wide span of different requirements is to be considered when designing and building such a dedicated facility.

Shielding

The traditional technology for building a bunker is mass concrete construction. Such wall sizes range from 2–7 metres, resulting in huge hydration heat during dry out. To keep the walls from cracking, a considerable amount of reinforcement steel needs to be applied. Furthermore, adding dry ice at around –70 degree Celsius to the liquid concrete and only pouring at night times are mandatory, but expensive, strategies to protect against unwanted radiation-permeable cracks.

At MedAustron an alternative and more cost-effective technology was applied for the first-time at a large-scale particle accelerator site: Forster Sandwich Construction® (http://www.forster-systemverbau.de/sandwich-construction.html) (Mayer et al. 2014). This method takes advantage of the fact that the radiation shielding property of concrete is solely determined by its mineral content, that is, gravel. Since the thickness of the wall is

TABLE 5.2   Key Infrastructural Parameter

System	Value
Heat supply	Approximately 5,000 kilowatts (depending on the site).
Cooling supply	2–5 megawatts (depending on the accelerator demand), closed circuits for activated accelerator-cooling system.
Electrical power	5–10 megavolt ampere (pulsed + base). UPS for IT infrastructure and safety relevant equipment.
Ventilation	Activated air handling, hygienic requirements in medical areas.
Water and sewage	Potable water, no activated discharge.
Control systems	For all Heating, Ventilation, and Air Conditioning (HVAC) and accelerator systems.
Safety management system (SMS)	Patrol control system (PCS), medical paging system, HVAC alarm systems.
Radiation protection	According to national standards, gamma and neutron detection and monitoring.
Communication systems	Fast internet connection for medical imaging exchange.
Medical gases	'Air medicalis', vacuum, oxygen.

not a structural requirement but a concern of radiation protection, the sandwich construction applies loose mineral material in 'boxes' made of prefabricated partition walls working as the load carrying elements. The loose mineral material resulting from excavation is compacted on-the-spot to the density of conventional concrete, thus resulting in similar shielding properties but at a much lower cost than concrete.

Technical Infrastructure

The technical infrastructure provides all required media for the different parts of the facility: medical examination and treatment, medical devices and apparatuses, the particle accelerator and the administration areas. Typical key figures and systems are shown in Table 5.2 (Mayer et al. 2014).

IONS FROM SOURCE TO TUMOUR

Accelerator Complex and High-Energy Beam Transport Line

The design of the accelerator building clearly depends strongly on the chosen accelerator technology. For synchrotron-based concepts, the system includes an injector, where ions from different ion sources are pre-accelerated by a linear accelerator (LINAC); a synchrotron and a high-energy beam transport (HEBT) system to deliver the beam to various beam ports at the irradiation rooms (Mayer et al. 2014). Cyclotron-based facilities are characterised by a much more compact design. The main components of a cyclotron are semi-circular hollow electrodes, a high-frequency power source and a powerful electromagnet. At maximum acceleration, protons released from the atom reach the inner edge of the cyclotron and are then directed outside the cyclotron by deflector magnets to a beam transport system. The strength of the alternating voltage magnetic field on the electrodes remains constant, and therefore the cyclotron produces a beam of particles with a constant speed and, most important, a constant energy. To adapt to the tumour-specific penetration depth, energy degraders are applied.

Irradiation Rooms: Equipment Choice and Integration

Several factors determine the number of treatment rooms for a facility (number of patients, type of beam delivery, treatment time, Quality Assurance (QA) and maintenance time, etc.). Dual particle facilities usually comprise several treatment rooms – one or two rooms are equipped with fixed beam lines (horizontal and vertical) and one room is equipped with a gantry. Until now, only two carbon gantries have been built (Haberer et al. 2004; Iwata et al. 2012), and from a practical point of view, only proton gantries will be available in new facilities until new advances in beam delivery technology are achieved. The standard realisation of a beam delivery system (BDS) in a fixed beam room of a dual particle facility is based on nozzles without movable snouts, and this arrangement is used for carbon ions as well as for protons. In order to preserve the ballistic capabilities of protons, patients need to be positioned as closely as possible to the exit window of the fixed beam nozzle, and a so-called non-isocentric treatment setup should be used. Implementation of this strategy is only possible if the patient alignment system (PAS) is supported by a comprehensive collision avoidance system. Integration of all this complicated and expensive equipment requires careful planning, staff training and effective interaction with the manufacturers. It is a task for the user's team to combine the recommendations from different suppliers into the logical sequence and develop a careful planning schedule. For example, the complete acceptance of the PAS would not be possible (testing of collision avoidance system) if the nozzle has not yet been installed and the anti-collision system has not yet been accepted.

Dosimetry

The implementation of radiotherapy equipment into clinical practice involves a variety of measurements and calibration procedures that require the use of different dosimetry equipment and associated phantoms for determination of various physics and beam data. The users should acquire not only standard devices and detectors like computerised water scanners, films, 2D array and different ionisation chambers, but also the equipment specifically devoted to the pencil ion beam scanning delivery technique, such as water column, 2D scintillating screen detector or multilayer ionisation chambers. Before dosimetry equipment use for the measurements, a series of acceptance tests must be conducted. Their aim is to make sure that the equipment meets the specifications previously agreed upon. Finally, the needed detectors, computerised beam scanners, multi-detector arrays, film scanners, phantoms and other measurement equipment should be commissioned when the user has to check the ability of the equipment to meet the requirements in terms of functionalities and accuracy (Moyers and Vatnitsky 2012). The user must also prepare different tools for mechanical and alignment checks.

Dose Delivery System

Currently two types of ion beam delivery systems are available. The first technique, called passive scattering, uses scatterers or wobblers spreading the beam laterally and varying the depth of penetration to cover the target sufficiently with dose while minimizing the dose delivered to normal tissues with patient-specific apertures. The depth of penetration is varied uniformly with off-axis position using rotating modulators or static ridge filters to create a region of uniform depth dose across the target. Protection of normal tissues distal to the target is performed by inserting a 3D bolus in the beam path just proximal to the patient's skin. The second technique –PBS deflects the pencil bean magnetically (quasi-discrete or raster scanning) while modulating both the ion flux and energy as a function of off-axis position, thus allowing better dose conformity and sparing of healthy tissues. This technique doesn't require patient-specific devices, eliminates the time for installing the devices and thereby reduces the treatment time for each fraction. The main hardware components of the scanning BDS are the nozzle, the scanning magnet power converters and connected scanning magnets. Included in the nozzle are dose and spot position monitors, and passive elements (range shifter and ripple filter). Compared to nozzles for dual particle delivery, nozzles in proton facilities are usually equipped with movable snouts. The spot position monitors provide information about the spot position during irradiation, and the dose monitors measure the dose delivery during radiation treatments. The ripple filter is a range modulator device that produces just enough variation in ion energies entering the patient so that a reduced number of accelerator energies may be used without producing ripples in the depth dose distribution. Separate ripple filters should be used for protons and carbon ions. The range shifter is a device that degrades the energy of the proton or carbon beam to treat superficial tumours. The user should be aware that, compared to the passive scattering delivery method, the modulated scanning technique takes slightly longer and increases the treatment time for each fraction. Using the modulated scanning technique also requires much more beam time for performing patient-specific QA activities, reduces the number of patients treated each day and increase the facility cost per patient.

Treatment Planning

When selecting a treatment planning system (TPS) the user should look for support for different delivery techniques, availability of Monte Carlo algorithms, the ability of dose summation with other modalities, robust planning and scripting. A primary difference in the treatment planning process for light ion beams and conventional beams is the need to account for the biological effects of light ion beams on tissue. Currently, the common solution for proton beams is to use a relative biological effectiveness (RBE) factor that is constant over the beam irradiation volume, and the required number of particles (NP) or monitor units (MUs) for each portal is calculated by reducing the planned dose by a value of 1.1. In contrast to protons, a TPS for ions heavier than protons should take into account the change of RBE as a function of ion species, ion energy, depth of penetration, type of tissue and so on. In addition to RBE aspects, other major development items for ion TPSs include multi-criteria optimisation focused on robust planning and Monte Carlo dose calculation algorithms. Due to the large amount of calculations for each of these development items, algorithms are being developed to take advantage of multiple processors and graphical processing units (GPUs) (Jia et al. 2012).

Patient Positioning and Positioning Verification

The optimal application of ion beam treatments requires high precision in aligning the patient with respect to the treatment beam in order to secure accurate and reproducible dose delivery according to the plan. To comply with this requirement, each treatment room of a particle facility is equipped with a PAS. The major components of a PAS consist of the patient positioner (PP), devices to register the patient to the PP, devices to immobilise the patient with respect to the PP, a collision avoidance and detection system (CADS) and the patient position verification system (PPVS) that itself consists of X-ray sources, image capture devices, image processing software and alignment software. Modern facilities and all new facilities are equipped only with robotic positioners. The robotic PPs are usually floor mounted; however, the ceiling-mounted robotic PP can improve the possibility of a non-isocentric setup (Stock et al. 2016). Automated collision detection should be integrated into the treatment planning phase and in the treatment room, robotic table movements, imaging and beam delivery should be supported by the record and verify system and checked upfront for collision. Finally, during robot movements the same checks are performed in real time for avoiding potential collision. The user should be aware that maximizing machine treatment time is essential to increase patient throughput because a complex patient setup and immobilisation consume valuable time. To improve the facility workflow, it was proposed to immobilise the patient on an interchangeable tabletop outside the treatment room and to bring the immobilised patient on the stretcher to the treatment room and to dock the tabletop to the PAS (Stock et al. 2017). However, the user should evaluate the cost/benefit ratio of the whole package (stretchers, transfer plates, docking couplings, etc.) by analysing the logistics of the workflow. It is unlikely that all patients at a hadron therapy facility will require stretcher transportation, but it is useful to support offline Positron Emission Tomography (PET) range verification, when the patient should be quickly transported to the PET imaging room.

The localisation and verification systems at particle facilities (stationary or robotic based) use kilo-voltage X-ray images to adjust the patient's position in the treatment room with respect to digitally reconstructed radiographs (DRRs) generated by the TPS. With the implementation of modulated scanning beams, planar imaging is in the process of being replaced by the beam's eye view (BEV) imaging when the X-Rays pass through the radiation head before passing through the patient (Pedroni, 2004), or by the cone-beam computed tomography (CBCT) systems in which an imaging ring is placed either on the treatment couch of the PPS, or on the C-arm (Deutschmann et al.,2013). In some cases, for patients treated in a vertical position, for example, when seated in a chair, a CT that extends down from the ceiling is used (Vatnitsky and Moyers 2013). Future developments will be probably focused on the ion computed tomography (ICT) using the same ion species used for treatment. The first prototype of a proton CT scanner is close to clinical implementation (Giacometti et al. 2017). This technique allows reducing uncertainties in beam penetration and can also be used for daily localisation imaging.

FACILITY COMMISSIONING

Acceptance testing is by definition verifying that the components of ion beam therapy equipment are functioning as specified and can be calibrated for clinical implementation. An accelerator, its control system and BDS with nozzles are the most important parts of the facility, and the acceptance and commissioning of these components requires close collaboration of the accelerator physics team and medical physics team at the facility. As a first step the particle beam is delivered to the medical irradiation room even if it is outside the clinical specification, and then in the next step it is tuned to comply with the agreed clinical specifications. It is obvious that the development of the acceptance testing procedure for the equipment supplied by different suppliers is a complicated task and requires understanding of the interaction of different components and coordination of the testing process.

Following the acceptance of an accelerator complex and of all medical equipment at a particle facility, a full characterisation of its performance for clinical use over the whole range of possible operation must be undertaken. This process is called medical commissioning and is differentiated from accelerator commissioning. The accelerator commissioning is performed by the accelerator physics group to bring the physical beam properties (number of ion spices, energies, beam and spot sizes, NP per spot or MU per portal and their correspondent position and dose stability etc.) to the specifications previously agreed upon. The major tasks for medical commissioning of a hadron therapy facility are to calibrate the various parameters of the BDS, PAS and medical software and test these systems under varying clinical conditions to determine appropriate intervention thresholds, acquire data for entry into the TPS and perform end-to-end tests for planned patient treatments. Medical commissioning involves a variety of different measurements that will be performed in order to characterise the complete irradiation system and, as a result, give confidence that not only the characteristics of the beam delivered to the patient comply with clinical requirements but also the resulting dose distribution is in agreement with the prescribed treatment plan. It is the responsibility of the user to implement periodic Quality Control (QC) checks to ensure that the performance of the systems remains consistent in clinical practice within accepted tolerances on a daily basis.

REFERENCES

Baron, M.H. et al. 2004. One-day survey: As a reliable estimation of the potential recruitment for proton- and carbon- ion therapy in France. *Radiother Oncol* 73 (S2): 18–20.

Deutschmann, H. et al. 2013. Robotic positioning and imaging. *Strahlenther Onkol* 189: 185–188.

Giacometti, V. et al. 2017. Software platform for simulation of a prototype proton CT scanner. *Med Phys* 44: 1002–1016.

Haberer, T. et al. 2004. The Heidelberg ion therapy center. *Radiother Oncol* 73 (S2): 186–190.

Iwata, Y. et al. 2012. Design of a superconducting rotating gantry for heavy-ion therapy. *Phys Rev ST Accel Beams* 15: 044701, 1–14.

Jia, X. et al. 2012. GPU-based fast Monte Carlo dose calculation for proton therapy. *Phys Med Biol* 57: 7783–7788.

Krengli, M., and Orecchia, R. 2004. Medical aspects of the national centre for oncological hadrontherapy (CNAO—Centro Nazionale Adroterapia Oncologica) in Italy. *Radiother Oncol* 73 (S2): 21–23.

Langendijk, J. et al. 2013. Selection of patients for radiotherapy with protons aiming at reduction of side effects: The model-based approach. *Radiother Oncol* 107: 267–273.

Mayer, R. et al. 2004. Epidemiological aspects of hadron therapy: A prospective nationwide study of the Austrian project MedAustron and the Austrian Society of Radiooncology (OEGRO). *Radiother Oncol* 73 (S2): 24–28.

Mayer, R., Magrin, G., and Schreiner, T. 2014. *Ion Beam Radiotherapy at MedAustron*. Vienna, Austria: Robitschek.

Moyers, M.F., and Vatnitsky, S. 2012. *Practical Implementation of Light Ion Beam Treatment.* Madison, WI: Medical Physics Publishing.

Orrecchia, R., and Krengli, M. 1998. Number of potential patients to be treated with proton therapy in Italy. *Tumori* 84: 205–208.

Pedroni, E. et al. 2004. The PSI Gantry 2: A second generation proton scanning gantry. *Z Med Phys* 14: 25–34.

Stock, M. et al. 2016. Development of clinical programs for carbon ion beam therapy at MedAustron. *Int J Part Ther* 2 (3): 474–477.

Stock, M. et al. 2017. Optimization of a dual particle facility for protons: acceptance and commissioning results of the whole treatment workflow at MedAustron. *PTCOG 56*, May 11–13, Yokohama, Japan.

Vatnitsky, S., and Moyers, M.F. 2013. Radiation therapy with light ions. In: *The Modern Technology in Radiation Oncology*, J. Van Dyke (Ed.). Madison, WI: Medical Physics Publishing.

CHAPTER 6

Imaging and Particle Therapy

Current Status and Future Perspectives

G. Landry, G. Dedes, M. Pinto and K. Parodi

CONTENTS

INTRODUCTION

Ion-beam therapy, hereafter also referred to as particle therapy, is a rapidly emerging treatment modality which makes use of the favourable interaction properties of swift ions in matter. In particular, ion beams offer the ability to confine the maximum energy deposition to a narrow region at an adjustable penetration depth – the so-called Bragg peak. This ion-specific characteristic opens new possibilities for tighter conformation of the dose to arbitrarily complex tumour shapes for superior dose escalation to the tumour and/or sparing of normal tissue and critical organs in comparison to conventional photon radiotherapy. Already in the early pioneering phase of ion-beam therapy, since the 1980s, the enhanced ballistic selectivity of the dose delivery triggered integration of, at that time, advanced imaging technologies such as orthogonal X-ray radiographies employed for patient positioning (Gragoudas et al., 1979; Verhey et al., 1982); however, for many decades, in-room image guidance for ion-beam therapy did not evolve much (Engelsman et al., 2013), eventually lagging behind the tremendous innovation of in-room and on-board volumetric imaging for photon therapy (Verellen et al., 2007). Fortunately, with the rapid clinical spread of facilities enabling even more accurate ion-dose delivery with state-of-the-art pencil beam scanning (PBS), several vendors started introducing innovative solutions for integration of volumetric in-room image guidance in ion-therapy treatment rooms (See Section 'In-Room X-ray-Based Imaging'). Along with the exploration of new opportunities of advanced X-ray imaging for treatment planning and adaptation (See Sections 'In-Room X-ray-Based Imaging' and 'Dual Energy-Computed Tomography') beyond the realm of solely patient positioning, several investigations are also ongoing to develop novel imaging modalities aimed at reducing the major challenge of range uncertainties in clinical ion-beam therapy (See Sections 'Ion-based Imaging' and 'Imaging for Treatment Verification'). In fact, the accuracy of the ion-dose delivery does not only depend on the initial anatomical patient model and its correct replication at the treatment site, but also on the knowledge of the tissue-stopping properties and the correct delivery of each individual Bragg peak building up the total dose delivery. While first attempts of novel means for probing patient tissue with ion beams and visualising treatment delivery with radioactive-implanted beams or by-products of irradiation date back to the 1960–1970s (Koehler, 1968; Tobias et al., 1977; Bennett et al., 1978; Sommer et al., 1978), current solutions being developed for clinical translation can rely on modern detector technologies and powerful computers for unprecedented performances. Moreover in the last few years, new ideas emerged (or regained interest) for *in vivo* treatment verification based on the detection of additional secondary emissions carrying information on the dose delivery, stimulating a wide range of research activities aimed at enabling full clinical utilisation of the theoretical advantages of ion beams for radiation therapy in the clinical practice.

This contribution reviews the evolution and ongoing research of modern imaging technologies which promise to play a crucial role in all stages of treatment planning and treatment delivery as well as *in vivo* treatment verification and adaptation for highly accurate ion-beam therapy.

IMAGING FOR PATIENT POSITIONING, TREATMENT PLANNING AND ADAPTIVE THERAPY

In-Room X-Ray-Based Imaging

Lost Ground to X-Ray Radiotherapy

The volumetric image-guidance implementation delay of particle therapy can partly be attributed to the small number of and significant heterogeneity in design between facilities, many of which were conceived before the widespread adoption of cone-beam computed tomography (CBCT) imaging in linear accelerator (linac)-based photon radiotherapy (Engelsman et al., 2013). Additionally, the room and gantry layouts of particle therapy

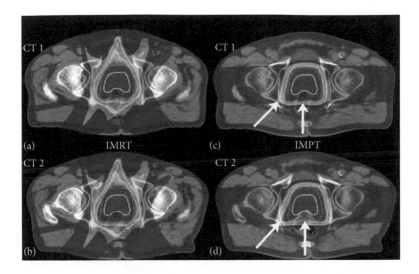

FIGURE 6.1   (a) Eight-field intensity-modulated radiotherapy (IMRT) plan optimized on planning CT (CT 1) and (b) recalculated on a control CT image (CT 2) aligned to CT 1 by matching the target which was delineated on both images. The IMRT dose distributions show little differences between CT 1 and CT 2. (c and d) A two-field full intensity-modulated proton therapy (IMPT) plan shows dose distribution degradation (marked by arrows) due to WET differences between (c) CT 1 and (d) CT 2. This illustrates the limitations of the shift invariance assumption for particle therapy. The clinical target volume (CTV) and planning target volume (PTV) are shown as green and blue contours. (From Landry, G. and Hua, C., *Med. Phys.*, 2018, in press. With permission.)

facilities can differ significantly from their linac counterparts, limiting the direct applicability of existing volumetric image-guidance solutions. A typical example would be the longer source to detector distances requiring higher output X-ray sources. Finally, in photon-based it is possible to obtain a reasonable estimate of the dose distribution resulting from a given patient position correction simply by displacing the original dose distribution calculated at treatment planning by the corresponding patient shift vector. This so-called shift invariance assumption for dose distributions (McCarter and Beckham, 2000; Booth and Zavgorodni, 2001; van Herk et al., 2002) is valid for X-ray radiotherapy at reasonable distances from the skin (Craig et al., 2003) and facilitates correcting for target displacements observed from volumetric imaging, since a new dose calculation can be avoided. As illustrated in Figure 6.1, this frequently breaks down for particle therapy due to the sensitivity to water-equivalent thickness (WET) changes. This entails that for particle therapy a new dose calculation may need to be performed to verify that the desired dosimetric improvement from a patient shift are realised; nonetheless, the past few years have seen particle therapy make up lost ground in terms of volumetric image guidance.

Current State of Volumetric-Image Guidance

Recently, X-ray-based volumetric-imaging capability has become commercially available and has been clinically implemented at new ion-therapy facilities. The main types of existing volumetric-imaging technologies found in modern particle therapy centres

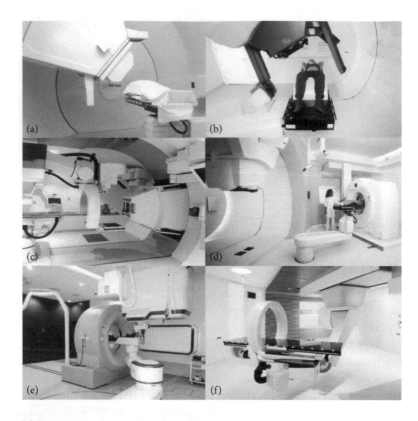

FIGURE 6.2 (a) Varian gantry-mounted CBCT at Scripps Proton Therapy Center (Only one of the two flat-panel detectors is used for CBCT); (b) Ion beam applications (IBA) nozzle-mounted CBCT; (c) C-arm-mounted CBCT at St. Jude Children Research Hospital; (d) Siemens in-room CT requiring couch rotation at IBA gantry-equipped facility; (e) i-ROCK (ion-beam Radiation Oncology Center in Kanagawa, medPhoton) Toshiba in-room CT with no couch rotation at the fixed-beam line facility of the ion-beam Radiation Oncology Center in Kanagawa, and (f) medPhoton couch-mounted CBCT. (From Landry, G. and Hua, C. *Med. Phys.*, 2018, in press. With permission.)

are shown in Figure 6.2. As seen in this figure, two main approaches have been adopted: gantry-, nozzle-, couch- or C-arm-mounted CBCT for imaging the patient directly in the treatment position at isocentre, and in-room CT on rails requiring the movement of the immobilised patient from the imaging to the treatment position. Key characteristics of each approach are summarised in Table 6.1 (Landry and Hua, 2018, in press).

Beyond Patient Positioning

The capability to perform WET or dose calculations on daily volumetric images is crucial for adaptive particle therapy for the reasons presented in Figure 6.1 which illustrates the sensitivity of particle therapy to inter-fractional anatomical changes. The calculation of WET is inherently simpler when using volumetric-imaging data acquired with a CT on rails than with CBCT (Kurz et al., 2015) due to the well-documented image-quality issues of the latter (Thing et al., 2016). CBCT image-correction techniques based on deformable image registration at the image level, possibly complemented by scatter corrections at the

TABLE 6.1 Comparison of Radiological Volumetric Image Guidance Solutions in Particle Therapy

Type	Currently Used for	Vendors	First Clinical Use	On- or Off-Isocentre Imaging	Advantages	Current Challenges
Gantry-mounted CBCT	360 gantry	Hitachi, IBA, Sumitomo, Varian	2014	On	Imaging at treatment isocentre; clearance with retracted imaging equipment; experience from more users	Large field of view (FOV) with half fan mode; CT number accuracy; difficulty in retrofitting to existing facilities
Nozzle-mounted CBCT	Partial gantry	IBA	2015	On	Viable solution for partial gantry and treatment room with limited space	Physical clearance; large FOV with half fan mode; CT number accuracy
Robotic C-arm CBCT	Partial gantry	Hitachi	2015	Both	Viable solution for partial gantry; multiple imaging positions; high accuracy and precision robots; easier to upgrade with decoupling imaging system	Size of robot for occupying ceiling or floor space; collision avoidance; robot movement time; CT number accuracy
In-room CT on rails	Both	Siemens, Toshiba	1997	Off[a]	Soft tissue image quality; larger radial and longitudinal FOV; 4D CT capability; automatic exposure control; dual energy CT potential	System integration; lack of imaging at treatment isocentre; longer overall image guidance time due to additional couch and CT movement
Couch-mounted CBCT	Both	medPhoton	2016	Both	Longitudinal FOV in helical mode; viable for both 360 and partial gantries; fan beam CT possible with collimation	Impact of equipment weight on couch position accuracy; collision avoidance; CT number accuracy

Source: Reproduced with permission from Landry, G. and Hua, C., *Med. Phys.*, 2018, in press.

[a] The majority of the in-room CT scanners were configured for off-treatment isocenter imaging except for those specially modified to descend from the ceiling for the seated patients.

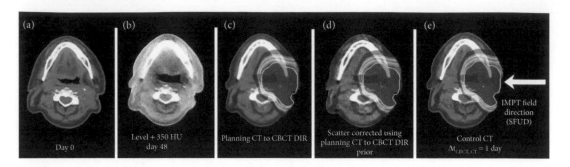

FIGURE 6.3 Example of in-room volumetric-imaging data for dose calculation. All images acquired in the context of photon IMRT and used for illustration purposes. (a) Planning CT scan; (b) late treatment in-room CBCT scan showing weight loss; (c–e) dose distributions for a single-field uniform dose (SFUD)-adapted plan using IMPT; (c, d) doses were recomputed on CBCT images corrected using (c) deformable image registration (DIR) of the planning CT to the daily CBCT image. (From Landry, G. et al., *Med. Phys.*, 42, 1354–1366, 2015.) and (d) using the result of the deformation as prior information used to estimate a scatter correction. (From Park, Y. K. et al., *Med. Phys.*, 42, 4449–4459, 2015.) In (e), an in-room equivalent control CT acquired one day after the CBCT was used to calculate the adapted plan and serves as reference. (From Kurz, C. et al., *Med. Phys.*, 43, 975–982, 2016. With permission.)

projection level, have recently shown potential for accurate WET and dose calculations in particle therapy. A few examples are summarised in Figure 6.3; however, regardless of the ability of CBCT intensity correction to restore accurate Hounsfield unit values, all WET and dose calculations relying on empirical or stoichiometric calibration of single energy-computed tomography (SECT) data into relative stopping power (RSP) bear uncertainties frequently quoted as high as 3.5% (Paganetti, 2012; Yang et al., 2012). This calls for advanced imaging which aims to improve the patient model for treatment planning and adaptation (See next sections).

Dual Energy-Computed Tomography

The first attempts of using dual energy-computed tomography (DECT) (Hounsfield, 1973; Alvarez and Macovski, 1976; Brooks, 1977) to reduce RSP estimation uncertainties were presented during 2009 in conference papers (Bazalova and Verhaegen, 2009; Beaulieu et al., 2009). These early attempts made use of the methods of Bazalova et al. (2008a, 2008b), an extension of monoenergetic synchrotron X-ray work (Torikoshi et al., 2003, 2005) to poly-energetic X-ray spectra which required knowledge of the X-ray spectra employed (Bazalova and Verhaegen, 2007). In those first attempts, the DECT estimated relative electron density (ρ_e) and effective atomic number (Z_{eff}) were used to segment images into materials for Monte Carlo (MC) simulation using Monte Carlo N-Particle extended, a general-purpose Monte Carlo radiation transport code (MCNPX), and the authors concluded that misassignment of materials caused large mass density (ρ) errors that did not support the use of DECT.

The concept was revisited in a theoretical paper (Yang et al., 2010) based on the same methods as Bazalova et al. (2008a) which bypassed material segmentation for direct use of ρ_e in the Bethe equation to calculate RSP:

$$RSP = \frac{\rho_e \left[\ln \left[2m_e c^2 \beta^2 / I(1-\beta^2) \right] - \beta^2 \right]}{\ln \left[2m_e c^2 \beta^2 / I_{water}(1-\beta^2) \right] - \beta^2} \qquad (6.1)$$

where:

m_e is the electron mass

β is the velocity relative to the speed of light c

I and I_{water} are the mean excitation potentials of the medium and water, respectively

Yang et al. realised that the DECT-extracted Z_{eff} could be converted into ln I by piecewise linear fits (Yang et al., 2010) for tabulated soft and bone tissues (Woodard and White, 1986), thus providing all ingredients for RSP calculation for so-called pencil beam dose calculation algorithms with an accuracy within 1%. The first experimental validation using a radiology dual-source scanner was presented in a pair of papers by Hünemohr et al. (2013, 2014a). While for tissue-mimicking plastics the work of Hünemohr et al. and others confirmed the expectations of Yang et al. compared to residual range measurements (Hünemohr et al., 2014a; Hudobivnik et al., 2016), an early attempt at validation using biological tissues proved inconclusive (Hünemohr et al., 2013). This was attributed by the authors to experimental uncertainties stemming from the measurement of residual range with a large area integrating detector behind tissue samples which are heterogeneous by nature (Hünemohr et al., 2013). Using a pencil-beam algorithm for treatment planning on surrogate radiology DECT scans of head trauma patients, Hudobivnik et al. showed range differences between SECT and DECT in the order of 1–2% of the proton range (See Figure 6.4 for an example case)

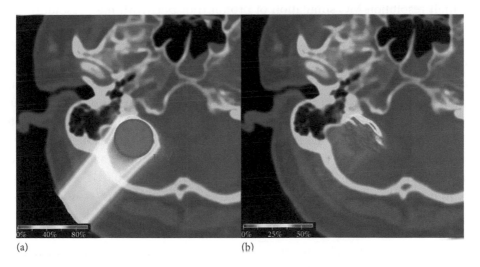

(a)   (b)

FIGURE 6.4 Comparison of head tumour proton therapy treatment plans optimized on the basis of (a) DECT images and SECT images (not shown) and (b) the absolute dose difference between the two plans was calculated to evaluate the range difference. For both panels, the colour bar is in percentage of the prescription dose. (From Hudobivnik, N. et al., *Med. Phys.*, 43, 495, 2016. With permission.)

(Hudobivnik et al., 2016). This would entail range differences of up to four millimetres for tumours at a 20-centimetre depth, and recent findings for prostate cancer patients using a clinical DECT scanner support this figure (Wohlfahrt et al., 2017a).

Since the seminal work of Yang et al., several groups have presented alternative methods to estimate RSP directly via DECT data (Bourque et al., 2014; Farace, 2014; Hünemohr et al., 2014a, 2014b; Hansen et al., 2015; Han et al., 2016; Hudobivnik et al., 2016; Lalonde and Bouchard, 2016; Mohler et al., 2016; Taasti et al., 2016; Zhu and Penfold, 2016; Bar et al., 2017) or via estimation of tissue composition (Landry et al., 2013; Hünemohr et al., 2014c; Lalonde and Bouchard, 2016; Berndt et al., 2017) where the latter allows replacing pencil beam algorithms by MC simulations for particle therapy. The early issues faced by Bazalova et al. when using DECT data for MC simulation have been circumvented by either converting ρ_e to ρ using a linear fit (Landry et al., 2011), or ensuring consistency between direct RSP estimation and the stopping power used in MC software by adjusting either ρ or I (Berndt et al., 2017).

While most of the seminal work on the use of DECT for particle therapy has been based on state-of-the-art dual-source radiology scanners (Flohr et al., 2006), the first clinical implementations for radiotherapy have relied on the so-called dual spiral technology where two scans are acquired back-to-back and registered (Wohlfahrt et al., 2017b). Although scanners have become available in the clinic, DECT import functionality is still lacking in clinical treatment planning systems (TPS), explaining why the first paper on the use of a DECT scanner for particle therapy made use of 120 peak kilovoltage equivalent – in terms of range – pseudo-monoenergetic images (Wohlfahrt et al., 2017b).

There remains the need for a conclusive experimental validation using biological tissues. The shortcomings mentioned earlier in this section will most likely require the combination of high resolution MC simulation of proton transport with high resolution residual range measurements behind samples (Xie et al., 2016).

Ion-Based Imaging

Ion-based imaging has recently seen increased research interest as a candidate modality for particle-therapy treatment planning, patient positioning, and dose recalculation. Instead of converting photon attenuation measured with X-rays (SECT or DECT) to RSP, the latter can be directly obtained by means of imaging with ion beams at the treatment position. The basic concept is that the RSP line integrals of ions traversing an object can be calculated by measuring the residual energy or range of the particles. A tomographic image can be reconstructed by acquiring several angular projections (radiographies) of the scanned object. The advantage would be more accurate RSP maps at lower imaging dose with respect to X-ray CT.

The utilisation of protons for imaging was already proposed in 1963 by Cormack (1963). The first particle radiographs and tomographies were published by Kohler (1968) and Goitein (1972) for protons and α particles, respectively. Cormack and Koehler (1976) demonstrated that proton computed tomography (pCT) can detect low-contrast features. In 1982, further work by Hanson et al. (1982) resulted in pCT images of human tissue specimens which tested two different detector concepts for measuring residual energy/range – calorimeter and range telescope – (Hanson et al., 1978, 1979, 1982) who also proposed the employment of proton exit angle selection for improved spatial resolution. That first era of ion-based imaging research was concluded in the 1980s with the work of Ito and Koyamaito (1987) and Takada et al. (1988).

In the past two decades, several different detector schemes for ion-based imaging have been proposed. They can be divided into two main concepts: (1) single particle tracking and (2) particle-integrating detectors.

Single particle-tracking detectors typically consist of tracker modules which detect the position of each particle before and after the scanned object and a residual energy/range detector after the object. An example of such a prototype scanner for pCT is shown in Figure 6.5a. Imaging systems based on range telescopes as residual range detector were developed in Paul Scherrer Institute (PSI) (Pemler et al., 1999), by the foundation Terapio con Radiazioni Adroniche (TERA) (Amaldi et al., 2011), and the Proton Radiotherapy Verification and Dosimetry Applications (PRaVDA) consortium (Poludniowski et al., 2014a). Imaging systems based on calorimetry were developed by the Loma Linda University (LLU) and the University of California Santa Cruz (UCSC) (Sadrozinski et al., 2013), the Proton IMAging (PRIMA) collaboration (Sipala et al., 2011; Civinini et al., 2013), and the Niigata University (Saraya et al., 2014). For ^{12}C ions, a prototype system based on calorimetry was developed at National Institute of Radiological Sciences (NIRS) (Shinoda et al., 2006). A hybrid system for protons combining the concepts of a range telescope and a calorimeter was presented by the LLU/UCSC collaboration (Johnson et al., 2014; Bashkirov et al., 2016).

Particle-integrating detectors do not register individual particles but rely on the formation of the signal by many particles, for example, a Bragg peak in a stacked detector after the scanned object. A tracking system might or might not be required depending on the ion species and beam delivery. An example of a particle-integrating scanner for ^{12}C tomography is shown in Figure 6.5b. Particle-integrating systems for proton imaging have been presented in (Zygmanski et al., 2000; Seco and Depauw, 2011; Testa et al., 2013; Bentefour et al., 2016).

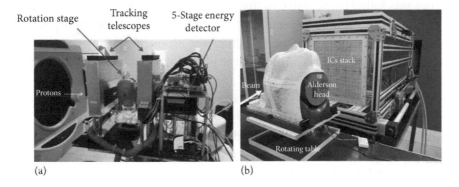

FIGURE 6.5 (a) Photograph of the LLU/UCSC single proton-tracking scanner indicating the scanner components and a paediatric head phantom as the scanned object. (From Sadrozinski, H. F. W. et al., *Nucl. Instrum. Meth. A*, 831, 394–399, 2016. With permission.) (b) Photograph of the HIT ^{12}C- integrating scanner indicating the residual energy detector and the Alderson head phantom. (From Rinaldi, I. et al., *Phys. Med. Biol.*, 59, 3041–3057, 2014. With permission.)

For ^{12}C ions, a first particle-integrating prototype was developed by Abe et al. (2002) while different setups based on a flat panel or a range telescope were investigated by the Heidelberg Ion-Beam Therapy Center (HIT) (Telsemeyer et al., 2012; Rinaldi et al., 2013). It should be noted that there is a growing list of variations of the particle-imaging prototypes named above, but it is beyond the scope of this chapter to comprehensively describe all of them.

The spreading of particle therapy facilities stimulated several studies on ion-based imaging by particularly focusing on the impact of proton radiography and pCT on proton therapy. In Schneider and Pedroni (1995), proton radiographies with a single particle-tracking detector were produced, verifying patient position with an accuracy better than two millimetres. Pemler et al. (1999) developed a detector with tracker modules before and after the object capable of a one megahertz single proton detection rate, achieving proton radiographs with spatial resolution of approximately one millimetre and confirming the limitations in spatial resolution due to multiple Coulomb scattering. With the same system, Schneider et al. (2004) achieved proton radiographies of a large animal patient with spatial resolution of one millimetre at an imaging dose of 0.03 milligray. In a recent systematic study (Saraya et al., 2014), the spatial resolution achievable in proton radiography was shown to be less than one millimetre for a 25-millimetre thick phantom and about 1.2 millimetres for a 200-millimetre thick phantom. Further improvements were proposed in Krah et al. (2015) where advanced data processing with prior X-ray information could enhance the spatial resolution of proton radiographies acquired with a particle-integrating detector system.

To complement the abovementioned experimental efforts, simulation studies of spatial resolution for ideal proton radiography (no detector effects) revealed a superior contrast-to-noise ratio when compared to X-rays at the same imaging dose with spatial resolution in the one to two millimetres range, assuming straight proton paths (Depauw and Seco, 2011). Image reconstruction efforts to improve spatial resolution in pCT introduced filtered back projection (FBP) along curved proton paths via the most likely path formalism (Rit et al., 2013) and could achieve better spatial resolution of 0.7–1.6 millimetres compared to 1.0–2.4 millimetres with a straight-line path assumption under ideal detection conditions. This performance was found comparable to that of three state-of-the-art iterative algorithms in Hansen et al. (2016). An alternative direct reconstruction method incorporating curved proton paths was developed in Poludniowski et al. (2014b).

Proton radiography and tomography has the potential of reducing uncertainties in proton range estimation. Farace et al. (2016) found range differences of less than three millimetres between proton radiography and TPS calculations when using a commercial dosimetric detector acting as range telescope. In Wang et al. (2016), the maximum error between the nominal WET in a homogeneous PMMA (polymethyl methacrylate) phantom and the calculated WET from its proton radiography was found to be 1.5 millimetres. Doolan et al. (2015) demonstrated that the WET accuracy estimated from an X-ray CT can be improved even with the use of a single ideal proton radiography. This resulted in a reduction of the WET errors from −2.1% to −0.2%. The main goal of pCT is to provide a highly accurate three-dimensional (3D) RSP map of the patient. Consequently, Arbor et al. (2015) showed that under ideal detection conditions, mean RSP absolute deviations obtained by pCT are at least three times smaller than those obtained with calibration of X-ray images. The state-of-the-art performance of pCT was summarised (Johnson et al., 2016; Sadrozinski et al., 2016); with an advanced single particle-tracking experimental setup, a full proton scan acquired in less than seven minutes was shown able to reach an RSP accuracy of 1% at an imaging dose of about one milligray and a spatial resolution between five and eight line pairs per cm at 10% of the modulation transfer function (Plautz et al., 2016). In Figure 6.6, a pCT of a paediatric head phantom obtained with the LLU/UCSC scanner is shown. The potential to further reduce imaging dose in proton imaging by obtaining a spatially varying image quality was shown in (Dedes et al., 2017). In theoretical comparisons, pCT yielded slightly better RSP estimates than DECT (Hansen et al., 2015). Finally, Bopp et al. (2013) and Quinones et al. (2016) concluded that under ideal detection conditions, proton angular deviation and transmission rate could bring complementary knowledge on tissue composition which in its turn can improve treatment planning with MC computational engines.

Heavy ion-based imaging is less limited than proton imaging by multiple Coulomb scattering, thus reducing the need of a particle tracking system although at the cost of a usually higher imaging dose. For ^{12}C ion beams, the particle-integrating prototype (Abe et al., 2002) exhibited a spatial resolution better than two millimetres and an electron density resolution better than 0.07 while the system of Rinaldi et al. (2014) demonstrated a WET precision up to 0.8 millimetre at HIT. In Meyer et al. (2017), single-particle (list mode) detection and particle-integrating approaches for ^{12}C ion tomography were directly compared in theoretical studies, resulting in average median RSP error below 1.0% and 1.8%, respectively. A comparison of ion-based imaging for different ion species was presented by Hansen et al. (2014) under ideal single particle-tracking detection conditions.

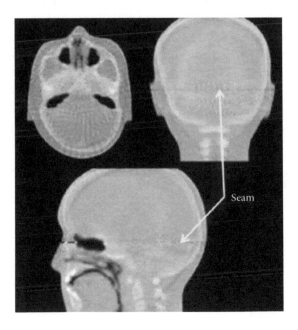

FIGURE 6.6 Tomographic reconstruction of a paediatric head phantom. Scanned acquired with the LLU/UCSC scanner. (From Sadrozinski, H. F. W. et al., *Nucl. Instrum. Meth. A*, 831, 394–399, 2016. With permission.)

The authors concluded that all ions yielded accurate stopping power estimates and the best spatial resolution was obtained with helium.

In the last decade, a small number of ion-based imaging prototype scanners have been developed. The state-of-the-art performance of those devices indicates that ion-based imaging is a promising candidate for particle therapy treatment planning, patient positioning, and dose recalculation. The next steps should be comprehensive preclinical comparisons between ion-based imaging systems and their X-ray counterparts currently in clinical particle therapy practice; therefore, the field has matured into a state where industrial support and testing in clinical facilities is essential.

IMAGING FOR TREATMENT VERIFICATION

While all the above-discussed imaging modalities can only provide information on the patient anatomy and tissue-stopping properties shortly before the start of treatment, the ultimate verification of the actual dose delivery should entail a signal directly correlated to the dose deposition. This is possible in photon therapy where the transmitted primary radiation can be detected with electronic portal imaging devices behind the patient and used for *in vivo* dose reconstruction (Mijnheer et al., 2013); however, in ion therapy, the primary beam particles are stopped in the Bragg peak placed within the tumour. Nevertheless, several secondary emissions are produced which can emerge from the patient and be detected to infer indirect information on the ion-dose delivery. Whereas the most extensively investigated methods rely on the original idea of using positron emission tomography (PET) to image the image the irradiation-induced tissue and, for $Z \geq 5$, beam β^+-activation (Tobias et al., 1977; Bennett et al., 1978), more attention has been recently given to the detection of prompt emissions of photons and charged particles from nuclear deexcitation processes as well as thermoacoustic waves. This chapter focuses particularly on the three methods of Figure 6.7 which already have been investigated in

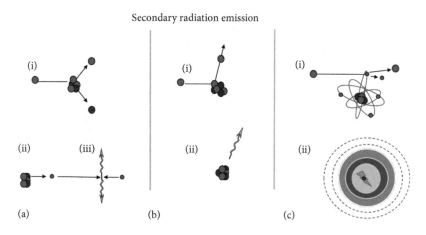

FIGURE 6.7 Secondary emissions induced by ion interaction in tissue (see text). (a) Positron annihilation gammas; (b) PGs; and (c) thermoacoustic. (Reprinted with permission from Polf, J. C. and Parodi, K., *Phys. Today*, 68, 28–33, 2015. Copyright 2015 by the American Institute of Physics.)

the clinical context although at very different levels of maturity. For ongoing research on the role of secondary protons, see Gaa et al. (2017) and citations therein.

Positron-Emission-Tomography

PET is so far the method which has been most extensively investigated clinically for verification of proton and carbon ion treatments (Parodi, 2015). The physical principle relies on the production of positron emitters – ^{15}O and ^{11}C with half-lives of approximately 2 and 20 minutes, respectively, in nuclear fragmentation reactions between the projectiles and the target nuclei of the traversed tissue (Figure 6.7i). The 511-kiloelectron volt photon pairs (Figure 6.7iii) from the annihilation of the positron emitted in β^+-decay (Figure 6.7ii) can be detected by either customised PET scanners integrated in the beam delivery (in-beam; Pawelke et al., 1996) and the treatment room (in-room; Zhu et al., 2011), or commercial nuclear medicine instrumentation (nowadays often combined with computed tomography [CT]) in a nearby room (offline; Vynckier et al., 1993). The strength of the measurable activity and its correlation with the dose delivery depend on the ion type and underlying projectile (for $Z > 1$) or target fragmentation mechanism as well as on the time elapsed between irradiation and imaging (Parodi, 2015). For example, activity induced by carbon-ion irradiation typically exhibits stronger spatial correlation but weaker signal than for protons at the same dose while offline PET imaging suffers from a low measurable signal due to physical decay and biological washout. The acquired PET image can be compared to a prediction based on a MC (Ponisch et al., 2004; Parodi et al., 2007a; Kraan, 2015) or an analytical (Miyatake et al., 2011; Frey et al., 2014) model of the expected treatment delivery, or a reference measurement acquired in a previous treatment fraction (Nishio et al., 2010). Despite promising clinical results showing the possibility to detect inter-, and in some cases, even intrafractional anatomical changes or positioning errors (Enghardt et al., 2004; Nishio et al., 2010; Bauer et al., 2013; Kurz et al., 2016; Handrack et al., 2017), along with (local) range verification accuracy and precision of few millimetres in cranial sites (Parodi ct al., 2007b; Nischwitz et al., 2015), several challenges were identified (Knopf and Lomax, 2013); however, limiting factors such as counting statistics, image quality, coregistration, and physiological washout are especially due to the current suboptimal workflows and detectors, which use or adapt instrumentation originally designed for higher statistics nuclear medicine or small animal imaging. Major improvements are thus expected from new generation scanners intentionally designed for in-beam PET application, relying either on limited angle dual-heads (Bisogni et al., 2017) or a special axially shifted full-ring design (Tashima et al., 2016) as shown in Figure 6.8. Along with hardware and software improvements, opening the perspective of on-the-fly reconstruction (Crespo et al., 2007), additional proposals aim to visualise irradiation-induced short-lived positron emitters (Buitenhuis et al., 2017) in order to overcome the major drawback of PET as an imaging technique intrinsically delayed with respect to the irradiation.

The just started or planned clinical trials with the new approaches will enable drawing more conclusive statements on the role of PET for *in vivo* treatment verification in comparison to the other emerging methods described in the following sections. This overview has been focused on the monitoring of clinical treatment with stable ion beams; however, it is

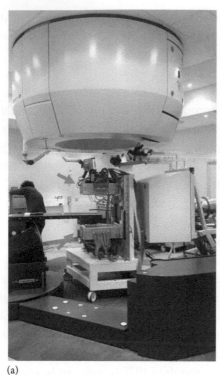

(a)   (b)

FIGURE 6.8   Example of state-of-the-art instrumentation for in-beam PET using a dual-head ((a), marked by the arrows) and an axially shifted full ring (so-called openPET) detector geometry in this example to verify the delivery of a horizontal ion-beam port. (Courtesy of [a] Giuseppina Bisogni, INFN, Pisa, Italy and CNAO, Pavia, Italy and [b] Taiga Yamaya, QST-NIRS, Chiba, Japan.)

worth mentioning that the original idea of using (low-dose exposure of) β^+-radioactive ion beams for direct localisation of their stopping position in tissue via PET imaging (Tobias et al., 1977; Llacer et al., 1984) has regained interest in Europe (Augusto et al., 2016) and Japan (Mohammadi et al., 2017). This renewed interest is mainly driven by the envisioned capabilities of facilities under construction or upgrade, opening new perspectives of post-acceleration of radioactive ion beams rather than inefficient production via nuclear fragmentation of stable beams on selected targets.

Prompt Gamma Monitoring

Another possibility for treatment verification is the prompt gamma (PG) monitoring. PG rays are emitted after the same type of nuclear fragmentation reactions as for the production of positron emitters (Figure 6.7); however, contrary to the case of PET monitoring, PG are emitted almost instantaneously, hence the measured spatial distribution is not affected by any type of physiological or physical processes taking place after emission. On the other hand, the need to monitor during irradiation and the broad energy spectrum of the emitted PG makes this a demanding technique to implement. A comparison between PET and PG distributions is shown in Figure 6.9.

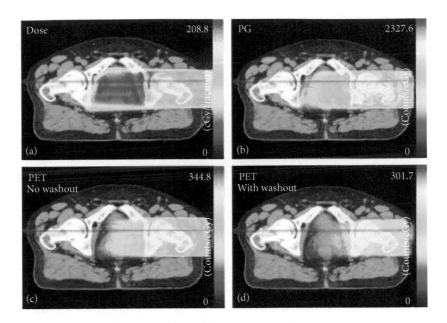

FIGURE 6.9   Comparison between PET (c: no washout modelling; d: with washout modelling) and PG (b) distributions estimated with MC simulations for a given treatment plan (prostate). The dose is also shown to illustrate the similarities and differences between dose (a) and monitoring (b–d) distributions. (Reprinted with permission from Moteabbed, M. et al., Monte Carlo patient study on the comparison of prompt gamma and PET imaging for range verification in proton therapy, *Phys. Med. Biol.*, 56, 1063–1082, 2011. Copyright 2011, Institute of Physics.) ©Institute of Physics and Engineering in Medicine. Reproduced by permission of IOP Publishing. All rights reserved.

Concerning detection methods, electronic collimation, mechanical collimation, and nonimaging methods are usually considered. Regarding electronic collimation, it comprises several designs of Compton cameras based on the Compton scattering effect. Each design tries to fully exploit different detection approaches using an arrangement of single or multiple scatter and absorber detectors in order to increase detection efficiency. Although the detection efficiency of Compton cameras is potentially higher than the ones using mechanical collimation, in reality, it is challenging to achieve high efficiencies due to the energy of the PG. The literature covers designs using scintillators and scintillators (Llosa et al., 2012, 2013; Taya et al., 2016), and semiconductors and scintillators (Kormoll et al., 2011; Richard et al., 2011; Roellinghoff et al., 2011; Hueso-Gonzalez et al., 2014; Thirolf et al., 2014, 2016) for scatter and absorbers detectors, respectively, and semiconductors only (Peterson et al., 2010; McCleskey et al., 2015; Polf et al., 2015). In addition to the detection of scattered photons, tracking the ejected electron after a Compton scattering can also be used in an attempt to increase image reconstruction efficiency (Thirolf et al., 2014). In addition, Compton cameras provide 3D data which other techniques cannot when using a single device. Several studies employing Compton cameras for PG monitoring can be found in the literature but the complexities in designing, building, testing, and commissioning such devices for range monitoring lead to a time-consuming process; therefore, most studies focus on the characterisation of the different camera components or proof-of-concept prototypes. Some developments, however, are noteworthy due to the use of clinical beams. For example, Polf et al. have tested a Compton camera at the University of Pennsylvania Roberts Proton Therapy Center. It was possible to detect three-millimetre proton range shifts with an accuracy of 1.5 millimetres while using clinical dose rates (Polf et al., 2015). Figure 6.10iii depicts the aforementioned Compton camera (semiconductors only) being used in the experimental campaign.

The mechanical collimation approaches can be divided into two categories according to whether they use: (1) spatial information (Min et al., 2006; Testa et al., 2008; Peterson et al., 2010; Roellinghoff et al., 2011; Bom et al., 2012; Smeets et al., 2012; Pinto et al., 2014); or (2) energy information (Verburg et al., 2013; Verburg and Seco, 2014). The spatial information of the PG emission is usually treated as a one-dimensional (1D) profile along the beam axis. This provides a strong indication of ion-range shifts along the beam axis but it may also hinder the detection of off-axis treatment deviations due to lateral heterogeneities (Gueth et al., 2013). The two most researched designs are the knife-edge single-slit collimator (Bom et al., 2012; Smeets et al., 2012; Richter et al., 2016; Xie et al., 2017) and the multislit collimator (Gueth et al., 2013; Pinto et al., 2014; Lin et al., 2017) cameras with the former being the one closer to be part of the clinical routine (Figure 6.10i) (Richter et al., 2016; Xie et al., 2017). By comparing the yields of several characteristic PG, a team at Massachusetts General Hospital proposed the use of energy information to assess the proton range using a single-slit camera (Figure 6.10ii) (Verburg et al., 2013; Verburg and Seco, 2014); however, this method requires detectors with high-energy resolution, a good knowledge of cross-sectional data for PG emission, and the ability to significantly reduce the neutron-induced background (such as via time-of-flight techniques which then implies the use of detectors with good time resolution). Nevertheless, this approach offers the possibility to monitor the entire irradiation field with a single measurement point, and since it relies on spectroscopic methods, it also leaves the door open for future developments on tissue and elemental composition identification (Polf et al., 2009, 2013).

Finally, the nonimaging techniques rely on the time information of the emitted PG (Golnik et al., 2014; Testa et al., 2014; Krimmer et al., 2017). The method initially presented by Golnik et al. (2014) provides the smallest footprint device suggested for PG range verification so far. It comprises a single noncollimated detector placed in a backward angle with respect to the beam direction (Figure 6.10iv). Based on the transit times of protons inside the patient and the PG emitted along their path, it is possible to estimate the proton range by measuring the PG time spectrum and calculating its average. An alternative method has been proposed by Krimmer et al. (2017) where instead of using the average value of the PG time spectrum, it utilises the integral of the PG peak in the time spectrum.

The complexity of PG monitoring has prevented a swift translation into a clinical setting. Even though the idea of exploiting PG has been discussed at least since 2003 (Jongen and Stichelbaut, 2003), the development of cameras able to efficiently cope with the clinical count rates has been a bottleneck. The main issue arises from the fact that the cameras must register the events during irradiation, thus dealing with both PG signal and background. In comparison, PET monitoring, when employed during the treatment, is usually only measured during the beam pauses. Only recently, a study of a PET scanner making use of the registered events

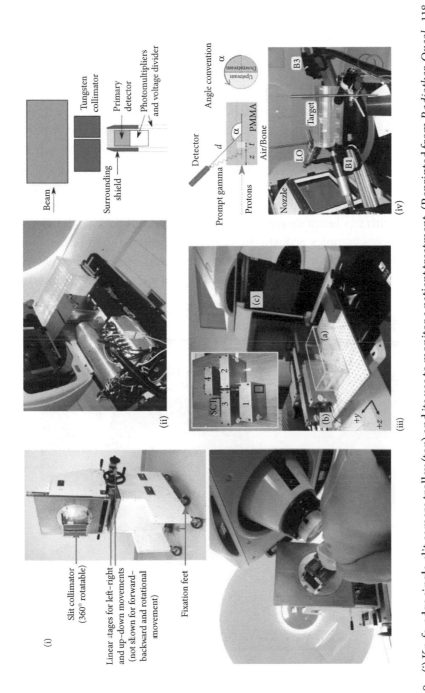

FIGURE 6.10 (i) Knife-edge single-slit camera trolley (top) and its use to monitor a patient treatment. (Reprinted from *Radiother. Oncol.*, 118, Richter, C. et al, First clinical application of a prompt gamma based in vivo proton range verification system, 232–237, Copyright 2016, with permission from Elsevier.) (ii) Experimental setup using a device aimed at exploiting the energy information of PG. (Reprinted with permission from Verburg, J. M. and Seco, J., Proton range verification through prompt gamma-ray spectroscopy. *J. Phys. Med. Biol.*, 59, 7089–7106, 2014. Copyright 2014, Institute of Physics.) (iii) Setup for measurements using a Compton camera. (Reprinted with permission from Polf, J. C. and Parodi, K., Imaging particle beams for cancer treatment, *Phys. Today*, 68, 28–33, 2015. Copyright 2015, Institute of Physics.) (iv) Setup of a PG-timing experiment. (Reprinted with permission from Huesc-Gonzalez, F. et al, First test of the prompt gamma ray timing method with heterogeneous targets at a clinical proton therapy facility, *Phys. Med. Biol.*, 60, 6247–6272, 2015. Copyright 2015, Institute of Physics.)

during beam pauses plus the ones during delivery has been reported (Sportelli et al., 2014). Nevertheless, clinical PG monitoring is becoming a reality. In 2016, the first ever PG monitoring clinical study of a proton treatment using passive scattered delivery (Richter et al., 2016) was presented and later in 2017, another breakthrough clinical study combined PG monitoring with a PBS (Xie et al., 2017). Such investigations and the prospective of large-scale clinical studies pave the way for a new level of treatment quality assurance and delivery precision.

Ionoacoustics

In contrast to secondary emissions arising from irradiation-induced nuclear reactions, thermoacoustic processes are generated by local tissue heating and subsequent expansion (Figure 6.7ii) due to electromagnetic energy deposition in tissue (Figure 6.7i); hence, this process is more directly related to the dose delivery and is naturally enhanced at the Bragg peak, thus theoretically offering a more straightforward correlation than nuclear-based approaches for range or dose monitoring. Pioneering attempts to visualise the acoustic waves associated with ion-beam delivery *in vivo* were reported back in the 1990s (Hayakawa et al., 1995) when a complex acoustic signature was detected with broadband hydrophones placed on the patient skin during passively scattered proton treatment of a liver tumour at a specially pulsed proton synchrotron (Figure 6.11); however, modern

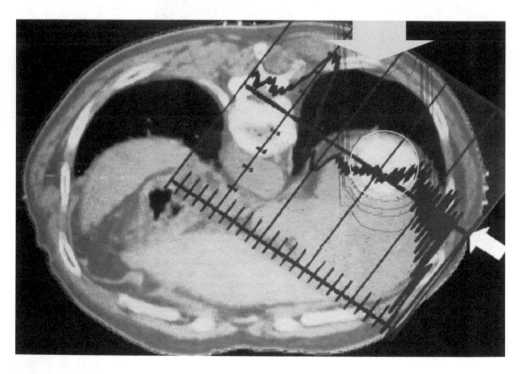

FIGURE 6.11 Example of acoustic signal (inserted graph) measured by a hydrophone placed on the patient skin (small arrow) during posterior-anterior passively scattered irradiation of a hepatic cancer patient (outlined approximate tumour location and treatment plan isodoses) by a pulsed proton beam (large arrow). (Adapted from Hayakawa, Y. et al., *Radiat. Oncol. Investig.*, 3, 42–45, 1995. With permission.)

beam scanning delivery with superposition of several well-localised energy deposition patterns indeed offers a better scenario for enhanced signal generation of more straightforward interpretation. Since the acoustic emission generated by the ion-energy deposition (hereafter called ionoacoustics) also depends on the time scale of the heating process with respect to the transit time of the sound wave across the Bragg peak, considerations on the beam time structure and related local current are crucial (Jones et al., 2016). Also in this respect, current trends of very compact synchrocyclotron accelerators for single-room proton therapy facilities offer almost ideal beam properties for ionoacoustics. In particular, pulses of approximately 2.5–4.0 microseconds length up to two pC (pico Coulomb) charge at kilohertz repetition result in very promising range verification performances with (sub) millimetre accuracy and precision in water (dose dependent) as recently experimentally shown in phantom experiments (Lehrack et al., 2017). Alternatively, the beam current can be artificially pulsed as recently demonstrated down to the not yet ideal length of approximately 17 microseconds at a clinical isochronous cyclotron, also enabling promising ionoacoustic detection of pristine Bragg peaks in water (Jones et al., 2015b). Although tissue heterogeneities are expected to increase signal attenuation and complicate the acoustic signal shape, encouraging simulation studies under realistic clinical conditions show the promise of the method to retain millimetre range accuracy in proton therapy when using multiple detectors and proper triangulation techniques (Jones et al., 2015a). Along with progress in instrumentation of enhanced sensitivity, this technology, although still at its infancy in comparison to the other methods reviewed in this chapter, can offer a very intriguing and cost-effective method for *in vivo* range (and eventually also dose) verification; moreover, its combination with anatomical ultrasound imaging (Kellnberger et al., 2016; Patch et al., 2016) opens the unique perspective of quasi real-time range (dose) verification coregistered to anatomy which is crucial for the envisioned application to challenging anatomical sites such as prostate, liver, and breast, subject to inter- and intra-fractional organ motion (Assmann et al., 2015).

CONCLUSION AND OUTLOOK

With the worldwide rapid spread of particle therapy facilities equipped with state-of-the-art PBS delivery and robotic patient positioning, the enhanced demands on treatment accuracy and precision have promoted a considerable evolution of imaging technologies tailored to the ion-beam therapy needs, catching up lost ground in advanced image guidance. Nowadays, volumetric X-ray-imaging based on CBCT or CT on rails is becoming a reality for patient positioning and an extensive subject of research for adaptive treatment strategies in most ion-therapy centres. Moreover, alternative approaches based on commercial DECT scanners (initially developed for radiology) and several prototypes of ion-transmission imaging are being intensively explored to enable improved RSP estimations for range and dose calculations in treatment planning besides low-dose image guidance with ion radiography and tomography. To complement the in-room knowledge of patient position, anatomy, and tissue-stopping properties prior to treatment delivery, several techniques based on the detection of secondary emissions induced by electromagnetic or nuclear interactions are being investigated at different levels of maturity and complexity for *in vivo* verification of the actual range (dose) delivery, ideally at an individual pencil beam level. While for the latter technologies, current efforts are still focused on clinical translation and integration in the clinical workflow, fruitful synergies among the different treatment modalities have already been recognised. For example, Krah et al. (2015) and Doolan et al. (2015) showed the positive impact of high-resolution X-ray data serving as prior information to ion-based imaging. In addition, the refined elemental tissue assignment enabled by DECT was shown by Berndt et al. (2017) to improve the prediction of secondary nuclear-based emission products in PET with a similar impact anticipated for PG monitoring. Finally, the potential benefit of hybrid approaches combining PET and PG (or charged particles) imaging was discussed, for example, in Parodi (2016) with first systems being under investigation and development in several projects such as the innovative solutions for in-beam dosimetry in hadron therapy (INSIDE) project (Marafini et al., 2015).

Hence, the next years will likely see widespread clinical integration and combination of advanced imaging solutions specifically tailored to ion beam therapy. These systems are expected to contribute to the entire chain of treatment planning, delivery as well as *in-vivo* verification and adaptation, with the ultimate goal to enable full exploitation of the ballistic advantages of ion beam therapy in clinical practice.

ACKNOWLEDGEMENTS

The authors would like to acknowledge several colleagues and collaborators for fruitful discussions, especially Walter Assmann from LMU, Frank Verhaegen from Maastricht, and Chia-ho Hua from St. Judes Children's Hospital.

REFERENCES

Abe S, Nishimura K, Sato H, Muraishi H, Inada T, Tomida T, Fujisaki T et al. 2002. Heavy ion CT system based on measurement of residual range distribution. *Igaku. Butsuri.* **22:** 39–47.

Alvarez R E and Macovski A. 1976. Energy-selective reconstructions in X-ray computerized tomography. *Phys. Med. Biol.* **21:** 733–744.

Amaldi U, Bianchi A, Chang Y H, Go A, Hajdas W, Malakhov N, Samarati J, Sauli F and Watts D. 2011. Construction, test and operation of a proton range radiography system. *Nucl. Instrum. Meth. A* **629:** 337–344.

Arbor N, Dauvergne D, Dedes G, Letang J M, Parodi K, Quinones C T, Testa E and Rit S. 2015. Monte Carlo comparison of x-ray and proton CT for range calculations of proton therapy beams. *Phys. Med. Biol.* **60:** 7585–7599.

Assmann W, Kellnberger S, Reinhardt S, Lehrack S, Edlich A, Thirolf P G, Moser M, Dollinger G et al. 2015. Ionoacoustic characterization of the proton Bragg peak with submillimeter accuracy. *Med. Phys.* **42:** 567–574.

Augusto R, Mendonca T, Wenander F, Penescu L, Orecchia R, Parodi K, Ferrari A and Stora T. 2016. New developments of 11C post-accelerated beams for hadron therapy and imaging. *Nucl. Instrum. Meth. Phys. Res. Sec. B* **376:** 374–378.

Bar E, Lalonde A, Royle G, Lu H M and Bouchard H. 2017. The potential of dual-energy CT to reduce proton beam range uncertainties. *Med. Phys.* **44:** 2332–2344.

Bashkirov V A, Schulte R W, Hurley R F, Johnson R P, Sadrozinski H F W, Zatserklyaniy A, Plautz T and Giacometti V. 2016. Novel scintillation detector design and performance for proton radiography and computed tomography. *Med. Phys.* **43:** 664–674.

Bauer J, Unholtz D, Sommerer F, Kurz C, Haberer T, Herfarth K, Welzel T, Combs S E, Debus J and Parodi K. 2013. Implementation and initial clinical experience of offline PET/CT-based verification of scanned carbon ion treatment. *Radiother. Oncol.* **107**: 218–226.

Bazalova M and Verhaegen F. 2007. Monte Carlo simulation of a computed tomography x-ray tube. *Phys. Med. Biol.* **52**: 5945–5955.

Bazalova M and Verhaegen F. 2009. *PTCOG 48 Conference,* 2009, vol. Series.

Bazalova M, Carrier J F, Beaulieu L and Verhaegen F. 2008a. Dual-energy CT-based material extraction for tissue segmentation in Monte Carlo dose calculations. *Phys. Med. Biol.* **53**: 2439–2456.

Bazalova M, Carrier J F, Beaulieu L and Verhaegen F. 2008b. Tissue segmentation in Monte Carlo treatment planning: A simulation study using dual-energy CT images. *Radiother. Oncol.* **86**: 93–98.

Beaulieu L, Bazalova M, Furstoss C and Verhaegen F. 2009. SU-FF-T-408: Tissue Inhomogeneities in Monte Carlo treatment planning for proton therapy. *Med. Phys.* **36**: 2616.

Bennett G W, Archambeau J O, Archambeau B E, Meltzer J I and Wingate C L. 1978. Visualization and transport of positron emission from proton activation in vivo. *Science* **200**: 1151–1153.

Bentefour E, Schnuerer R and Lu H M. 2016. Concept of proton radiography using energy resolved dose measurement. *Phys. Med. Biol.* **61**: N386–N393.

Berndt B, Landry G, Schwarz F, Tessonnier T, Kamp F, Dedes G, Thieke C et al. 2017. Application of single- and dual-energy CT brain tissue segmentation to PET monitoring of proton therapy. *Phys. Med. Biol.* **62**: 2427–2448.

Bisogni M G, Attili A, Battistoni G, Belcari N, Camarlinghi N, Cerello P, Coli S et al. 2017. INSIDE in-beam positron emission tomography system for particle range monitoring in hadrontherapy. *J. Med. Imag.* **4**: 011005.

Bom V, Joulaeizadeh L and Beekman F. 2012. Real-time prompt gamma monitoring in spot-scanning proton therapy using imaging through a knife-edge-shaped slit. *Phys. Med. Biol.* **57**: 297–308.

Booth J T and Zavgorodni S F. 2001. Modelling the dosimetric consequences of organ motion at CT imaging on radiotherapy treatment planning. *Phys. Med. Biol.* **46**: 1369–1377.

Bopp C, Colin J, Cussol D, Finck C, Labalme M, Rousseau M and Brasse D. 2013. Proton computed tomography from multiple physics processes. *Phys. Med. Biol.* **58**: 7261–7276.

Bourque A E, Carrier J F and Bouchard H. 2014. A stoichiometric calibration method for dual energy computed tomography. *Phys. Med. Biol.* **59**: 2059–2088.

Brooks R A. 1977. A quantitative theory of the hounsfield unit and its application to dual energy scanning. *J. Comput. Assist. Tomogr.* **1**: 487–493.

Buitenhuis H J T, Diblen F, Brzezinski K W, Brandenburg S and Dendooven P. 2017. Beam-on imaging of short-lived positron emitters during proton therapy. *Phys. Med. Biol.* **62**: 4654–4672.

Civinini C, Bruzzi M, Bucciolini M, Carpinelli M, Cirrone G A P, Cuttone G, Lo Presti D et al. 2013. Recent results on the development of a proton computed tomography system. *Nucl. Instrum. Meth. A* **732**: 573–576.

Cormack A M. 1963. Representation of a function by its line integrals with some radiological applications. *J. Appl. Phys.* **34**: 2722.

Cormack A M and Koehler A M. 1976. Quantitative proton tomography: Preliminary experiments. *Phys. Med. Biol.* **21**: 560–569.

Craig T, Battista J and Van Dyk J. 2003. Limitations of a convolution method for modeling geometric uncertainties in radiation therapy. I. The effect of shift invariance. *Med. Phys.* **30**: 2001–2011.

Crespo P, Shakirin G, Fiedler F, Enghardt W and Wagner A. 2007. Direct time-of-flight for quantitative, real-time in-beam PET: A concept and feasibility study. *Phys. Med. Biol.* **52**: 6795–6811.

Dedes G, De Angelis L, Rit S, Hansen D, Belka C, Bashkirov V, Johnson R P et al. 2017. Application of fluence field modulation to proton computed tomography for proton therapy imaging. *Phys. Med. Biol.* **62**: 6026–6043.

Depauw N and Seco J. 2011. Sensitivity study of proton radiography and comparison with kV and MV x-ray imaging using GEANT4 Monte Carlo simulations. *Phys. Med. Biol.* **56:** 2407–2421.

Doolan P J, Testa M, Sharp G, Bentefour E H, Royle G and Lu H M. 2015. Patient-specific stopping power calibration for proton therapy planning based on single-detector proton radiography. *Phys. Med. Biol.* **60:** 1901–1917.

Engelsman M, Schwarz M and Dong L. 2013. Physics controversies in proton therapy. *Semin. Radiat. Oncol.* **23:** 88–96.

Enghardt W, Parodi K, Crespo P, Fiedler F, Pawelke J and Ponisch F. 2004. Dose quantification from in-beam positron emission tomography. *Radiother. Oncol.* 73(Suppl 2): S96–S98.

Farace P. 2014. Experimental verification of ion stopping power prediction from dual energy CT data in tissue surrogates. *Phys. Med. Biol.* **59:** 7081–7084.

Farace P, Righetto R and Meijers A. 2016. Pencil beam proton radiography using a multilayer ionization chamber. *Phys. Med. Biol.* **61:** 4078–4087.

Flohr T G, McCollough C H, Bruder H, Petersilka M, Gruber K, Suss C, Grasruck M et al. 2006. First performance evaluation of a dual-source CT (DSCT) system. *Eur. Radiol.* **16:** 256–268.

Frey K, Bauer J, Unholtz D, Kurz C, Kramer M, Bortfeld T and Parodi K. 2014. TPSPET-A TPS-based approach for in vivo dose verification with PET in proton therapy. *Phys. Med. Biol.* **59:** 1–21.

Gaa T, Reinhart M, Hartmann B, Jakubek J, Soukup P, Jakel O and Martisikova M. 2017. Visualization of air and metal inhomogeneities in phantoms irradiated by carbon ion beams using prompt secondary ions. *Phys. Med.* **38:** 140–147.

Goitein M. 1972. Three-dimensional density reconstruction from a series of two-dimensional projections. *Nucl. Instrum. Meth.* **101:** 509–518.

Golnik C, Hueso-Gonzalez F, Muller A, Dendooven P, Enghardt W, Fiedler F, Kormoll T et al. 2014. Range assessment in particle therapy based on prompt gamma-ray timing measurements. *Phys. Med. Biol.* **59:** 5399–5422.

Gragoudas E S, Goitein M, Koehler A, Wagner M S, Verhey L, Tepper J, Suit H D, Schneider R J and Johnson K N. 1979. Proton irradiation of malignant melanoma of the ciliary body. *Br. J. Ophthalmol.* **63:** 135–139.

Gueth P, Dauvergne D, Freud N, Letang J M, Ray C, Testa E and Sarrut D. 2013. Machine learning-based patient specific prompt-gamma dose monitoring in proton therapy. *Phys. Med. Biol.* **58:** 4563–4577.

Han D, Siebers J V and Williamson J F. 2016. A linear, separable two-parameter model for dual energy CT imaging of proton stopping power computation. *Med. Phys.* **43:** 600.

Handrack J, Tessonnier T, Chen W, Debus J, Bauer J and Parodi K. 2017. Sensitivity of post treatment positron-emission-tomography/computed-tomography to detect inter-fractional range variations in scanned ion beam therapy. *Acta Oncol.* 56(11): 1451–1458.

Hansen D C, Bassler N, Sorensen T S and Seco J. 2014. The image quality of ion computed tomography at clinical imaging dose levels. *Med. Phys.* **41:** 111908.

Hansen D C, Seco J, Sorensen T S, Petersen J B B, Wildberger J E, Verhaegen F and Landry G. 2015. A simulation study on proton computed tomography (CT) stopping power accuracy using dual energy CT scans as benchmark. *Acta Oncol.* **54:** 1638–1642.

Hansen D C, Sorensen T S and Rit S. 2016. Fast reconstruction of low dose proton CT by sinogram interpolation. *Phys. Med. Biol.* **61:** 5868–5882.

Hanson K M. 1979. Proton computed tomography. *IEEE Trans. Nucl. Sci.* **26:** 1635–1640.

Hanson K M, Bradbury J N, Koeppe R A, Macek R J, Machen D R, Morgado R, Paciotti M A, Sandford S A and Steward V W. 1982. Proton computed tomography of human specimens. *Phys. Med. Biol.* **27:** 25–36.

Hanson K, Bradbury J, Cannon T, Hutson R, Laubacher D, Macek R, Paciotti M and Taylor C. 1978. The application of protons to computed tomography. *J. Comput. Assist. Tomogr.* **2:** 671–674.

Hayakawa Y, Tada J, Arai N, Hosono K, Sato M, Wagai T, Tsuji H and Tsujii H. 1995. Acoustic pulse generated in a patient during treatment by pulsed proton radiation beam. *Radiat. Oncol. Investig.* **3:** 42–45.

Hounsfield G N. 1973. Computerized transverse axial scanning (tomography). 1. Description of system. *Br. J. Radiol.* **46:** 1016–1022.

Hudobivnik N, Schwarz F, Johnson T, Agolli L, Dedes G, Tessonnier T, Verhaegen F et al. 2016. Comparison of proton therapy treatment planning for head tumors with a pencil beam algorithm on dual and single energy CT images. *Med. Phys.* **43:** 495.

Hueso-Gonzalez F, Enghardt W, Fiedler F, Golnik C, Janssens G, Petzoldt J, Prieels D et al. 2015. First test of the prompt gamma ray timing method with heterogeneous targets at a clinical proton therapy facility. *Phys. Med. Biol.* **60:** 6247–6272.

Hueso-Gonzalez F, Golnik C, Berthel M, Dreyer A, Enghardt W, Fiedler F, Heidel K et al. 2014. Test of Compton camera components for prompt gamma imaging at the ELBE bremsstrahlung beam. *J. Instrum.* **9:** P05002.

Hünemohr N, Krauss B, Dinkel J, Gillmann C, Ackermann B, Jakel O and Greilich S. 2013. Ion range estimation by using dual energy computed tomography. *Z. Med. Phys.* **23:** 300–313.

Hunemohr N, Krauss B, Tremmel C, Ackermann B, Jakel O and Greilich S. 2014a. Experimental verification of ion stopping power prediction from dual energy CT data in tissue surrogates. *Phys. Med. Biol.* **59:** 83–96.

Hunemohr N, Niebuhr N and Greilich S. 2014b. Reply to Comment on Experimental verification of ion stopping power prediction from dual energy CT data in tissue surrogates. *Phys. Med. Biol.* **59:** 7085–7087.

Hunemohr N, Paganetti H, Greilich S, Jakel O and Seco J. 2014c. Tissue decomposition from dual energy CT data for MC based dose calculation in particle therapy. *Med. Phys.* **41:** 061714.

Ito A and Koyamaito H. 1987. Proton computed-tomography applied to small biomedical samples. *Biol. Trace Elem. Res.* **13:** 423.

Johnson R P, Bashkirov V, DeWitt L, Giacometti V, Hurley R F, Piersimoni P, Plautz T E et al. 2016. A Fast experimental scanner for proton CT: Technical performance and first experience with phantom scans. *IEEE Trans. Nucl. Sci.* **63:** 52–60.

Johnson R P, Bashkirov V, Giacometti V, Hurley R F, Piersimoni P, Plautz T E, Sadrozinski H F W et al. 2014. Results from a pre-clinical head scanner for proton CT. *2014 IEEE Nuclear Science Symposium and Medical Imaging Conference (Nss/Mic)*.

Jones K, Sehgal C and Avery S. 2015a. MO-F-CAMPUS-J-**01**: Acoustic range verification of proton beams: Simulation of heterogeneity and clinical proton pulses. *Med. Phys.* **42:** 3582.

Jones K C, Seghal C M and Avery S. 2016. How proton pulse characteristics influence protoacoustic determination of proton-beam range: Simulation studies. *Phys. Med. Biol.* **61:** 2213–2242.

Jones K C, Stappen F V, Bawiec C R, Janssens G, Lewin P A, Prieels D, Solberg T D, Sehgal C M and Avery S. 2015b. Experimental observation of acoustic emissions generated by a pulsed proton beam from a hospital-based clinical cyclotron. *Med. Phys.* **42:** 7090–7097.

Jongen Y and Stichelbaut F. 2003. Verification of the proton beam position in the patient by the detection of prompt gamma-rays emission. *Presentation at the 39th Particle Therapy Co-Operative Group (PTCOG)*.

Kellnberger S, Assmann W, Lehrack S, Reinhardt S, Thirolf P, Queiros D, Sergiadis G, Dollinger G, Parodi K and Ntziachristos V. 2016. Ionoacoustic tomography of the proton Bragg peak in combination with ultrasound and optoacoustic imaging. *Sci. Rep.* **6:** 29305.

Knopf A C and Lomax A. 2013. In vivo proton range verification: A review. *Phys. Med. Biol.* **58:** R131–R160.

Koehler A M. 1968. Proton radiography. *Science* **160:** 303–304.

Kormoll T, Fiedler F, Schone S, Wustemann J, Zuber K and Enghardt W. 2011. A compton imager for in-vivo dosimetry of proton beams-A design study. *Nucl. Instrum. Meth. A* **626:** 114–119.

Kraan A C. 2015. Range verification methods in particle therapy: underlying physics and Monte Carlo modelling. *Front. Oncol.* **5:** 150.

Krah N, Testa M, Brons S, Jakel O, Parodi K, Voss B and Rinaldi I. 2015. An advanced image processing method to improve the spatial resolution of ion radiographies. *Phys. Med. Biol.* **60:** 8525–8547.

Krimmer J, Angellier G, Balleyguier L, Dauvergne D, Freud N, Herault J, Letang J M et al. 2017. A cost-effective monitoring technique in particle therapy via uncollimated prompt gamma peak integration. *Appl. Phys. Lett.* **110:** 154102.

Kurz C, Bauer J, Unholtz D, Richter D, Herfarth K, Debus J and Parodi K. 2016. Initial clinical evaluation of PET-based ion beam therapy monitoring under consideration of organ motion. *Med. Phys.* **43:** 975–82.

Kurz C, Dedes G, Resch A, Reiner M, Ganswindt U, Nijhuis R, Thieke C, Belka C, Parodi K and Landry G. 2015. Comparing cone-beam CT intensity correction methods for dose recalculation in adaptive intensity-modulated photon and proton therapy for head and neck cancer. *Acta Oncol.* **54:** 1651–1657.

Lalonde A and Bouchard H. 2016. A general method to derive tissue parameters for Monte Carlo dose calculation with multi-energy CT. *Phys. Med. Biol.* **61:** 8044.

Landry G and Hua C. 2018, in press. Current state and future applications of radiological image guidance for particle therapy. *Med. Phys.* doi: https://doi.org/10.1002/mp.12744.

Landry G, Granton P V, Reniers B, Ollers M C, Beaulieu L, Wildberger J E and Verhaegen F. 2011. Simulation study on potential accuracy gains from dual energy CT tissue segmentation for low-energy brachytherapy Monte Carlo dose calculations. *Phys. Med. Biol.* **56:** 6257–6278.

Landry G, Nijhuis R, Dedes G, Handrack J, Thieke C, Janssens G, Orban de Xivry J et al. 2015. Investigating CT to CBCT image registration for head and neck proton therapy as a tool for daily dose recalculation. *Med. Phys.* **42:** 1354–1366.

Landry G, Parodi K, Wildberger J E and Verhaegen F. 2013. Deriving concentrations of oxygen and carbon in human tissues using single- and dual-energy CT for ion therapy applications. *Phys. Med. Biol.* **58:** 5029–5048.

Lehrack S, Assmann W, Bertrand D, Henrotin S, Herault J, Heymans V, Vander Stappen F, Thirolf P G, Vidal M and Van de Walle J. 2017. Submillimeter ionoacoustic range determination for protons in water at a clinical synchrocyclotron. *Phys. Med. Biol.* **62:** L20.

Lin H H, Chang H T, Chao T C and Chuang K S. 2017. A comparison of two prompt gamma imaging techniques with collimator-based cameras for range verification in proton therapy. *Radiat. Phys. Chem.* **137:** 144–150.

Llacer J, Chatterjee A, Alpen E L, Saunders W, Andreae S and Jackson H C. 1984. Imaging by injection of accelerated radioactive particle beams. *IEEE Trans. Med. Imaging* 3: 80–90.

Llosa G, Barrio J, Cabello J, Crespo A, Lacasta C, Rafecas M, Callier S, de la Taille C and Raux L. 2012. Detector characterization and first coincidence tests of a Compton telescope based on LaBr3 crystals and SiPMs. *Nucl. Instrum. Meth. A* **695:** 105–108.

Llosa G, Cabello J, Callier S, Gillam J E, Lacasta C, Rafecas M, Raux L et al. 2013. First compton telescope prototype based on continuous LaBr3-SiPM detectors. *Nucl. Instrum. Meth. A* **718:** 130–133.

Marafini M, Attili A, Battistoni G, Belcari N and Bisogni M G. 2015. Innovative solutions for in-beam dosimetry in hadrontherapy. *Acta Phys. Pol.* **127:** 1465–1467.

McCarter S D and Beckham W A. 2000. Evaluation of the validity of a convolution method for incorporating tumour movement and set-up variations into the radiotherapy treatment planning system. *Phys. Med. Biol.* **45:** 923–931.

McCleskey M, Kaye W, Mackin D S, Beddar S, He Z and Polf J C. 2015. Evaluation of a multistage CdZnTe Compton camera for prompt gamma imaging for proton therapy. *Nucl. Instrum. Meth. A* **785:** 163–169.

Meyer S, Gianoli C, Magallanes L, Kopp B, Tessonnier T, Landry G, Dedes G, Voss B and Parodi K. 2017. Comparative Monte Carlo study on the performance of integration-and list-mode detector configurations for carbon ion computed tomography. *Phys. Med. Biol.* **62:** 1096–1112.

Mijnheer B, Olaciregui-Ruiz I, Rozendaal R, Sonke J, Spreeuw H, Tielenburg R, van Herk M, Vijlbrief R and Mans A. 2013. 3D EPID-based in vivo dosimetry for IMRT and VMAT. *7th International Conference on 3d Radiation Dosimetry (Ic3ddose)*, Vol. **444,** Sydney, Australia.

Min C H, Kim C H, Youn M Y and Kim J W. 2006. Prompt gamma measurements for locating the dose falloff region in the proton therapy. *Appl. Phys. Lett.* **89:** 183517.

Miyatake A, Nishio T and Ogino T. 2011. Development of activity pencil beam algorithm using measured distribution data of positron emitter nuclei generated by proton irradiation of targets containing (12)C, (16)O, and (40)Ca nuclei in preparation of clinical application. *Med. Phys.* **38:** 5818–5829.

Mohammadi A, Yoshida E, Tashima H, Nishikido F, Inaniwa T, Kitagawa A and Yamaya T. 2017. Production of an ^{15}O beam using a stable oxygen ion beam for in-beam PET imaging. *Nucl. Instrum. Meth. A* **849:** 76–82.

Mohler C, Wohlfahrt P, Richter C and Greilich S. 2016. Range prediction for tissue mixtures based on dual-energy CT. *Phys. Med. Biol.* **61:** N268–N275.

Moteabbed M, Espana S and Paganetti H. 2011. Monte Carlo patient study on the comparison of prompt gamma and PET imaging for range verification in proton therapy. *Phys. Med. Biol.* **56:** 1063–1082.

Nischwitz S P, Bauer J, Welzel T, Rief H, Jakel O, Haberer T, Frey K, Debus J, Parodi K, Combs S E and Rieken S. 2015. Clinical implementation and range evaluation of in vivo PET dosimetry for particle irradiation in patients with primary glioma. *Radiother. Oncol.* **115:** 179–185.

Nishio T, Miyatake A, Ogino T, Nakagawa K, Saijo N and Esumi H. 2010. The development and clinical use of a beam on-line pet system mounted on a rotating gantry port in proton therapy. *Int. J. Radiat. Oncol. Biol. Phys.* **76:** 277–286.

Paganetti H. 2012. Range uncertainties in proton therapy and the role of Monte Carlo simulations. *Phys. Med. Biol.* **57:** R99–R117.

Park Y K, Sharp G C, Phillips J and Winey B A. 2015. Proton dose calculation on scatter-corrected CBCT image: Feasibility study for adaptive proton therapy. *Med. Phys.* **42:** 4449–4459.

Parodi K. 2015. Vision 20/20: Positron emission tomography in radiation therapy planning, delivery, and monitoring. *Med. Phys.* **42:** 7153–7168.

Parodi K. 2016. Unconventional imaging in ion beam therapy: Status and perspectives. *Acta Phys. Pol.* **47:** 447–452.

Parodi K, Ferrari A, Sommerer F and Paganetti H. 2007a. Clinical CT-based calculations of dose and positron emitter distributions in proton therapy using the FLUKA Monte Carlo code. *Phys. Med. Biol.* **52:** 3369–3387.

Parodi K, Paganetti H, Shih H A, Michaud S, Loeffler J S, DeLaney T F, Liebsch N J et al. 2007b. Patient study of in vivo verification of beam delivery and range, using positron emission tomography and computed tomography imaging after proton therapy. *Int. J. Radiat. Oncol. Biol. Phys.* **68:** 920–934.

Patch S K, Covo M K, Jackson A, Qadadha Y M, Campbell K S, Albright R A, Bloemhard P et al. 2016. Thermoacoustic range verification using a clinical ultrasound array provides perfectly co-registered overlay of the Bragg peak onto an ultrasound image. *Phys. Med. Biol.* **61:** 5621–5638.

Pawelke J, Byars L, Enghardt W, Fromm W D, Geissel H, Hasch B G, Lauckner K, Manfrass P, Schardt D and Sobiella M. 1996. The investigation of different cameras for in-beam PET imaging. *Phys. Med. Biol.* **41:** 279–296.

Pemler P, Besserer J, de Boer J, Dellert M, Gahn C, Moosburger M, Schneider U, Pedroni E and Stauble H 1999 A detector system for proton radiography on the gantry of the Paul-Scherrer-Institute. *Nucl. Instrum. Meth. A* **432:** 483–495.

Peterson S W, Robertson D and Polf J. 2010. Optimizing a three-stage Compton camera for measuring prompt gamma rays emitted during proton radiotherapy. *Phys. Med. Biol.* **55**: 6841–6856.

Pinto M, Dauvergne D, Freud N, Krimmer J, Letang J M, Ray C, Roellinghoff F and Testa E. 2014. Design optimisation of a TOF-based collimated camera prototype for online hadrontherapy monitoring. *Phys. Med. Biol.* **59**: 7653–7674.

Plautz T E, Bashkirov V, Giacometti V, Hurley R F, Johnson R P, Piersimoni P, Sadrozinski H F W, Schulte R W and Zatserklyaniy A. 2016. An evaluation of spatial resolution of a prototype proton CT scanner. *Med. Phys.* **43**: 6291–6300.

Polf J C and Parodi K. 2015. Imaging particle beams for cancer treatment. *Phys. Today* **68**: 28–33.

Polf J C, Avery S, Mackin D S and Beddar S. 2015. Imaging of prompt gamma rays emitted during delivery of clinical proton beams with a Compton camera: Feasibility studies for range verification. *Phys. Med. Biol.* **60**: 7085–7099.

Polf J C, Panthi R, Mackin D S, McCleskey M, Saastamoinen A, Roeder B T and Beddar S. 2013. Measurement of characteristic prompt gamma rays emitted from oxygen and carbon in tissue-equivalent samples during proton beam irradiation. *Phys. Med. Biol.* **58**: 5821–5831.

Polf J C, Peterson S, Ciangaru G, Gillin M and Beddar S. 2009. Prompt gamma-ray emission from biological tissues during proton irradiation: A preliminary study. *Phys. Med. Biol.* **54**: 731–743.

Poludniowski G, Allinson N M and Evans P M. 2014b. Proton computed tomography reconstruction using a backprojection-then-filtering approach. *Phys. Med. Biol.* **59**: 7905–7918.

Poludniowski G, Allinson N M, Anaxagoras T, Esposito M, Green S, Manolopoulos S, Nieto-Camero J, Parker D J, Price T and Evans P M. 2014a. Proton-counting radiography for proton therapy: A proof of principle using CMOS APS technology. *Phys. Med. Biol.* **59**: 2569–2581.

Ponisch F, Parodi K, Hasch B G and Enghardt W. 2004. The modelling of positron emitter production and PET imaging during carbon ion therapy. *Phys. Med. Biol.* **49**: 5217–5232.

Quinones C T, Letang J M and Rit S. 2016. Filtered back-projection reconstruction for attenuation proton CT along most likely paths. *Phys. Med. Biol.* **61**: 3258–3278.

Richard M H, Dahoumane M, Dauvergne D, Dedes G, De Rydt M, Freud N, Letang J M et al. 2011. Design study of the absorber detector of a compton camera for on-line control in ion beam therapy. *IEEE Nucl. Sci. Conf. R* 3496–3500. http://ieeexplore.ieee.org/document/6152642/.

Richter C, Pausch G, Barczyk S, Priegnitz M, Keitz I, Thiele J, Smeets J et al. 2016. First clinical application of a prompt gamma based in vivo proton range verification system. *Radiother. Oncol.* **118**: 232–237.

Rinaldi I, Brons S, Gordon J, Panse R, Voss B, Jakel O and Parodi K. 2013. Experimental characterization of a prototype detector system for carbon ion radiography and tomography. *Phys. Med. Biol.* **58**: 413–427.

Rinaldi I, Brons S, Jakel O, Voss B and Parodi K. 2014. Experimental investigations on carbon ion scanning radiography using a range telescope. *Phys. Med. Biol.* **59**: 3041–3057.

Rit S, Dedes G, Freud N, Sarrut D and Letang J M. 2013. Filtered backprojection proton CT reconstruction along most likely paths. *Med. Phys.* **40**: 031103.

Roellinghoff F, Richard M H, Chevallier M, Constanzo J, Dauvergne D, Freud N, Henriquet P et al. 2011. Design of a compton camera for 3D prompt-gamma imaging during ion beam therapy. *Nucl. Instrum. Meth. A* **648**: S20–S23.

Sadrozinski H F W, Geoghegan T, Harvey E, Johnson R P, Plautz T E, Zatserklyaniy A, Bashkirov V et al. 2016. Operation of the preclinical head scanner for proton CT. *Nucl. Instrum. Meth. A* **831**: 394–399.

Sadrozinski H F W, Johnson R P, Macafee S, Plumb A, Steinberg D, Zatserklyaniy A, Bashkirov V A, Hurley R F and Schulte R W. 2013. Development of a head scanner for proton CT. *Nucl. Instrum. Meth. A* **699**: 205–210.

Saraya Y, Izumikawa T, Goto J, Kawasaki T and Kimura T. 2014. Study of spatial resolution of proton computed tomography using a silicon strip detector. *Nucl. Instrum. Meth. A* **735**: 485–489.

Schneider U and Pedroni E. 1995. Proton radiography as a tool for quality-control in proton therapy. *Med. Phys.* **22**: 353–363.

Schneider U, Besserer J, Pemler P, Dellert M, Moosburger M, Pedroni E and Kaser-Hotz B. 2004. First proton radiography of an animal patient. *Med. Phys.* **31**: 1046–1051.

Seco J and Depauw N. 2011. Proof of principle study of the use of a CMOS active pixel sensor for proton radiography. *Med. Phys.* **38**: 622–623.

Shinoda H, Kanai T and Kohno T. 2006. Application of heavy-ion CT. *Phys. Med. Biol.* **51**: 4073–4081.

Sipala V, Brianzi M, Bruzzi M, Bucciolini M, Cirrone G A P, Civinini C, Cuttone G et al. 2011. PRIMA: An apparatus for medical application. *Nucl. Instrum. Meth. A* **658**: 73–77.

Smeets J, Roellinghoff F, Prieels D, Stichelbaut F, Benilov A, Busca P, Fiorini C et al. 2012. Prompt gamma imaging with a slit camera for real-time range control in proton therapy. *Phys. Med. Biol.* **57**: 3371–3405.

Sommer F G, Tobias C A, Benton E V, Woodruff K H, Henke R P, Holly W and Genant H K. 1978. Heavy-ion radiography: Density resolution and specimen radiography. *Invest. Radiol.* **13**: 163–170.

Sportelli G, Belcari N, Camarlinghi N, Cirrone G A, Cuttone G, Ferretti S, Kraan A et al. 2014. First full-beam PET acquisitions in proton therapy with a modular dual-head dedicated system. *Phys. Med. Biol.* **59**: 43–60.

Taasti V T, Petersen J B, Muren L P, Thygesen J and Hansen D C. 2016. A robust empirical parametrization of proton stopping power using dual energy CT. *Med. Phys.* **43**: 5547.

Takada Y, Kondo K, Marume T, Nagayoshi K, Okada I and Takikawa K. 1988. Proton computed-tomography with a 250 Mev pulsed-beam. *Nucl. Instrum. Meth. A* **273**: 410–422.

Tashima H, Yoshida E, Inadama N, Nishikido F, Nakajima Y, Wakizaka H, Shinaji T et al. 2016. Development of a small single-ring OpenPET prototype with a novel transformable architecture. *Phys. Med. Biol.* **61**: 1795–1809.

Taya T, Kataoka J, Kishimoto A, Iwamoto Y, Koide A, Nishio T, Kabuki S and Inaniwa T. 2016. First demonstration of real-time gamma imaging by using a handheld Compton camera for particle therapy. *Nucl. Instrum. Meth. A* **831**: 355–361.

Telsemeyer J, Jaakel O and Martisikova M. 2012. Quantitative carbon ion beam radiography and tomography with a flat-panel detector. *Phys. Med. Biol.* **57**: 7957–7971.

Testa E, Bajard M, Chevallier M, Dauvergne D, Le Foulher F, Freud N, Letang J M, Poizat J C, Ray C and Testa M. 2008. Monitoring the bragg peak location of 73 MeV/u carbon ions by means of prompt gamma-ray measurements. *Appl. Phys. Lett.* **93**: 093506.

Testa M, Min C H, Verburg J M, Schumann J, Lu H M and Paganetti H. 2014. Range verification of passively scattered proton beams based on prompt gamma time patterns. *Phys. Med. Biol.* **59**: 4181–4195.

Testa M, Verburg J M, Rose M, Min C H, Tang S K, Bentefour E, Paganetti H and Lu H M. 2013. Proton radiography and proton computed tomography based on time-resolved dose measurements. *Phys. Med. Biol.* **58**: 8215–8233.

Thing R S, Bernchou U, Mainegra-Hing E, Hansen O and Brink C. 2016. Hounsfield unit recovery in clinical cone beam CT images of the thorax acquired for image guided radiation therapy. *Phys. Med. Biol.* **61**: 5781–5802.

Thirolf P G, Aldawood S, Bohmer M, Bortfeldt J, Castelhano I, Dedes G, Fiedler F et al. 2016. A Compton camera prototype for prompt gamma medical imaging. *12th International Conference on Nucleus-Nucleus Collisions 2015* **117**: 05005.

Thirolf P G, Lang C, Aldawood S, Von der Kolff H G V, Maier L, Schaart D R and Parodi K. 2014. Development of a compton camera for online range monitoring of laser-accelerated proton beams via prompt-gamma detection. *EPJ Web Conf.* **66**: 11036.

Tobias C A, Benton E V, Capp M P, Chatterjee A, Cruty M R and Henke R P. 1977. Particle radiography and autoactivation. *Int. J. Radiat. Oncol. Biol. Phys.* **3**: 35–44.

Torikoshi M, Tsunoo T, Ohno Y, Endo M, Natsuhori M, Kakizaki T, Ito N, Uesugi K and Yagi N. 2005. Features of dual-energy X-ray computed tomography. *Nucl. Instrum. Methods Phys. Res. A* **548:** 99–105.

Torikoshi M, Tsunoo T, Sasaki M, Endo M, Noda Y, Ohno Y, Kohno T, Hyodo K, Uesugi K and Yagi N. 2003. Electron density measurement with dual-energy x-ray CT using synchrotron radiation. *Phys. Med. Biol.* **48:** 673–685.

van Herk M, Remeijer P and Lebesque J V. 2002. Inclusion of geometric uncertainties in treatment plan evaluation. *Int. J. Radiat. Oncol. Biol. Phys.* **52:** 1407–1422.

Verburg J M and Seco J. 2014. Proton range verification through prompt gamma-ray spectroscopy. *Phys. Med. Biol.* **59.** 7089 7106.

Verburg J M, Riley K, Bortfeld T and Seco J. 2013. Energy- and time-resolved detection of prompt gamma-rays for proton range verification. *Phys. Med. Biol.* **58:** L37–L49.

Verellen D, De Ridder M, Linthout N, Tournel K, Soete G and Storme G. 2007. Innovations in image-guided radiotherapy. *Nat. Rev. Cancer* **7:** 949–960.

Verhey L J, Goitein M, McNulty P, Munzenrider J E and Suit H D. 1982. Precise positioning of patients for radiation therapy. *Int. J. Radiat. Oncol. Biol. Phys.* **8:** 289–294.

Vynckier S, Derreumaux S, Richard F, Bol A, Michel C and Wambersie A. 1993. Is it possible to verify directly a proton-treatment plan using positron emission tomography? *Radiother. Oncol.* **26:** 275–277.

Wang P, Cammin J, Bisello F, Solberg T D, McDonough J E, Zhu T C, Menichelli D and Teo B K K. 2016. Proton computed tomography using a 1D silicon diode array. *Med. Phys.* **43:** 5758–5766.

Wohlfahrt P, Möhler C, Enghardt W, Greilich S and Richter C. 2017a. OC-0150: Dual-energy CT-based proton treatment planning to assess patient-specific range uncertainties. *Radiother. Oncol.* **123:** S73–S75.

Wohlfahrt P, Möhler C, Hietschold V, Menkel S, Greilich S, Krause M, Baumann M, Enghardt W and Richter C. 2017b. Clinical implementation of dual-energy CT for proton treatment planning on pseudo-monoenergetic CT scans. *Int. J. Radiat. Oncol. Biol. Phys.* **97:** 427–434.

Woodard H Q and White D R. 1986. The composition of body tissues. *Br. J. Radiol.* **59:** 1209–1218.

Xie Y, Bentefour E H, Janssens G, Smeets J, Vander Stappen F, Hotoiu L, Yin L et al. 2017. Prompt gamma imaging for in vivo range verification of pencil beam scanning proton therapy. *Int. J. Radiat. Oncol. Biol. Phys.* **99:** 210–218.

Xie Y, Yin L, Ainsley C, McDonough J, Solberg T, Lin A and Teo B. 2016. TU-FG-BRB-01: Dual energy CT proton stopping power ratio calibration and validation with animal tissues. *Med. Phys.* **43:** 3756.

Yang M, Virshup G, Clayton J, Zhu X R, Mohan R and Dong L. 2010. Theoretical variance analysis of single- and dual-energy computed tomography methods for calculating proton stopping power ratios of biological tissues. *Phys. Med. Biol.* **55:** 1343–1362.

Yang M, Zhu X R, Park P C, Titt U, Mohan R, Virshup G, Clayton J E and Dong L. 2012. Comprehensive analysis of proton range uncertainties related to patient stopping-power-ratio estimation using the stoichiometric calibration. *Phys. Med. Biol.* **57:** 4095–4115.

Zhu J and Penfold S N. 2016. Dosimetric comparison of stopping power calibration with dual-energy CT and single-energy CT in proton therapy treatment planning. *Med. Phys.* **43:** 2845–2854.

Zhu X, Espana S, Daartz J, Liebsch N, Ouyang J, Paganetti H, Bortfeld T R and El Fakhri G. 2011. Monitoring proton radiation therapy with in-room PET imaging. *Phys. Med. Biol.* **56:** 4041–4057.

Zygmanski P, Gall K P, Rabin M S Z and Rosenthal S J. 2000. The measurement of proton stopping power using proton-cone-beam computed tomography. *Phys. Med. Biol.* **45:** 511–528.

CHAPTER 7

Why Particle Therapy Rather than Photon Therapy or How to Integrate the Decision into Multimodal Management

Joachim Widder and Richard Pötter

CONTENTS

COMPARING DEPTH-DOSE PROFILES OF photons with particles – protons or heavier ions – yields a clear advantage for any particles compared with photons. This is the essence of the indubitable theoretical advantage of particle therapy over photon therapy. So why is there still debate about the clinical advantage of particles, and why does this bear on recommending particle versus photon therapy in present multidisciplinary oncology (Ruysscher et al., 2012; Mitin and Zietman, 2014)?

PHOTONS OR PARTICLES?

Deciding to use particles rather than photons for treatment resembles employing surgical approach A (e.g., video-assisted endoscopic surgery) versus approach B (e.g., open surgery) more than it resembles giving drug A rather than drug B. Frequently, different drugs that are compared in the treatment of a given disease target different pathogenic pathways. This may or may not imply different adverse effects when the disease is treated using drug A rather than drug B. Most important, however, it means that the active agents' mechanisms addressing the pathology, the mechanisms of action leading to desired effects, may differ significantly. For instance, comparing cytotoxic chemotherapy with antibodies blocking specific tumour-growth factors in the treatment of certain malignant tumours entails a wide range of unknowns that can only be compared in a head-to-head comparison, which in turn is best performed as a double-blind randomised clinical trial to limit bias as much as possible (Howick et al., 2009). In contrast, changing the surgical approach is unlikely to lead to better tumour-specific outcome, as the essence – completely removing tumour tissue – remains unchanged when altering the approach. The main purpose of recent advances in surgical technology – for example, laparoscopic, thoracoscopic, robot-assisted surgery – is reduction of surgical collateral damage by reducing the surgical trauma, thus by reducing the normal tissue complication probability (NTCP) (Howington et al., 2013). As a by-product, increasing the degrees of freedom by using robots to direct surgical instruments to their target independent of restrictions of human hands or their three-dimensional extensions (endoscopic video-assisted approaches) could theoretically yield higher rates of complete resections of tumours. But still, showing equivalence or non-inferiority of any novel surgical approach compared with classical surgery is expected to be used more often than superiority trials, because superiority is likely to be marginal at best, given that complete resection rates of about 90%–95% or higher are statistically hard to ameliorate. As *employing* certain complex technology for therapeutic ends always means *applying* the respective technology, which implies handling, using machines, it will remain more or less operator dependent (Fleshman et al., 2015; Stevenson et al., 2015). This implies that eventual suboptimal results achieved will therefore constitute a strong stimulus for improving the technology, as long as the underlying principle triggering the idea of applying that very technology for therapeutic purposes – that is, reducing toxicity – has not been falsified (Strong and Soper, 2015).

This certainly applies for any particle therapy as it has been utilised for decennia using technology that is arguably not optimal (especially, passive scattering; no on-board soft-tissue imaging; no beam flexibility using gantries). Therefore, 'proton therapy' or 'carbon ion therapy' are insufficiently described entities; more technical details need to be known before either therapy can be compared with photon therapy: for example, image guidance, intensity modulation and beam flexibility (Widder et al., 2015). This has important consequences for the methodology to gain or increase evidence using any health technology (McCulloch et al., 2009). First and typically, technology is developed while being used, and thus it comes in versions. Improvements may happen as small gradual changes, very rarely as quantum leaps. Second, when increasing precision or accuracy (or both) is the aim of using health technology, in general terms this means that collateral damage caused by therapeutic interventions will be reduced proportionally with attaining this aim. Minimally invasive and organ-sparing surgery, both crucial developments in contemporary surgery, are typical examples. Particle therapy fits nicely into this line, its main purpose being reduction of radiation toxicity by reducing dose to tissues surrounding malignant tumours. Employing particle therapy, in particular proton therapy, thus encounters comparable issues as surgery when having to demonstrate its added value in terms of superiority regarding toxic effects of radiation, or its non-inferiority or equivalence regarding tumour outcome. However, in sharp contrast to surgery, side effects of radiotherapy critically depend on dose and dose distributions – dose–volume parameters – delivered to critical tissues and organs. These parameters are quantifiable and in turn lend themselves to modelling. Toxic outcomes after radiotherapy are a function of dose–volume parameters that are modulated more or less by clinical, genetic, molecular or other patient- and tumour-related factors. Any dose-planning of curative radiotherapy has to navigate between delivering sufficient dose at the tumour and dose-limits, dose-constraints and dose-objectives at unaffected organs and tissues surrounding the tumour in order to limit toxicity as much as possible.

Predictability of adverse radiation effects is regarded certain enough in daily radiation oncology practice to form absolute contraindications for escalating tumour doses in several situations, even if tumour control rates are suboptimal with doses generally regarded as standard treatment. In addition, there are numerous examples for detrimental and contrary-to-expectation outcomes – that is, worse overall outcome in the experimental arm – when dose administered to tumours was escalated in randomised controlled trials (Minsky et al., 2002; Bradley et al., 2015). The most likely explanation for these observations is increased toxicity – partly hidden and not readily traceable to distinct organs or tissues – due to higher doses at critical tissues. Advancing the field thus requires meticulous analysis of outcomes – tumour control, survival as well as adverse effects of treatment – as a function of dose-volume data to critical organs and tissues. Contemporary treatment planning both for photons – intensity modulation including volumetric modulation of arc therapy or intensity modulated brachytherapy – and for particles – intensity modulation using multi-field optimisation of scanned beams – opens tremendously more degrees of freedom for distributing the dose than could be rationally constrained based on evidence regarding radiation toxicity. In other words, there is a considerable gap of knowledge regarding optimal dose distributions, especially when dose–volume parameters to several organs have to be weighed against each other. Given that particle therapy yields considerably different dose distributions than photon therapy, this physical fact renders an exceptional opportunity for gaining much more insight than could possibly be gained when limiting normal tissue effects modelling to photon treatment only.

FOUR PRINCIPAL CLASSES OF INDICATIONS

There are thus four principal classes of indications for particle therapy, when treatment is delivered within a framework where optimising indications for particle therapy is a main objective.

1. *NTCP-model–based indications*: Making the indication for particle treatment based on predicting lower risks of certain clinically relevant adverse events of radiotherapy for the individual patient; that is: reducing toxicity by applying NTCP models (Langendijk et al., 2013; Widder et al., 2015).

2. *Dose–volume histogram-based indications*: Making the indication for particle treatment based on large reductions of dose to critical organs even in the absence of validated NTCP models, under the precept that sparing these organs would be regarded highly desirable by every radiation oncologist, and when this can only be achieved by using particles (van de Sande et al., 2016). In addition, the reasoning to considerably reduce normal tissue dose-volume parameters, equally applies when prevention of secondary tumours forms the indication. Tumours induced secondary to radiation therapy form a special and serious class of radiation toxicity, where the latency period is longer than for most other adverse effects. Consequently, young age at primary radiotherapy will always be a strong factor when prevention of secondary tumours forms the indication for particle therapy. Resembling NTCP-model-based indications, a judgement about an eventual dose-volume histogram (DVH)-based indication can only be substantiated employing comparative planning on an individual-patient basis or, with growing experience, based on models predicting dose-gains (not yet toxicity gains) with the use of particles in place of photons on the basis of baseline tumour extension and patient anatomy alone, or combined with photon-planning reference data.

3. *Standard indications*: Making the indication for particle treatment because the disease and/or clinical situation is regarded as classical standard indication for treatment with particles, for example, proton treatment of children, skull-base tumours, or ocular melanomas (Mitin and Zietman, 2014) and carbon ion treatment, for example, for inoperable sarcoma and adenoidcystic carcinoma in the head and neck region (Kamada et al., 2015). Recording dose–volume parameters from both photon and particle plans is key for the first two classes of indications – to validate, improve or even newly construct not yet available NTCP models – while particle treatment indications classified as standard do not strictly require comparative photon-particle treatment planning. Still, they might also benefit from planning comparison studies, which might further clarify the definition of standard indications, addressing the question whether there would be some patients within this cohort who would not or only marginally benefit from particle therapy (Ruysscher et al., 2012).

4. *Comparative photon-particle treatment studies testing TCP/NTCP relations*: Usually, one physical Gray administered with protons is regarded to have the same biological effect as 1.1 Gray administered with photons, and for carbon ions this value is two or three times higher (Paganetti, 2014; Ebner and Kamada, 2016). However, biological effects per dose unit may be different for tumour- as compared with non-tumour tissue when using particles – especially heavier particles such as carbon ions. A fourth – and most interesting – class of indications therefore

TABLE 7.1   Indication Categories

Indication Category	Tumour Control Probability (TCP)	Complication Probability (NTCP)
NTCP-model-based	Equal at equal tumour dose	Less due to less normal tissue (NT) dose
DVH-based	Required tumour dose feasible	Less due to large dose reduction at critical organs even in absence of validated NTCP models
Classical standard	Required tumour dose (only) feasible with particles	Dose to NT acceptable even though fewer complications not yet clearly demonstrated
Comparative particle-photon studies	Potentially higher per tumour-dose unit due to favourable (higher) RBE in tumour	Potentially lower per NT-dose unit due to favourable (lower) RBE[a] in NT

[a] RBE – relative biological effectiveness.

derives from clinical studies investigating eventually different therapeutic ratios of tumour-to-normal tissue biological effects of (heavy) particles compared with photons. Here, biology-driven concepts of radiation effect in different tissues would form the rationale of treating patients with, for example, soft-tissue sarcomas. The assumption put to the test in those studies rests on quantitatively different biological effects of a dose unit of particles as compared to photons resulting in a more favourable tumour-control probability (TCP)-to-NTCP relationship. Phenomena to be studied in this context include, but are not limited to, different oxygen-enhancement ratios with different particles or dependence on the linear energy transfer (LET) along the beam-path of biological effects in different normal or neoplastic tissues and on fraction size (Suit et al., 2010). Extremely little evidence for *clinical* applicability of the hypothesis of differing biological effects along the beam path is presently available, so any indication made under this presumption must be regraded experimental (Mitin and Zietman, 2014) (Table 7.1).

REASONS FOR LACK OF EVIDENCE

In the early 1960s, protons were first used for radiotherapy purposes, and some 20 years later, a few facilities started treating highly selected patients in the United States, Europe, Japan and Russia. Well in advance of the introduction of 3D-image–based dose planning and calculation in conventional photon-based radiotherapy, tumours technically amenable for proton therapy were selected. Patients with locally highly aggressive tumours with relatively low metastatic propensity were treated, for example, skull-base chordomas and chondrosarcomas, or ocular melanoma, enabling eye-preserving treatment with high local control rates even for patients whose tumours were too large for contact brachytherapy (Potter et al., 1997). Similar reasoning assuming a favourable TCP-to-NTCP ratio was used first for neutrons and then for carbon ions (Kamada et al., 2015), in particular for 'radioresistant' tumours, such as inoperable sarcoma. In these patients, dose distributions were achieved enabling administration of very high biologically effective doses to tumours located near critical organs, which had been impossible to achieve with photon technology of the time. These tumours did not exhibit respiratory movements, and usually anatomically abutted non-moving bony structures, enabling orthogonal X-ray–based position verification with high accuracy. In addition, patients with those rare malignancies were enrolled in particle therapy programs based on their readiness to be referred to the very few treatment facilities often located far from their homes, sometimes even on a different continent. This rendered any comparisons of outcomes achieved with proton or carbon ion therapy as opposed to photon therapy almost impossible due to an excessive selection bias. Therefore, during this period, little evidence in terms of superiority of particles over photons was attained, but there were some promising signs for neutron radiotherapy (Laramore et al., 1993a, 1993b). Only in the beginning of the twenty-first century the number of proton facilities around the world rapidly increased. But still, proton therapy – and even more so, therapy utilising heavier ions – remains an extremely scarce resource, entailing a high degree of patient selection that had not typically been based on demonstrated dosimetric advantages in individual cases.

There is unanimous agreement that the proton (or heavy ion) dose profile enabling reduction of dose outside the target volume (together with eventually increasing the tumour effect within the target) is why ions might be the better beams to deliver therapeutic radiation than photons. Why has this not been taken seriously enough to actually demonstrate instead of continually announcing superiority based on quite weak proof? The reasons why there is still a painful lack of evidence for ion beam therapy are many.

Proton therapy has been; is; and, for the foreseeable future, will remain an extremely scarce resource (Bekelman et al., 2014). At the same time, it still has the reputation of being the technology capable of delivering better radiotherapy than is possible with photon therapy: it is often suggested – and communicated sometimes even by radiation oncologists – that protons permit delivery of higher dose to the tumour while surrounding tissues would be entirely spared. This largely unsubstantiated statement – most proton treatments actually delivered do not administer higher doses to tumours than photon treatments do – combined with the scarcity of the resource at least partly explain why performing comparative research or research based on technically selecting patients who are most likely to benefit from particles remains challenging. There may be patients, sometimes supported by physicians, who simply 'want' protons and are ready to pay for it. There may also be administrators of proton facilities whose primary concern is to generate revenues. Moreover, manufacturers of particle facilities do not have any incentive to put their product to the test: quite to the contrary. Selling particle facilities and offering the respective treatment does not require demonstration of its superior value compared with standard photon therapy. Claiming, instead of showing, its superiority is therefore more economically rational from the producers' perspective (Zietman et al., 2010). Arguably, this is a considerable impediment for defining situations where proton therapy should be indicated, because it demonstrably leads to better outcomes for patients. The pharmaceutical industry would not sponsor a single clinical trial, if registration and distribution of drugs were possible without demonstrating benefit in a trial comparing the new drug with the best current standard. At the other extreme, some radiation oncologists recommend not even considering particle therapy for any patient because they seem to know – absent proof, however – that ion therapy will not exhibit any clinically relevant advantages over photon therapy despite its dosimetric advantages.

INTEGRATING PARTICLES IN MULTIDISCIPLINARY TUMOUR BOARD DECISIONS

This environment of scarce evidence together with relevant interests against evidence generation has significant bearings on how to integrate particle therapy into tumour board deliberations and discussions. Without a clear methodological framework and defined aims of considering particle therapy at tumour boards, it is nearly certain that subjective inclinations of involved persons will determine the course of recommendations and therewith the composition of cohorts undergoing particle therapy.

As long as there is no clear proof that any tumour clinically fares demonstrably better with particle rather than photon irradiation, considering particle therapy for a given patient comes down to the question of whether a dose distribution achievable with particles is *likely to result in favourable outcome* for the patient; in other words, whether the dose distribution achievable with particles will be relevantly different in critical normal tissues, organs or parts of organs at a given tumour dose. The elements of rational patient selection for particle therapy at tumour boards should rest on predictable effects of treatment, just as with any other treatment reasonably recommended at tumour boards. Any recommendation for treating cancer rests on two elements: given the baseline prognosis of a given tumour at a given stage in a given patient, treatment recommendations have to take into account the probability of improving the prognosis with respect to the tumour as well as probabilities of adverse effects. There is nothing special here regarding a recommendation for particle therapy compared with any other tumour-directed treatment. Considering radiotherapy, however, and when it comes to deciding for particle versus photon treatment, the recommendation should be based on expected dose distributions of the respective treatment modalities. At a given tumour dose – implying a certain tumour control probability in turn – the patient's risk of suffering adverse effects due to receiving particle therapy will decrease as a function of dose in relevant normal tissues. And this can – and should, therefore – be shown by comparative *in silico* photon-particle treatment planning comparisons. The situation is even more challenging for carbon ion radiotherapy, when an improved therapeutic ratio (TCP/NTCP) due to improved tumour control with heavy particles is assumed. Here, local control will have to be a primary endpoint (with meticulous monitoring of adverse effects) and validly comparing results with photon-achieved local control is unavoidable.

It may at times be difficult to impossible at tumour boards to estimate the amount of dose-reduction in critical organs *en face*, and it will be generally impossible to quantify the amount of risk reduction for adverse effects with sufficient precision.

Practical Compromise: Planning on Diagnostic Imaging

A practical solution for this need to compare dose distributions among modalities may be to perform a planning comparison study for photons versus particles on diagnostic imaging, because an estimate based on such a comparative endeavour will be much closer to the actual treatment situation than even the best guess of an experienced photon-and-particle radiation oncologist. It is clear, however, that such a comparative planning will again only approximate the actual treatment situation, but it has the great advantage that the patient will be spared undergoing treatment preparation and planning-CTs or -MRIs for particle therapy when there still is a chance of actually not receiving a recommendation for particle therapy. It will be very useful to develop tumour-location-specific pre-selection models for these situations. Planning on diagnostic imaging will partly render better-than-actually-achievable dose distributions for particles due to a number of factors: lack of any motion management, diagnostic as opposed to treatment positioning of the patient at imaging, lack of dual-energy CT-scanning at diagnostic imaging, absence of positioning aids and treatment couches, eventual contrast media perturbing the dose and other smaller or larger differences introducing location-dependent deviations from actually achievable treatment plans. In addition, these given differences may influence particle planning differently than photon planning, which might in turn be predictable using planning-pre-assessment models. However, in terms of feasibility, performing *in-silico* comparative treatment planning on diagnostic imaging may be the optimal approach – mainly due to feasibility – for selecting patients who might eventually benefit from particle therapy. It will render more reliable estimates of expected benefit than deciding on indications just by looking at images at tumour boards.

The Need for Standardisation

Definitions of gross tumour and clinical target volumes as well as contouring of organs-at-risk for specific tumour sites should be standardised and automatised as much as possible, before a reasonable assessment of presumptive advantages of particle therapy can be made. Also, generation of treatment plans using uncertainty settings as actually achievable in a given department (planning target volume margins for photons and robustness setting for protons and ions, respectively) in turn critically depend on numerous factors, such as machine parameters and respective imaging employed for treatment guidance. Standardisation and automatisation are not only required in order to achieve more objective photon-particle treatment plan comparisons, but are also highly desirable for reasons of feasibility and efficiency, as such planning comparison exercises entail a considerable extra burden on treatment planning divisions.

REQUIREMENTS: THE BACKBONE

Meticulous databases should be established containing both multi-factorial baseline and outcome assessments regarding tumour, patient and treatment parameters and characteristics (Niezink et al., 2015). Such databases are principally open for any kind of data, including molecular biology of tumour as well as normal tissues; always, full dosimetric data from photon as well as particle treatment must necessarily be included (Kessel et al., 2012; Combs et al., 2013). Patient- as well as physician- or health-care-professional-rated measures of disease burden and treatment-related toxicity along with tumour control, relapse and survival data accompanied by relevant imaging data should form the backbone for both the development and validation of predictive models needed to estimate the benefit of particle treatment. Especially toxicity-associated data need to be attained strictly prospectively, as retrospective scoring is unavoidably entrenched with bias, severely reducing data reliability. Full dose–volume parameters need to be routinely retrieved from patients treated with particles, but also from planning comparisons and (ideally) also from patients with comparable indications undergoing photon therapy. Outcome measures as well as NTCP models require continuous updating and recalibration in order to maintain or improve their accuracy.

Who Should Decide for Particle Therapy?

As the decision for particle therapy rather than photon treatment is a decision about radiotherapy technology and technique, it should clearly be made only by radiation oncologists, just like deciding about surgical approach or selecting a certain antineoplastic drug should be the task of oncologic surgeons or medical oncologists, respectively. Sometimes, patients themselves make up their minds and self-refer themselves to proton centres. Also, non-radiation-oncologists (surgeons, medical oncologists and even non-oncology specialists) might recommend proton or even carbon ion therapy to patients at tumour boards or even bypass multidisciplinary discussions. Such practice should be discouraged, as it does not reflect reasonable patient self-determination, but rather subjecting patients to decisions based on likely unbalanced information. Patients in need of radiotherapy will understandably be receptive for an option suggesting increased efficacy of treatment by delivering higher dose to the tumour along with improved sparing of healthy tissues and they may be less likely to request proof for such claims made.

Enriched Populations

The population eligible for particle treatment should therefore be enriched with patients likely to benefit from such treatment. As long as there is no clear indication for any tumour entity or molecular signature to be more susceptible for particles rather than photons, potential benefit can only be determined by dosimetric advantages: a higher tumour-control-to-normal-tissue complication probability (TCP/NTCP) (Suit et al., 2010). This approach highly resembles molecularly targeted therapies, that – in order to be efficacious – require molecular features in tumour or immune cells they can target. Molecularly targeted agents are typically ineffective if given to patients not exhibiting the critical (molecular) signature. Particle therapy will predictably be ineffective when given to patients lacking the critical dosimetric and/or biological signature. Such enrichment will be necessary both for regular treatment and for eventual comparative clinical studies. At regular treatment, patient, tumour and treatment data are collected in order to use them to validate and update models predicting normal tissue complications and outcome in various dimensions. Even and especially for comparative clinical studies, a dosimetric *in-silico* difference between treating with photons versus particles needs to be an inclusion criterion, without which a meaningful outcome of the study will be highly unlikely a priori (Widder et al., 2015). There is already clinical evidence for the detrimental effect of bypassing dosimetrically enriching study populations: a phase II prospective randomised trial of photons versus protons for locally advanced non-small-cell lung cancer was negative for both toxicity as well as for tumour control (Liao et al. 2018). This was perfectly explained by complete absence of any dosimetric differences between photons and protons. Consequentially, and unsurprisingly, there were no outcome differences between the treatment arms in any respect. A retrospective investigation of prostate cancer treated with passively scattered protons versus IMRT was also negative, and patient eligibility had again not been based on any dosimetrically predictable advantage (Sheets et al., 2012).

CONCLUSION

In summary, integrating particle beam therapy into multidisciplinary treatment decisions should follow generally accepted principles of evidence-based medicine: just like using a new drug or a novel surgical technique does not require different reasoning at multidisciplinary tumour boards, neither does particle therapy. Any rational recommendation of one treatment option over another is based on a model predicting favourable outcome as, for example, a favourable hazard-ratio of treatment A versus treatment B in terms of a clinically relevant outcome. Tumour control as well as clinically relevant toxicity are the outcome measures at stake. This principle retains validity for making justified recommendations for particle therapy.

REFERENCES

Bekelman, J E, D A Asch, Z Tochner, J Friedberg, D J Vaughn, E Rash, K Raksowski, and S M Hahn. 2014. Principles and reality of proton therapy treatment allocation. *International Journal of Radiation Oncology, Biology, Physics* 89 (3): 499–508. doi:10.1016/j.ijrobp.2014.03.023.

Bradley, J D, R Paulus, R Komaki, G Masters, G Blumenschein, S Schild, J Bogart et al. 2015. Standard-dose versus high-dose conformal radiotherapy with concurrent and consolidation carboplatin plus paclitaxel with or without cetuximab for patients with stage IIIA or IIIB non-small-cell lung cancer (RTOG 0617): A randomised, two-by-two factorial P. *The Lancet Oncology* 16 (2): 187–199. doi:10.1016/S1470-2045(14)71207-0.

Combs, S E., M Djosanjh, R Potter, R Orrechia, T Haberer, M Durante, P Fossati et al. 2013. Towards clinical evidence in particle therapy: ENLIGHT, PARTNER, ULICE and beyond. *Journal of Radiation Research* 54 (suppl 1): i6–i12. doi:10.1093/jrr/rrt039.

Ebner, D K, and T Kamada. 2016. The emerging role of carbon-ion radiotherapy. *Frontiers in Oncology* 6: 140. doi:10.3389/fonc.2016.00140.

Fleshman, J, M Branda, D J Sargent, A M Boller, V George, M Abbas, W R Peters et al. 2015. Effect of laparoscopic-assisted resection versus open resection of stage II or III rectal cancer on pathologic outcomes: The ACOSOG Z6051 randomized clinical trial. *JAMA* 314 (13): 1346–1355. doi:10.1001/jama.2015.10529.

Howick, J, P Glasziou, and J K Aronson. 2009. The evolution of evidence hierarchies: What can bradford hill's 'guidelines for causation' contribute? *Journal of the Royal Society of Medicine* 102 (5): 186–194. doi:10.1258/jrsm.2009.090020.

Howington, J A, M G Blum, A C Chang, A A Balekian, and S C Murthy. 2013. Treatment of stage I and II non-small cell lung cancer: Diagnosis and management of lung cancer, 3rd ed: American college of chest physicians evidence-based clinical practice guidelines. *Chest* 143 (suppl 5): e278S–e313S. doi:10.1378/chest.12-2359.

Kamada, T, H Tsujii, E A Blakely, J Debus, W De Neve, M Durante, O Jäkel et al. 2015. Carbon ion radiotherapy in Japan: An assessment of 20 years of clinical experience. *The Lancet Oncology* 16 (2): e93–e100. doi:10.1016/S1470-2045(14)70412-7.

Kessel, K A, N Bougatf, C Bohn, D Habermehl, D Oetzel, R Bendl, U Engelmann et al. 2012. Connection of European particle therapy centers and generation of a common particle database system within the European ULICE-framework. *Radiation Oncology* 7 (1): 115. doi:10.1186/1748-717X-7-115.

Langendijk, J A, P Lambin, D De Ruysscher, J Widder, M Bos, and M Verheij. 2013. Selection of patients for radiotherapy with protons aiming at reduction of side effects: The model-based approach. *Radiotherapy and Oncology.* doi:10.1016/j.radonc.2013.05.007.

Laramore, G E, J M Krall, F J Thomas, K J Russell, M H Maor, F R Hendrickson, K L Martz, T W Griffin, and L W Davis. 1993a. Fast neutron radiotherapy for locally advanced prostate cancer. Final report of radiation therapy oncology group randomized clinical trial. *American Journal of Clinical Oncology* 16 (2): 164–167.

Laramore, G E, J M Krall, T W Griffin, W Duncan, M P Richter, K R Saroja, M H Maor, and L W Davis. 1993b. Neutron versus photon irradiation for unresectable salivary gland tumors: Final report of an RTOG-MRC randomized clinical trial. Radiation therapy oncology group. Medical research council. *International Journal of Radiation Oncology, Biology, Physics* 27 (2): 235–240.

Liao, Z, J Jack Lee, R Komaki, D R Gomez, M S O'Reilly, F V Fossella, G R Blumenschein, et al. 2018. Bayesian adaptive randomization trial of passive scattering proton therapy and intensity-modulated photon radiotherapy for locally advanced non-small-cell lung cancer. *Journal of Clinical Oncology.* doi:10.1200/JCO.2017.74.0720.

McCulloch, P, D G Altman, W B Campbell, D R Flum, P Glasziou, J C Marshall, J Nicholl et al. 2009. No surgical innovation without evaluation: The IDEAL recommendations. *The Lancet* 374 (9695): 1105–1112. doi:10.1016/S0140-6736(09)61116-8.

Minsky, B D, T F Pajak, R J Ginsberg, T M Pisansky, J Martenson, R Komaki, G Okawara, S A Rosenthal, and D P Kelsen. 2002. INT 0123 (Radiation Therapy Oncology Group 94-05) Phase III trial of combined-modality therapy for esophageal cancer: High-dose versus standard-dose radiation therapy. *Journal of Clinical Oncology: Official Journal of the American Society of Clinical Oncology* 20 (5): 1167–1174.

Mitin, T, and A L Zietman. 2014. Promise and pitfalls of heavy-particle therapy. *Journal of Clinical Oncology: Official Journal of the American Society of Clinical Oncology* 32 (26): 2855–2863. doi:10.1200/JCO.2014.55.1945.

Niezink, A G H, N J Dollekamp, H J Elzinga, D Borger, E J H Boer, J F Ubbels, M Woltman-van Lersel et al. 2015. An instrument dedicated for modelling of pulmonary radiotherapy. *Radiotherapy and Oncology.* doi:10.1016/j.radonc.2015.03.020.

Paganetti, H. 2014. Relative biological effectiveness (RBE) values for proton beam therapy. Variations as a function of biological endpoint, dose, and linear energy transfer. *Physics in Medicine and Biology* 59 (22): R419–R472. doi:10.1088/0031-9155/59/22/R419.

Potter, R, K Janssen, F J Prott, J Widder, U Haverkamp, H Busse, and R P Muller. 1997. Ruthenium-106 eye plaque brachytherapy in the conservative treatment of uveal melanoma: Evaluation of 175 patients treated with 150 Gy from 1981–1989. *Frontiers of Radiation Therapy and Oncology.* http://gateway.webofknowledge.com/gateway/Gateway. cgi?GWVersion=2&SrcAuth=ORCID&SrcApp=OrcidOrg&DestLinkType=FullRecord&DestApp=MEDLINE&KeyUT=MEDLINE:9205894&KeyUID=MEDLINE:9205894.

Ruysscher, D De, M M Lodge, B Jones, M Brada, A Munro, T Jefferson, and M Pijls-Johannesma. 2012. Charged particles in radiotherapy: A 5-year update of a systematic review. *Radiotherapy and Oncology: Journal of the European Society for Therapeutic Radiology and Oncology* 103 (1): 5–7. doi:10.1016/j.radonc.2012.01.003.

Sheets, N C, G H Goldin, A M Meyer, Y Wu, Y Chang, T Sturmer, J A Holmes et al. 2012. Intensity-modulated radiation therapy, proton therapy, or conformal radiation therapy and morbidity and disease control in localized prostate cancer. *JAMA* 307 (15): 1611–1620. doi:10.1001/jama.2012.460.

Stevenson, A R L, M J Solomon, J W Lumley, P Hewett, A D Clouston, V J Gebski, L Davies, K Wilson, W Hague, and J Simes. 2015. Effect of laparoscopic-assisted resection versus open resection on pathological outcomes in rectal cancer. *JAMA* 314 (13): 1356. doi:10.1001/jama.2015.12009.

Strong, S A., and N J. Soper. 2015. Minimally invasive approaches to rectal cancer and diverticulitis. *JAMA* 314 (13): 1343. doi:10.1001/jama.2015.11454.

Suit, H, T DeLaney, S Goldberg, H Paganetti, B Clasie, L Gerweck, A Niemierko et al. 2010. Proton versus carbon ion beams in the definitive radiation treatment of cancer patients. *Radiotherapy and Oncology* 95 (1): 3–22. doi:10.1016/j.radonc.2010.01.015.

van de Sande, M AE, C L Creutzberg, S van de Water, A W Sharfo, and M S Hoogeman. 2016. Which cervical and endometrial cancer patients will benefit most from intensity-modulated proton therapy? *Radiotherapy and Oncology* 120 (3): 397–403. doi:10.1016/j.radonc.2016.06.016.

Widder, J, A van der Schaaf, P Lambin, C A Marijnen, J P Pignol, C R Rasch, B J Slotman, M Verheij, and J A Langendijk. 2015. The quest for evidence for proton therapy: Model-based approach and precision medicine. *International Journal of Radiation Oncology, Biology, Physics.* doi:S0360-3016(15)26569-8.

Zietman, A, M Goitein, and J E Tepper. 2010. Technology evolution: Is it survival of the fittest? *Journal of Clinical Oncology: Official Journal of the American Society of Clinical Oncology* 28 (27): 4275–4279. doi:10.1200/JCO.2010.29.4645.

CHAPTER 8

Role of Multidisciplinary Collaborative Network for Advancing Cancer Therapy

Manjit Dosanjh and Jacques Bernier

THE EUROPEAN NETWORK FOR Light Ion Hadron Therapy (ENLIGHT) had its inaugural meeting at European Organization for Nuclear Research (CERN) in February 2002. About 70 specialists from different disciplines, including radiation biology, oncology, physics and engineering attended this first gathering [1]. At that time, 'multidisciplinarity' was not yet a buzzword and the network was a real pioneer in the field [2].

Clinicians, physicists, biologists and engineers with experience and interest in particle therapy were gathering for the first time under the network's umbrella. Started with the support of the EU Commission, ENLIGHT itself has run four other EU-funded projects – ULICE, PARTNER, ENVISION and ENTERVISION [3–6]. In fact, the network has worked as an open collaborative tool and served as a common multidisciplinary platform for all the communities involved. Since its foundation, ENLIGHT has relied on the variety of skills of its members to be able to identify and tackle the technical challenges, train young researchers, support innovation and lobby for funding.

The idea of creating a multi-disciplinary and transnational platform for researchers and experts involved with radiation therapy, including hadron therapy, was born in 2001, when the Proton–Ion Medical Machine Study (PIMMS) had been presented at MedAustron and the whole idea of setting up specialised centres providing multiple radiation modalities was taking off in Europe [7]. That was the time when European Society for Radiotherapy & Oncology (ESTRO) was also starting to see the importance of considering other radiation options; at the same time, at CERN, Ugo Amaldi was pushing to get the organisation more heavily involved with hadron therapy and applications of accelerators advances in medical physics. The creation of ENLIGHT was, indeed, the result of the work of a few visionary people who could see the power of collaboration and knowledge sharing.

One of the most enlightening initiatives that the network supported was the organization of conferences devoted to blending scientific backgrounds and expertise with the aim of creating a new culture of collaboration and sharing. The first of such conferences was Physics for Health in Europe (PHE), held in 2010 at CERN, the temple of fundamental physics. Although, at first sight, large accelerators and giant detectors do not seem to have much in common with sharp tools that medicine needs, physics is not new to producing applications for life sciences. Several detection techniques are used in diagnosis instruments, and radiation and hadron therapy were born in physics labs. The idea of using protons for cancer treatment was first proposed in 1946 by the physicist Robert Wilson, who later became the founder and first director of the Fermi National Accelerator Laboratory (Fermilab) near Chicago [8]. The first patients were treated in the 1950s in nuclear physics research facilities by means of non-dedicated accelerators. Initially, the clinical applications were limited to a few parts of the body, as accelerators were not powerful enough to allow protons to penetrate deep in the tissues.

In the late 1970s improvements in accelerator technology, coupled with advances in medical imaging and computing, made proton therapy a viable option for routine medical applications. However, it has only been since the beginning of the 1990s that proton facilities have been established in clinical settings, the first one being in Loma Linda, California, United States. At the end of 2016, nearly 70 centres were in operation worldwide, and another 63 are under construction or in the planning stage (Figure 8.1). Most of these are proton centres, 25 in the United States (protons only), 19 in Europe (3 dual centres), 15 in Japan (4 carbon and 1 dual), 3 (1 carbon and 1 dual) in China and 4 in other parts of the world. Globally there is a huge momentum in particle therapy, especially treatment with protons. Sixty-three new centres are under construction – so that by 2021, there will be 130

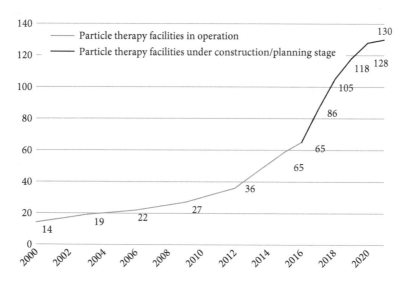

FIGURE 8.1 Hadron therapy facilities in operation worldwide, under construction and in the planning stage at the end of 2016. (From https://www.ptcog.ch/index.php/facilities-in-operation; https://www.ptcog.ch/index.php/facilities-under-construction. Particle Therapy Co-Operative Group (PTCOG)).

centres operating in nearly 30 different countries. We can proudly say that the ENLIGHT efforts definitely helped hadron therapy and high-tech medicine gain momentum by making scientists and doctors plan common and synchronised actions.

The first gatherings and conferences supported by ENLIGHT focused on the intrinsic physics processes that make hadron therapy a promising way of treating tumours which cannot be treated readily with conventional irradiation techniques since they are either radio-resistant or located very close to critical organs. They also discussed radiobiology, accelerators, radioisotope production, detectors and use of information technology. The seeds were there to rapidly extend and set more ambitious goals for the network. In a couple of years, the PHE conference had already united with the purely medical 'International Conference on Translational Research in Radio-Oncology' (ICTR) to become the interdisciplinary ICTR-PHE conference, held every second year since 2012. With its 10-year history and it's already established network of participants, ICTR injected new power into the newly created melting pot. Just like ENLIGHT itself, ICTR-PHE was established as a multidisciplinary conference since its beginning as it not only attracts experts in particle physics, hadron therapy, radiation therapy, but also nuclear medicine, immunotherapy and so on. PHE alone could only deal with topics such as detectors and techniques to detect the effects of specific particles in the body. On the other hand, translational research was already in ICTR but only within the limits of X-rays. The two together created a new blend. Over the years, the physicists learnt to orient their efforts in order to best meet doctors' needs, and the medical environment learnt methodologies and approaches that came from the fundamental science labs. Sharing and discussing the results of the most recent research and addressing the challenges and possible developments to indicate the subjects with the highest priority for further studies in diagnosis and therapy on the European scale became just ordinary business for members of the community. A new way of thinking, working and collaborating was born, with radiochemists, nuclear-medicine physicians, biologists, software developers, accelerators experts, radio-oncologists and detector and medical physicists being asked to make innovative proposals to boost further the comprehensive approach of cancer management. Adaptive radiotherapy driven by the tumour biological response to treatment, clinical applications of drug-radiation interactions as chemo- and bio-radiation, modulation of the host's immune system by radiotherapy as well as accurate simulation tools for hadron therapy are among the most significant innovations in treatment optimization retrieved from these 'bridging the gap' strategies.

Today, what used to be a pioneering vision has become evidence: complex diseases, including cancer, show that the best treatment can only come as a result of the collaboration of experts coming from many different fields. ENLIGHT has demonstrated the advantages of regular and organised exchange of data, information and best practice. The more we understand about cancer, the more we realise that it's not a single disease but rather a multiplicity of different health issues that can evolve in different ways and have different behaviours. If we want to fight it, we have to invest all our efforts in developing a personalised approach rather than searching for the single Holy Grail that will rescue all of humanity. In personalised medicine, we are looking for single signatures of specific issues that can be treated in a very targeted way. When we look at cancerous cells today, we search

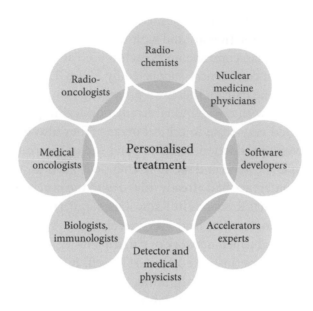

FIGURE 8.2   Collaborative approach to personalised treatment.

for specific mutations to understand the radiosensitivity we should expect; we evaluate the whole situation with the help of many experts; we take into account the information coming from a multitude of studies and contributions that are available in literature.

Personalised treatments are holding centre stage in the scientific debate on hadron therapy (Figure 8.2). Technology is not dormant: developments are crucial to reduce the costs; to provide treatments tailored to each specific case; and to reach the necessary level of sophistication in beam delivery to treat complex cases such as tumours inside, or close to, moving organs. In this context, imaging is key. Today, it is becoming obvious that the optimal imaging tool will necessarily have to combine different imaging modalities, for example, positron emission tomography (PET) and prompt photons. PET is, of course, a mainstay for dose imaging, but a well-known issue in its application to in-beam real-time monitoring for hadron therapy comes from having to allow room for the beam nozzle: partial-ring PET scanners cannot provide full angular sampling and therefore introduce artefacts in the reconstructed images. The time-of-flight (TOF) technique is often used to improve the image-reconstruction process. An innovative concept, a jagiellonian positron emission tomograph (J-PET) scanner detects back-to-back photons in plastic scintillators, and applies compressive sensing theory to obtain a better signal normalisation, and therefore improves the TOF resolution. The new EXPLORER full-body PET, which is being developed in the United States, promises to bring a sixfold improvement with respect to the current situation. The powerful machine is expected to allow clinicians to reconstruct images at a higher spatial resolution, detect smaller lesions and low-grade disease, and provide better statistics for kinetic modelling. EXPLORER will also deliver an expected fortyfold reduction in the radiation dose, which will allow more patients to be accurately diagnosed with a much lower risk and at an earlier stage.

Thirty years ago, we were not aware that all these developments would have been possible. We didn't have the technologies and the instrumentation; we didn't know how to look at the cellular behaviour and evolution. Molecular biology was not born; the CT scanner was not even a dream; PET and MRI didn't exist. For example, in order to evaluate the radioresistance of cancerous cells we could just bombard the sample of cells taken from the tumour and count the surviving fraction after the radiation treatment. Today, we find all this very primitive and inaccurate. If we look at history, we realize that the quantum leap was actually done by cell and molecular biology, which has resulted in radiobiology being able to make rapid unexpected progress. We have improved our comprehension of molecular tumour response to irradiation with both ions and photons, and of the biological consequences of the complex, less repairable DNA damage caused specifically by ions. Understanding the cell signalling mechanisms affected by hadron therapy will lead to improvements in therapeutic efficacy. A particularly thorny issue is the relative biological effectiveness (RBE) of protons and carbon with respect to photons. More extensive and systematic radiobiology studies with different ions, under standardised dosimetry and laboratory conditions, are needed to clarify this and other open issues.

If we look at particle therapy today, we see that the next step will be the broad use of carbon and other ions. These have some clear advantages even over protons in providing both a local control of very aggressive tumours and a lower acute or late toxicity, thus enhancing the quality of life during and after cancer treatment. Since the birth of hadron therapy, more than 160,000 patients have been treated globally with hadrons, including about 20,000 with carbon ions.

A key element for such a quick evolution was training as particle therapy centres require highly trained staff, and yet few experts exist in this rapidly expanding field. ENLIGHT demonstrated its visionary approach also in this field as training of the young generations was one of the main objectives of the network since the beginning. In addition, regular training sessions have been added to the ENLIGHT annual meetings as of 2016.

Hadron therapy is facing a dilemma when it comes to designing clinical trials. In fact, from a clinical standpoint, the ever-increasing number of hadron therapy patients would allow randomised trials to be performed – that is, systematic clinical studies in which patients are treated with comparative methods to determine which is the most effective curative protocol.

However, several considerations add layers of complexity to the clinical-trials landscape: the need to compare standard photon radiotherapy not only with protons but also with carbon ions, the positive results of hadron therapy treatments for main indications, and the non-negligible fact that most of the patients who contact a hadron therapy centre are well informed about the technique and will not accept being treated with conventional radiotherapy. Nevertheless, progress in clinical trials is being made. For the first time, in 2015, at the ENLIGHT meeting in Kraków, the two dual-ion (proton and carbon) centres in Europe – HIT, in Heidelberg (Germany) and CNAO, in Pavia (Italy) – presented patient numbers and dose-distribution studies carried out at their facilities [9–11]. The data were collected mainly in cohort studies carried out within a single institution, and the results often highlighted the need for larger statistics and a unified database. More data from patients treated with carbon ions are becoming available from the MedAustron hadron therapy centre in Wiener Neustadt (Austria). Clinical trials are also a major focus outside Europe: in the United States, several randomised and non-randomised trials have been set up to compare protons with photons, and to investigate either the survival improvement (for glioblastoma, non-small-cell lung cancer, hepatocellular carcinoma and oesophageal cancer) or the decrease of adverse effects (low-grade glioma, oropharyngeal cancer, naso-pharyngeal cancer, prostate cancer and post-mastectomy radiotherapy in breast cancer). The National Cancer Institute in the United States funded a trial comparing conventional radiation therapy and carbon ions for pancreatic cancer.

Fifteen years after the birth of ENLIGHT, we can be proud of our achievements, but we can't rest yet and simply enjoy the results of our efforts [12]. The current situation calls for more and more inclusivity in the various disciplines that can help get the best results for the patients. One recent add-on is immunology. We don't know yet where a deeper knowledge of the human immune system will bring us, but we can already see that this is becoming a key element that we have to take into account when we fight cancer. As a matter of fact, the interactions of radiation with the host's immune system are nowadays at the centre of many investigations. In addition, increasing attention is being paid to the association of various forms of immunotherapies with radiotherapy, in order to boost cancer cell killing. In this domain, current translational studies are investigating both the effects of immune checkpoint blockade strategies and adoptive immunotherapies in combination with radiation. We also need to involve all the professionals developing new technologies and make them focus on what the clinical field really needs. We need to closely monitor all the advances that experts are working on in their labs and rapidly bring them to the patient. We need to increase our effectiveness; we need to create new ways for the various disciplines to exchange best practice and to be willing to share what they have learnt in their respective research environments.

Several challenges still lie ahead, including securing funding and succeeding in harmonising data, which is key to sharing information and best practices within the various communities.

Medical imaging has a key role as it allows medical doctors to deliver an effective dose to the target tumour site while minimising the side effects on healthy tissues: the volume and position of the tumour has to be assessed before, after and during treatment whenever possible by using a whole range of imaging tools, such as PET, magnetic resonance imaging (MRI) and computed tomography (CT) scans – alone or in combination. Moving organs such as lungs presents a challenging task for medical imaging, since the position of the tumour has to be monitored while it is being treated. The integration of MRI with a linear accelerator, for example, can provide image guidance that is simultaneous with treatment, thus reducing the patient's exposure to the additional ionising radiation of CT scans.

In terms of health economics, evidence-based rationale for hadron therapy is of critical importance, especially when considering the necessity of balancing absolute costs with the therapeutic index of this treatment modality. In a time of healthcare reforms in most countries, evidence-based justification of new technologies is indeed an imperative step to support their utility and efficacy. This, together with the fact that hadron therapy cannot be uniformly the most sustainable option for all cancers, requires a precise identification of those patients and malignancies subgroups which will most likely benefit from the use of hadrons.

Emerging topics in all forms of radiation therapy are the collection, transfer and sharing of medical data, and the implementation of big data-analytics tools to inspect them. These tools will be crucial in implementing decision support systems, allowing treatment to be tailored to each individual patient. The flow of information in healthcare, and in particular in radiation therapy, is overwhelming not only in terms of data volume but also in terms of the diversity of data types involved. Indeed, experts need to analyse patient and tumour data, as well as complex physical dose arrays, and to correlate these with clinical outcomes that also have genetic determinants.

ENLIGHT has positioned itself as 'THE' network where all this can really take shape and grow. We can rely on world experts who are aware of the forefront technologies and latest breakthroughs. Over the years, the ENLIGHT community has shown a remarkable ability to reinvent itself, while maintaining its cornerstones of multi-disciplinarity, integration, openness and attention to future generations [12]. All the communities involved will have to draw a new list of priorities, which will allow them to tackle the latest challenges of a frontier discipline such as hadron therapy in the most effective way. Collaboration, inter-disciplinarity also add translational research are the models to follow. Making these models effective is key.

REFERENCES

1. http://cern.ch/enlight, viewed December 2017.
2. M. Dosanjh, April 2002. http://cerncourier.com/cws/article/cern/28632.
3. http://cern.ch/envision, viewed December 2017.
4. http://cern.ch/partner, viewed December 2017.
5. http://cern.ch/entervision, viewed December 2017.
6. http://cern.ch/ULICE, viewed December 2017.
7. Badano L, Benedikt M, Bryant PJ, Crescenti M, Holy P, Knaus P, Maier A, Pullia M, and Rossi S. 1999. Proton-ion medical machine study (PIMMS)—Part I: CERN/PS 99-010 DI, Geneva, Switzerland, March 1999 and with additional authors; G. Borri G, and S. Reimoser S, and contributors F. Grammatica F, M. Pavlovic M, and L. Weisser L., 2000. Part II: CERN/PS 00-007 DR, Geneva, Switzerland, July 2000. A CD with drawings and other data including software is available on request, Yellow Report number CERN 2000-2006.
8. Wilson RR. Radiological use of fast protons. *Radiology* (1946) 47(5):487–491. doi:10.1148/47.5.487.
9. Debus J, Gross KD, and Pavlovic M (Eds.). 1998. *Proposal for a Dedicated Ion Beam Facility for Cancer Therapy* (Darmstadt, Germany: GSI).
10. Durante M, Orecchia R, Loeffler JS. Charged-particle therapy in cancer: Clinical uses and future perspectives. *Nat Rev Clin Oncol* (2017) 14:483–495.
11. ENLIGHT Annual Meeting 2015, Kraków, Poland: http://indico.cern.ch/event/392790/.
12. Dosanjh M and Cirilli M. Networking against cancer, CERN Courier. http://cerncourier.com/cws/article/cern/63701, January 15, 2016.

CHAPTER 9

Long-Term Outcomes and Prognostic Factors in Patients with Indications for Particle Therapy in Sarcomas

Beate Timmermann and Stephanie E. Combs

CONTENTS

INTRODUCTION

Sarcomas encompass a rare and heterogeneous group of solid malignant neoplasms which originate from soft tissues (84%) or bones (14%) [1]. Multidisciplinary management of sarcomas that are often contiguous to critical organs at risk is challenging and varies widely according to the respective site, histopathology and stage. Even though improvements of surgical techniques were achieved over the last decades, complete resection is often impossible without risking major impairments. In these cases, surgical procedure is ideally complemented or replaced by high-precision radiotherapy (RT). However, sensitive structures in close proximity to the tumour may limit the delivery of sufficient dose required for effective local therapy in sarcomas. In this context, protons and other charged particles (e.g., carbon ions) are of increasing interest due to their physical and radiobiological properties. Protons and other charged particles have the advantage of restricting the irradiated volume and improving sparing of normal tissue compared to standard photon techniques and intensity-modulated radiotherapy (IMRT). However, evidence of the clinical benefits of particle therapy for the treatment of sarcomas, especially for long-term outcomes and their prognostic factors, is still limited. Nevertheless, early data in small cohorts suggest good feasibility as well as high efficacy. For rhabdomyosarcoma (RMS) and Ewing's sarcoma, data on treatment with proton beam therapy (PBT) are predominantly available, whereas carbon-ion radiotherapy (CIRT) was introduced predominantly for osteosarcoma and chordoma (CH) and chondrosarcoma (CS).

RHABDOMYOSARCOMAS

Epidemiology

RMS is the most common childhood soft-tissue sarcoma and accounts for 3.1% of all paediatric tumours, with an incidence of 4.8 per million [2,3]. The most common site of presentation in paediatric RMS is the head and neck, with the majority being in a parameningeal location. The five-year overall survival (OS) rate in the paediatric population has improved from 49% in 1975–1979 to 64% in 2003–2009 [3]. RMS in adults is less common, accounting for 2%–5% of adult soft-tissue sarcomas [4]. Experience in this population is limited, and data is predominantly derived from retrospective case series [5,6]. Compared to the paediatric population, adult RMS patients have a poorer prognosis, with a five-year OS of 31%–44% [5,7–9].

Risk Stratification

Risk stratification of patients with RMS is based on a number of factors such as extent of residual tumour after surgery with consideration of regional lymph node involvement, tumour size, invasiveness, nodal status and site of primary tumour. These factors are incorporated into the Children's Oncology Group (COG) risk stratification system (low, intermediate and high risk). Disease site alone was also proven to be a strong predictor for survival and disease control. While paediatric patients with orbital RMS have favourable 10-year OS and event-free survival (EFS) of 87% and 77% [10], tumours in parameningeal site are known to have a poorer prognosis, with a 10-year OS and EFS of 66.1% and 62.6%, respectively [11]. Besides very young age, large tumour size and histology other than embryonal, particularly intracranial extensions are associated with a poorer outcome in this unfavourable site [11–14].

Therapy

Chemotherapy is administered according to risk stratification. Local control (LC) is typically achieved by a combination of resection with or without RT. The goal of surgical resection is complete tumour removal while preserving organs and functional tissue. However, surgical approach depends highly on tumour site and feasibility of complete surgery. In parameningeal sites, LC is usually managed by CTx (chemotherapy) and RT alone [15].

According to Intergroup Rhabdomyosarcoma Studies (IRS) recommendation, all patients with group II-IV RMS will receive RT. Dose, duration and timing of RT depends on the clinical group and disease site [16]. RT has proven to be an important component of the combined-modality treatment in RMS [17,18]. Particularly in parameningeal RMS, analyses reinforce the necessity for RT. Significant differences in 10-year OS and EFS for paediatric patients with (68.5% and 66%) and without initial RT (40.8% and 25.1%) are reported [11]. PBT is considered highly conformal and dosimetric studies have confirmed that PBT, when compared to conventional photon-based RT and IMRT, may offer considerable dosimetric advantages in parameningeal [19,20], orbital [20,21] and genitourinary sites [20,22]. However, clinical comparative data on long-term outcomes after PBT is still sparse in paediatric RMS and does not exist in the adult population.

Long-Term Outcomes and Prognostic Factors

Five-year OS, LC and EFS for PBT in localised RMS or metastatic embryonal RMS are 78%, 81% and 69%, respectively [23]. Outcomes for parameningeal RMS are notably lower, with five-year OS, LC, EFS, PFS (progression-free survival) and FFS (failure-free survival) of 64%–73%, 67.5%–77%, 60%, 72% and 59%, respectively [23–26]. Analyses on prognostic factors after PBT support these data. Parameningeal site, compared to other sites, is a significant risk factor for experiencing local failure [24]. Furthermore, intracranial extensions in parameningeal RMS seem to strongly predict local failure with a hazard ratio (HR) of 3.78 ($p = 0.009$) [24].

COG risk group (high versus low/intermediate) with a HR of 4.86 ($p = 0.09$) as well as IRS stage (≥ 3) with an HR of 7.01 ($p = 0.003$) strongly predict LC [24]. Five-year EFS, OS and LC are higher in the low-risk group (93%, 100% and 93%) compared to intermediate-risk patients with a five-year EFS, OS and LC of 61% ($p = 0.04$), 70% ($p = 0.04$) and 77% ($p = 0.20$), respectively [23]. These outcomes for intermediate-risk patients treated with PBT were poorer than those in a COG trial using conventional RT including a larger cohort of intermediate-risk patients, reporting an FFS of 68%–73% and an OS of 79% [13].

Age of the patient also influences tumour control. A trend is seen towards improved LC for patients aged 2–10 years (88%) compared with patients younger than two years or older than 10 years (64%) ($p = 0.07$) [23]. In addition, a primary tumour size of more than five centimetres adversely affects OS ($p = 0.14$) [23] and local failure (HR = 3.13, $p = 0.04$) [24]. Another significant positive predictor for PFS is a shorter time interval between the start of neoadjuvant CTx (chemotherapy) and start of PT (cutoff at 13 weeks) with a PFS of 57% compared to 89% ($p = 0.04$). Particularly for the high-risk group with intracranial extensions, these data suggest that a delay in radiation should be avoided [24,25].

Late Toxicities

Depending on the site of treatment, RT can result in adverse events such as decreased statural growth, neuroendocrine dysfunction or visual impairments [27,28]. The incidence of late toxicities in paediatric RMS populations treated with PBT is promising. Late toxicities of any grade (18%–35%) in paediatric cohorts [23,24] were favourable compared to long-term toxicity data for head and neck RMS treated with IMRT ranging from 32% to 47% [29–31]. In several studies on PBT for RMS, late higher-grade toxicities were seen in 7% [23], 18% [24] and 8% [25] of the patients. These late sequelae occurred as endocrine abnormalities, facial hypoplasia and dry eye [23] as well as cataracts and hearing impairments [23–25]. So far, only a single case of radiation-induced secondary malignancy was reported [24].

CHORDOMAS AND CHONDROSARCOMAS

Epidemiology

CH as a primary bone tumour arises from remnants of the notochord. It occurs within the axial skeleton, two-thirds at the ends (sacrum and clivus) and one-third at the spine [32]. CS is a heterogeneous group of malignant, cartilage-forming bone tumours and is the second leading primary sarcoma after osteosarcoma [33].

Risk Stratification

Important prognostic factors are tumour volume and age [34,35]. A common staging system for CH and CS is the American Joint Committee on Cancer (AJCC) System for bone sarcomas [36,37]. It is based on tumour size, grade, presence and locations of metastases. Stage I bone sarcomas are low grade and stage II bone sarcomas are high grade. Patients with stage III present with synchronous tumours of the regional bone. Stage IV is subdivided in the presence of either pulmonary (A) or nonpulmonary metastases (B).

Therapy

Gross total resection (GTR) is rarely achievable due to the typical location of CH and CS near critical structures. Therefore, additional RT is indicated to achieve better LC rates and to improve the prognosis [38,39]. As CH and CS are highly radio-resistant, doses up to 70–74 Gray are necessary to be effective (Figure 9.1) [40,41]. However, sensitive structures in close proximity to the tumour may limit the delivery of sufficient dose. Only few series

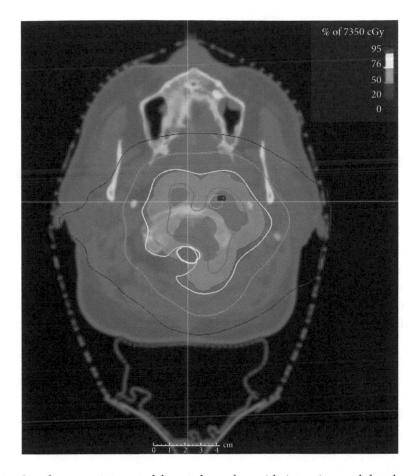

FIGURE 9.1 Simultaneous integrated boost dose plan with intensity modulated proton beam therapy for a clivus chordoma in a 50-year-old patient with sparing of myelon. Purple area: gross tumour volume, orange area: clinical target volume, red line: 95% isodose, yellow line: 76% isodose, light blue line: 50% isodose, dark blue line: 20% isodose.

with photons have delivered such high doses as required for long-term tumour control [42]. Particle therapy has been shown to have an important role since local doses beyond 70 Gray are essential for long-term LC. The physical properties of the particle beam, together with the intricate anatomy with close proximity to sensitive organs at risk such as the brain stem, optic chiasm or cranial nerves, which have relevant physiological functions, are central arguments for particle therapy.

Long-Term Outcomes and Prognostic Factors

Several centres have reported their results from photon RT and have shown rather disappointing long-term LC, ranging from 15%–65% for CH, and 80%–100% for CS [43–50]. Early data from proton centres have demonstrated superior results; therefore, currently, proton therapy should be considered the gold standard [51]. Early studies from Boston, Massachusetts, United States, have reported local failure rate of 31% in 204 CH patients, of which 95% were local recurrences [51].

French data published by Noel et al. reported on 67 patients with CH and CS with photons and a proton boost [35]; the three-year LC rates were 71% and 85% for CH and CS. Updated results for CH confirmed the LC rates at 86% at two and 54% at four years, respectively [34]. In that study, dose was limited to 67 Gray, which is lower than in most series treating the protons therapy only. Recent data included only patients treated with protons, no photons. While overall tolerability of such high doses is reported, including tumours of varying histologies, only a few series on pure CH or CS are available with long-term follow-up and providing long-term evidence [52].

The Paul Scherrer Institute (PSI) published their results after proton therapy delivered with spot-scanning. They treated 151 (68%) CH and 71 (32%) CS patients, and the median dose was 72.5 ± 2.2 Gray RBE. LC at seven years for CH was 71% and 94% for CS patients. Prognostic factors were optic apparatus and/or brainstem compression, histology and tumour volume. High-grade toxicities were observed in 13% of the patients at seven years. Detailed analyses on toxicity, predominantly temporal lobe toxicity, have shown no differences in dose-response relationships between photons and protons. Mainly tissue volume included into high-dose regions was predictive for side effects [53,54]; treatment planning recommendations stated that not only maximal doses but particularly volume relationships (D_{max}, V − 1 cm^3) were considered to be most relevant.

Recently, the Heidelberg group has reported long-term outcome after CIRT in skull base CHs, demonstrating high LC rates for CH of 54% at 10 years, and for CS of 91.5% [55,56]. Currently, randomised trials comparing protons and carbon ions in skull base CHs are recruiting in order to show which particle type leads to superior outcome. However, a treatment arm with modern, advanced photons is missing in both studies [57,58].

Late Toxicities

A PBT series reported an eight-year high-grade (≥ 3) toxicity-free survival (TFS) of 90.8% for skull base CS. Though not statistically significant, TFS was influenced by the number of weekly fractions, age and tumour volume. Observed late high-grade toxicities for base of skull tumours were hearing loss, cerebellum or spinal cord necrosis and optic neuropathy [40]. In a study on PBT for bone sarcomas of the skull base but also spine, late sequelae grade 3 or 4 toxicities occurred in 9.4% of the patients. Grade 3 late toxicities mainly affected musculoskeletal and connective tissue. However, grade 4 late toxicities included tissue necrosis and brainstem infarction as well [59].

OSTEOSARCOMAS

Epidemiology

Osteosarcomas predominantly arise in the metaphysis of long bones [60,61] and are the most common bone tumours amongst adolescents and young adults [62,63]. Like Ewing's sarcoma, osteosarcoma mainly occur in patients younger than 20 years with about 35% of all cases reported for patients aged 10–19 years [33]. Most common localisation for osteosarcoma amongst all age groups is lower long bones, though pelvic localisation is getting more important in higher age groups [63].

Risk Stratification

Initial tumour size, tumour location, response to chemotherapy, surgical remission and primary metastases are understood to be prognostic factors in osteosarcomas [64–66]. As in CH and CS, the AJCC staging system is used for osteosarcomas [67].

Therapy

The gold standard for achieving LC is complete surgery whenever amenable [68,69]. Wide surgical margins should be attempted [69]. However, wide surgery is difficult to achieve for lesions of the axial skeleton and the pelvis or base of the skull. Increasingly, multimodal treatment approaches including surgery and chemotherapy are used [69]. Though RT is predominantly used for patients with non-resectable osteosarcomas or in situations with positive margins, studies indicate that patients with some characteristics benefit from the addition of RT in their treatments [70,71]. If radiation therapy is considered for unresectable osteosarcomas, very high doses have to be applied due to low radiosensitivity of this entity [68]. Therefore, highly conformal radiation techniques like proton and carbon therapy have come into focus [69].

Long-Term Outcomes and Prognostic Factors

Five-year OS, LC and disease-free survival (DFS) rates after PBT of 65.5%–67%, 68%–72% and 65%, respectively, were reported [72,73]. Influencing factors for LC were grade ≥2 disease and prolonged treatment length. Even though not significant, craniofacial osteosarcomas were detected to be an indicator for local failure (HR = 2.6). However, presentation for primary or relapsed disease did not adversely influence LC. In addition, radiation treatment volumes as well as the absence of surgery did not interact with treatment failure [72]. Five-year LC and OS as well as DFS for patients with GTR or subtotal tumour resection (STR) were significantly better compared to those undergoing biopsy only. In addition, survival was significantly superior for patients treated for primary tumour than if treated for salvage. However, the anatomic site did not significantly influence LC [73].

For patients with unresectable spinal osteosarcomas, a five-year LC, OS and PFS rate after carbon ion irradiation of 79%, 52% and 48%, respectively, was reported [74]. Patients with carbon irradiation doses of <64 gray-equivalent showed significantly more recurrences than those who received ≥64 GyE. A tumour volume ≤100 cubic centimetres and a vertical tumour size larger than 40 millimetres do significantly tend to show more local recurrences than tumours with volumes >100 cubic centimetres and ≥40 millimetres. Lower survival and tumour control rates are seen for unresectable osteosarcomas of the trunk treated with CIRT. Five-year OS, disease specific survival (DSS), PFS and LC rate were 33%, 34%, 23% and 62%, respectively [75]. Eastern Cooperative Oncology Group (ECOG) performance status of 1, CTV <500 cubic centimetres, normal alkaline phosphatase (ALP) and C-reactive protein (CRP) level were detected as significant prognostic factors positively influencing OS. LC was significantly superior in patients with performance status of 1 as well as with smaller clinical target volumes (CTV) (<500 vs. ≥500 cubic centimetres). Five-year OS and LC for unresectable osteosarcoma of the head and neck were 44.4% and 85.7%, respectively [76]. The results for the entire cohort demonstrated a significant difference in survival for gross tumour volume (GTV) (≥100 millilitres vs. <100 millilitres). LC was significantly higher for patients after irradiation with 70.4 GyE compared to lower total doses.

Late Toxicities

Late higher-grade (3 and 4) toxicities after PBT for osteosarcoma in various sites were 30.1%, with patients showing grade 3 toxicities which predominantly consisted of pain and immobility of limb. Late grade 4 toxicities predominantly resulted in enucleation. However, two patients developed second malignancies (one acute lymphocytic leukaemia and one secondary squamous cell carcinoma) [72]. Late grade 3 and 4 toxicities after CIRT of unresectable osteosarcoma of the trunk were reported as skin/soft-tissue reactions and permanent neurologic and bone toxicities [75].

EWING'S SARCOMAS

Epidemiology

Ewing's sarcomas, peripheral primitive neuroectodermal tumours (PNET) and Askin tumours are assigned to the family of Ewing's tumours [77]. Typical localisations are lower extremities (18% tibia, 20% femur), pelvis (26%), chest wall (16%) and upper extremity (9%) [33,78]. About 40% of all Ewing's sarcomas appear in the age group 10–19 years [33]. Consequently, the highest incidence of Ewing's sarcomas is seen in the second decade of life [33].

Risk Stratification

Metastases at initial diagnosis, primary site and age [79] as well as tumour stage and size [80] are known to influence prognosis. For localised Ewing's sarcomas, strong prognostic factors are initial tumour size or volume in general and histologic response to chemotherapy [81]. In metastatic Ewing's sarcomas, age older than 14 years at diagnosis, a primary tumour volume more than 200 millilitres, presence of two or more bone metastases, presence of bone marrow metastases and additional pulmonary metastases are known to be negative prognostic factors [82].

Therapy

The implementation of multimodal treatment regimens consisting of local therapy as well as chemotherapy has improved outcomes for patients of Ewings's sarcomas over the last decades [81]. RT is used when complete surgery is difficult (Figure 9.2) [83]. If RT is used in definitive settings, doses between 55 and 60 Gray are applied. For RT in combination with surgery, lower doses (45–55 Gray) seem to be appropriate [84]. Typical sites for RT are the pelvis, extremities and spine [83]. For pelvic, spinal and paraspinal Ewing's sarcomas, a benefit is likely for the adjuvant RT [85,86]. Analyses on RT for Ewing's sarcomas support the importance of RT as part of multimodal treatment approaches. In a retrospective series on the comparison of treatment regimens of Ewing's tumours of the spine, five-year

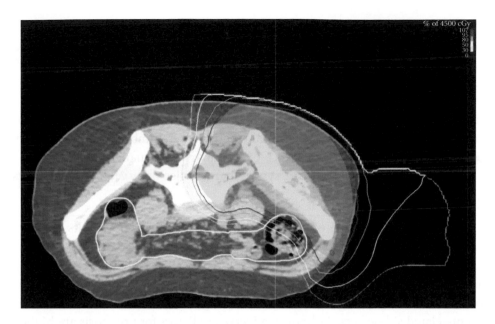

FIGURE 9.2   Proton beam therapy dose plan for a pelvic Ewing's sarcoma in a 27-year-old patient. Blue area: clinical target volume, red line: 95% isodose, orange line: 80% isodose, yellow line: 50% isodose, light blue line: 30% isodose.

LC was 50% for surgery alone, 74% for RT alone and 83% for surgery in combination with RT [87]. For the treatment of Ewing's sarcoma, protons have been infrequently used so far [88]. Hence, evidence of long-term outcomes for the treatment of Ewing's sarcoma with particle therapy is still scarce.

Long-Term Outcomes and Prognostic Factors

In a recent clinical trial on PBT for Ewing's sarcoma, an estimated five-year OS, LC and distant-metastasis-free survival (DMFS) of 83.0%, 81.5% and 76.4%, respectively, was reported. Metastatic status at diagnosis was found to be a significant prognostic factor for five-year LC ($p = 0.003$). A tumour volume larger than 200 millilitres was associated with decreased DMFS ($p = 0.03$) [89]. These outcomes after PBT are comparable with analyses on treatment with RT for Ewing's sarcoma reporting a three-year LC of 61% for patients presenting with metastatic disease and 84% for patients without metastases [90]. A higher risk of local relapse for patients with tumours larger than 200 millilitres compared to small tumours (subdistribution HR of 1.8) was also reported for patients treated with conventional RT [91].

Age greater than 10 years was a significant predictor for local ($p = 0.05$) and distant failure ($p = 0.003$) and a decreased OS ($p = 0.05$) after PBT [89]. For paediatric patients treated with PBT for Ewing's sarcoma, Japanese data reported a five-year OS rate of 56.8% [92] which is substantially lower when compared to the report by Weber et al. [89]. OS was inferior for patients with recurrent disease and superior for patients who were not previously irradiated. In addition, superior OS for patients who could have received photon RT instead of PT was reported.

Late Toxicity

A reported toxicity profile for a cohort treated with PBT for Ewing's sarcomas seemed to be favourable, with an estimated high-grade toxicity-free survival after five years of 90.9% [89]. Grade 1 and 2 late toxicities predominantly occurred as residual alopecia. Higher-grade (grade 3) effects were observed as kyphoscoliosis and endocrine dysfunction. No grade 4 or 5 late toxicity was observed. Scolioses/kyphoses was also seen in other reports on PBT for paediatric Ewing's sarcoma. Other mild reactions were limb length discrepancy, skin reactions and telangiectasia. Appearance of secondary malignancies of the blood (acute myeloid leukemia and myelodysplastic syndrome) was also observed in four patients [93].

SUMMARY AND CONCLUSION

Though there is still a dearth of coherent published data, current results permit a preliminary interpretation of long-term outcomes and prognostic factors in particle therapy for sarcomas. Long-term survival and disease outcomes for paediatric RMS after PBT are promising when using similar dose levels comparable to reports of photon-treated populations. However, due to the dosimetric advantages of PBT, there is a chance to limit treatment-related side effects. As known, adverse predictors for prognosis are higher COG group and higher IRS stage. Young age, tumour size of more than five centimetres, parameningeal primary disease site and intracranial extensions are also associated with a poorer prognosis. Furthermore, findings suggest that a delay in radiation should be avoided in the high-risk RMS group and particular in patients presenting with intracranial tumour extension. For CH and CS, no randomised large series on proton therapy are available. However, all data from proton therapy confirm that even high local doses beyond 74 Gray can be applied safely with protons; direct comparison with historical photon data confirms the dose-response relationship for CH. Clear data on any potential benefit of heavier ions over protons are still in the future. For osteosarcomas, particle therapy appears to be an effective and safe treatment even when escalating local doses. It is predominantly used in locations where radical surgical resection seems infeasible. After particle therapy, early data on survival, disease control and late adverse events are promising. Studies revealed various parameters like tumour size, irradiation dose and resection status to have an impact on prognosis. Also in Ewing's sarcomas, the use of proton and heavier charged particles resulted in promising survival and disease outcomes while late toxicities were tolerable. Local and distant control rates were adversely affected by later age ($\geq$10 years) and metastatic status, whereas survival was negatively influenced by tumour volume, higher stage and previous irradiation.

Up to now, studies on particle therapy in sarcomatous tumours were predominantly retrospectively conducted and non-randomised. Relatively small cohorts and short follow-up periods, usually not exceeding five years, are limiting the evidence for PBT and other charged particles. However, despite high doses, large volume and critical sites, particle therapy was proven to be both effective and feasible. Therefore, particle therapy will continue to be an important tool in the arsenal of multidisciplinary treatment concepts for difficult sarcoma cases. Future research with large prospective clinical trials and international registries will help to gain more evidence.

REFERENCES

1. Stiller, C.A. et al., Descriptive epidemiology of sarcomas in Europe: Report from the RARECARE project. *Eur J Cancer*, 2013. **49**(3): 684–695.
2. Kaatsch, P., and Spix, C. *German Childhood Cancer Registry—Annual Report 2015 (1980–2014)*. Institute of Medical Biostatistics, Epidemiology, and Informatics (IMBEI) at the University Medical Center of the Johannes Gutenberg University Mainz, 2015. Available from: http://www.kinderkrebsregister.de/dkkr-gb/latest-publications/annual-reports/annual-report-201314.html?L=1 (accessed June 11, 2017).
3. Ward, E. et al., Childhood and adolescent cancer statistics, 2014. *CA Cancer J Clin*, 2014. **64**(2): 83–103.
4. Ogilvie, C.M. et al., Treatment of adult rhabdomyosarcoma. *Am J Clin Oncol*, 2010. **33**(2): 128–131.
5. Ferrari, A. et al., Rhabdomyosarcoma in adults. A retrospective analysis of 171 patients treated at a single institution. *Cancer*, 2003. **98**(3): 571–580.
6. Khosla, D. et al., Adult rhabdomyosarcoma: Clinical presentation, treatment, and outcome. *J Cancer Res Ther*, 2015. **11**(4): 830–834.
7. Little, D.J. et al., Adult rhabdomyosarcoma: Outcome following multimodality treatment. *Cancer*, 2002. **95**(2): 377–388.
8. Esnaola, N.F. et al., Response to chemotherapy and predictors of survival in adult rhabdomyosarcoma. *Ann Surg*, 2001. **234**(2): 215–223.
9. Dumont, S.N. et al., Management and outcome of 239 adolescent and adult rhabdomyosarcoma patients. *Cancer Med*, 2013. **2**(4): 553–563.
10. Oberlin, O. et al., Treatment of orbital rhabdomyosarcoma: Survival and late effects of treatment--results of an international workshop. *J Clin Oncol*, 2001. **19**(1): 197–204.
11. Merks, J.H. et al., Parameningeal rhabdomyosarcoma in pediatric age: Results of a pooled analysis from North American and European cooperative groups. *Ann Oncol*, 2014. **25**(1): 231–236.
12. Spalding, A.C. et al., The effect of radiation timing on patients with high-risk features of parameningeal rhabdomyosarcoma: An analysis of IRS-IV and D9803. *Int J Radiat Oncol Biol Phys*, 2013. **87**(3): 512–516.
13. Arndt, C.A. et al., Vincristine, actinomycin, and cyclophosphamide compared with vincristine, actinomycin, and cyclophosphamide alternating with vincristine, topotecan, and cyclophosphamide for intermediate-risk rhabdomyosarcoma: Children's oncology group study D9803. *J Clin Oncol*, 2009. **27**(31): 5182–5188.
14. Yang, J.C. et al., Parameningeal rhabdomyosarcoma: Outcomes and opportunities. *Int J Radiat Oncol Biol Phys*, 2013. **85**(1): e61–e66.
15. Dasgupta, R., J. Fuchs, and D. Rodeberg, Rhabdomyosarcoma. *Semin Pediatr Surg*, 2016. **25**(5): 276–283.
16. Terezakis, S.A., and M.D. Wharam, Radiotherapy for rhabdomyosarcoma: Indications and outcome. *Clin Oncol (R Coll Radiol)*, 2013. **25**(1): 27–35.
17. Hiniker, S.M., and S.S. Donaldson, Recent advances in understanding and managing rhabdomyosarcoma. *F1000Prime Rep*, 2015. **7**: 59.
18. Yang, L., T. Takimoto, and J. Fujimoto, Prognostic model for predicting overall survival in children and adolescents with rhabdomyosarcoma. *BMC Cancer*, 2014. **14**: 654.
19. Kozak, K.R. et al., A dosimetric comparison of proton and intensity-modulated photon radiotherapy for pediatric parameningeal rhabdomyosarcomas. *Int J Radiat Oncol Biol Phys*, 2009. **74**(1): 179–186.
20. Ladra, M.M. et al., A dosimetric comparison of proton and intensity modulated radiation therapy in pediatric rhabdomyosarcoma patients enrolled on a prospective phase II proton study. *Radiother Oncol*, 2014. **113**(1): 77–83.

21. Yock, T. et al., Proton radiotherapy for orbital rhabdomyosarcoma: Clinical outcome and a dosimetric comparison with photons. *Int J Radiat Oncol Biol Phys*, 2005. **63**(4): 1161–1168.

22. Cotter, S.E. et al., Proton radiotherapy for pediatric bladder/prostate rhabdomyosarcoma: Clinical outcomes and dosimetry compared to intensity-modulated radiation therapy. *Int J Radiat Oncol Biol Phys*, 2011. **81**(5): 1367–1373.

23. Ladra, M.M. et al., Preliminary results of a phase II trial of proton radiotherapy for pediatric rhabdomyosarcoma. *J Clin Oncol*, 2014. **32**(33): 3762–3770.

24. Leiser, D. et al., Tumour control and quality of life in children with rhabdomyosarcoma treated with pencil beam scanning proton therapy. *Radiother Oncol*, 2016. **120**: 163–168.

25. Weber, D.C. et al., Pencil beam scanning proton therapy for pediatric parameningeal rhabdomyosarcomas: Clinical outcome of patients treated at the Paul Scherrer Institute. *Pediatr Blood Cancer*, 2016. **63**(10): 1731–1736.

26. Childs, S.K. et al., Proton radiotherapy for parameningeal rhabdomyosarcoma: Clinical outcomes and late effects. *Int J Radiat Oncol Biol Phys*, 2012. **82**(2): 635–642.

27. Paulino, A.C. et al., Long-term effects in children treated with radiotherapy for head and neck rhabdomyosarcoma. *Int J Radiat Oncol Biol Phys*, 2000. **48**(5): 1489–1495.

28. Raney, R.B. et al., Late complications of therapy in 213 children with localized, nonorbital soft-tissue sarcoma of the head and neck: A descriptive report from the Intergroup Rhabdomyosarcoma Studies (IRS)-II and—III. IRS group of the children's cancer group and the pediatric oncology group. *Med Pediatr Oncol*, 1999. **33**(4): 362–371.

29. Curtis, A.E. et al., Local control after intensity-modulated radiotherapy for head-and-neck rhabdomyosarcoma. *Int J Radiat Oncol Biol Phys*, 2009. **73**(1): 173–177.

30. Wolden, S.L. et al., Intensity-modulated radiotherapy for head-and-neck rhabdomyosarcoma. *Int J Radiat Oncol Biol Phys*, 2005. **61**(5): 1432–1438.

31. Combs, S.E. et al., Intensity modulated radiotherapy (IMRT) and fractionated stereotactic radiotherapy (FSRT) for children with head-and-neck-rhabdomyosarcoma. *BMC Cancer*, 2007. **7**: 177.

32. McMaster, M.L. et al., Chordoma: Incidence and survival patterns in the United States, 1973–1995. *Cancer Causes Cont*, 2001. **12**(1): 1–11.

33. Damron, T.A., W.G. Ward, and A. Stewart, Osteosarcoma, chondrosarcoma, and Ewing's sarcoma: National cancer data base report. *Clin Orthop Relat Res*, 2007. **459**: 40–47.

34. Noel, G. et al., Chordomas of the base of the skull and upper cervical spine. One hundred patients irradiated by a 3D conformal technique combining photon and proton beams. *Acta Oncol*, 2005. **44**(7): 700–708.

35. Noel, G. et al., Radiation therapy for chordoma and chondrosarcoma of the skull base and the cervical spine. Prognostic factors and patterns of failure. *Strahlenther Onkol*, 2003. **179**(4): 241–248.

36. Mendenhall, W.M. et al., Skull base chordoma. *Head Neck*, 2005. **27**(2): 159–165.

37. Coca-Pelaz, A. et al., Chondrosarcomas of the head and neck. *Eur Arch Otorhinolaryngol*, 2014. **271**(10): 2601–2609.

38. De Amorim Bernstein, K., and T. DeLaney, Chordomas and chondrosarcomas-The role of radiation therapy. *J Surg Oncol*, 2016. **114**(5): 564–569.

39. Jian, B.J. et al., Adjuvant radiation therapy and chondroid chordoma subtype are associated with a lower tumor recurrence rate of cranial chordoma. *J Neurooncol*, 2010. **98**(1): 101–108.

40. Weber, D.C. et al., Long term outcomes of patients with skull-base low-grade chondrosarcoma and chordoma patients treated with pencil beam scanning proton therapy. *Radiother Oncol*, 2016. **120**(1): 169–174.

41. Indelicato, D.J. et al., A prospective outcomes study of proton therapy for chordomas and chondrosarcomas of the spine. *Int J Radiat Oncol Biol Phys*, 2016. **95**(1): 297–303.

42. Schulz-Ertner, D. et al., Effectiveness of carbon ion radiotherapy in the treatment of skull-base chordomas. *Int J Radiat Oncol Biol Phys*, 2007. **68**(2): 449–457.

43. Zorlu, F. et al., Conventional external radiotherapy in the management of clivus chordomas with overt residual disease. *Neurol Sci*, 2000. **21**(4): 203–207.

44. Chang, S.D. et al., Stereotactic radiosurgery and hypofractionated stereotactic radiotherapy for residual or recurrent cranial base and cervical chordomas. *Neurosurg Focus*, 2001. **10**(3): E5.

45. Krishnan, S. et al., Radiosurgery for cranial base chordomas and chondrosarcomas. *Neurosurgery*, 2005. **56**(4): 777–784.

46. Koga, T., M. Shin, and N. Saito, Treatment with high marginal dose is mandatory to achieve long-term control of skull base chordomas and chondrosarcomas by means of stereotactic radiosurgery. *J Neurooncol*, 2010. **98**(2): 233–238.

47. Kano, H. et al., Stereotactic radiosurgery for chordoma: A report from the North American gamma knife consortium. *Neurosurgery*, 2011. **68**(2): 379–389.

48. Potluri, S. et al., Residual postoperative tumour volume predicts outcome after high-dose radiotherapy for chordoma and chondrosarcoma of the skull base and spine. *Clin Oncol (R Coll Radiol)*, 2011. **23**(3): 199–208.

49. Hauptman, J.S. et al., Challenges in linear accelerator radiotherapy for chordomas and chondrosarcomas of the skull base: Focus on complications. *Int J Radiat Oncol Biol Phys*, 2012. **83**(2): 542–551.

50. Sahgal, A. et al., Image-guided, intensity-modulated radiation therapy (IG-IMRT) for skull base chordoma and chondrosarcoma: Preliminary outcomes. *Neuro Oncol*, 2015. **17**(6): 889–894.

51. Fagundes, M.A. et al., Radiation therapy for chordomas of the base of skull and cervical spine: Patterns of failure and outcome after relapse. *Int J Radiat Oncol Biol Phys*, 1995. **33**(3): 579–584.

52. Grosshans, D.R. et al., Spot scanning proton therapy for malignancies of the base of skull: Treatment planning, acute toxicities, and preliminary clinical outcomes. *Int J Radiat Oncol Biol Phys*, 2014. **90**(3): 540–546.

53. Schlampp, I. et al., Temporal lobe reactions after radiotherapy with carbon ions: Incidence and estimation of the relative biological effectiveness by the local effect model. *Int J Radiat Oncol Biol Phys*, 2011. **80**(3): 815–823.

54. Pehlivan, B. et al., Temporal lobe toxicity analysis after proton radiation therapy for skull base tumors. *Int J Radiat Oncol Biol Phys*, 2012. **83**(5): 1432–1440.

55. Uhl, M. et al., Highly effective treatment of skull base chordoma with carbon ion irradiation using a raster scan technique in 155 patients: First long-term results. *Cancer*, 2014. **120**(21): 3410–3417.

56. Uhl, M. et al., High control rate in patients with chondrosarcoma of the skull base after carbon ion therapy: First report of long-term results. *Cancer*, 2014. **120**(10): 1579–1585.

57. Nikoghosyan, A.V. et al., Randomised trial of proton versus carbon ion radiation therapy in patients with chordoma of the skull base, clinical phase III study HIT-1-Study. *BMC Cancer*, 2010. **10**: 607.

58. Nikoghosyan, A.V. et al., Randomised trial of proton versus carbon ion radiation therapy in patients with low and intermediate grade chondrosarcoma of the skull base, clinical phase III study. *BMC Cancer*, 2010. **10**: 606.

59. Demizu, Y. et al., Proton beam therapy for bone sarcomas of the skull base and spine: A retrospective nationwide multicenter study in Japan. *Cancer Sci*, 2017. **108**: 972–977.

60. Lahat, G., A. Lazar, and D. Lev, Sarcoma epidemiology and etiology: Potential environmental and genetic factors. *Surg Clin North Am*, 2008. **88**(3): 451–481.

61. Picci, P., Osteosarcoma (osteogenic sarcoma). *Orphanet J Rare Dis*, 2007. **2**: 6.

62. Herzog, C.E., Overview of sarcomas in the adolescent and young adult population. *J Pediatr Hematol Oncol*, 2005. **27**(4): 215–218.

63. Mirabello, L., R.J. Troisi, and S.A. Savage, Osteosarcoma incidence and survival rates from 1973 to 2004: Data from the surveillance, epidemiology, and end results program. *Cancer*, 2009. **115**(7): 1531–1543.

64. Bielack, S.S. et al., Prognostic factors in high-grade osteosarcoma of the extremities or trunk: An analysis of 1,702 patients treated on neoadjuvant cooperative osteosarcoma study group protocols. *J Clin Oncol*, 2002. **20**(3): 776–790.

65. Grimer, R.J. et al., Osteosarcoma over the age of forty. *Eur J Cancer*, 2003. **39**(2): 157–163.

66. Kaste, S.C. et al., Tumor size as a predictor of outcome in pediatric non-metastatic osteosarcoma of the extremity. *Pediatr Blood Cancer*, 2004. **43**(7): 723–728.

67. Kundu, Z.S., Classification, imaging, biopsy and staging of osteosarcoma. *Indian J Orthop*, 2014. **48**(3): 238–246.

68. Bolling, T., J. Hardes, and U. Dirksen, Management of bone tumours in paediatric oncology. *Clin Oncol (R Coll Radiol)*, 2013. **25**(1): 19–26.

69. Bielack, S.S. et al., Advances in the management of osteosarcoma. *F1000Res*, 2016. **5**: 2767.

70. Schwarz, R. et al., The role of radiotherapy in oseosarcoma. *Cancer Treat Res*, 2009. **152**: 147–164.

71. Ozaki, T. et al., Osteosarcoma of the pelvis: Experience of the cooperative osteosarcoma study group. *J Clin Oncol*, 2003. **21**(2): 334–341.

72. Ciernik, I.F. et al., Proton-based radiotherapy for unresectable or incompletely resected osteosarcoma. *Cancer*, 2011. **117**(19): 4522–4530.

73. DeLaney, T.F. et al., Radiotherapy for local control of osteosarcoma. *Int J Radiat Oncol Biol Phys*, 2005. **61**(2): 492–498.

74. Matsumoto, K. et al., Impact of carbon ion radiotherapy for primary spinal sarcoma. *Cancer*, 2013. **119**(19): 3496–3503.

75. Matsunobu, A. et al., Impact of carbon ion radiotherapy for unresectable osteosarcoma of the trunk. *Cancer*, 2012. **118**(18): 4555–4563.

76. Jingu, K. et al., Carbon ion radiation therapy improves the prognosis of unresectable adult bone and soft-tissue sarcoma of the head and neck. *Int J Radiat Oncol Biol Phys*, 2012. **82**(5): 2125–2131.

77. Biswas, B., and S. Bakhshi, Management of Ewing sarcoma family of tumors: Current scenario and unmet need. *World J Orthop*, 2016. **7**(9): 527–538.

78. Ludwig, J.A., Ewing sarcoma: Historical perspectives, current state-of-the-art, and opportunities for targeted therapy in the future. *Curr Opin Oncol*, 2008. **20**(4): 412–418.

79. Cotterill, S.J. et al., Prognostic factors in Ewing's tumor of bone: Analysis of 975 patients from the European intergroup cooperative Ewing's sarcoma study group. *J Clin Oncol*, 2000. **18**(17): 3108–3114.

80. Rodriguez-Galindo, C. et al., Analysis of prognostic factors in ewing sarcoma family of tumors: Review of St. Jude children's research hospital studies. *Cancer*, 2007. **110**(2): 375–384.

81. Gaspar, N. et al., Ewing sarcoma: Current management and future approaches through collaboration. *J Clin Oncol*, 2015. **33**(27): 3036–3046.

82. Ladenstein, R. et al., Primary disseminated multifocal Ewing sarcoma: Results of the Euro-EWING 99 trial. *J Clin Oncol*, 2010. **28**(20): 3284–3291.

83. McGovern, S.L., and A. Mahajan, Progress in radiotherapy for pediatric sarcomas. *Curr Oncol Rep*, 2012. **14**(4): 320–326.

84. Bernstein, M. et al., Ewing's sarcoma family of tumors: Current management. *Oncologist*, 2006. **11**(5): 503–519.

85. Indelicato, D.J. et al., Spinal and paraspinal Ewing tumors. *Int J Radiat Oncol Biol Phys*, 2010. **76**(5): 1463–1471.

86. Casey, D.L. et al., Ewing sarcoma in adults treated with modern radiotherapy techniques. *Radiother Oncol*, 2014. **113**(2): 248–253.

87. Vogin, G. et al., Local control and sequelae in localised Ewing tumours of the spine: A French retrospective study. *Eur J Cancer*, 2013. **49**(6): 1314–1323.

88. Subbiah, V. et al., Ewing's sarcoma: Standard and experimental treatment options. *Curr Treat Options Oncol*, 2009. **10**(1–2): 126–140.

89. Weber, D.C. et al., Pencil beam scanned protons for the treatment of patients with Ewing sarcoma. *Pediatr Blood Cancer*, 2017. **64**: e26688.

90. La, T.H. et al., Radiation therapy for Ewing's sarcoma: Results from Memorial Sloan-Kettering in the modern era. *Int J Radiat Oncol Biol Phys*, 2006. **64**(2): 544–550.

91. Foulon, S. et al., Can postoperative radiotherapy be omitted in localised standard-risk Ewing sarcoma? An observational study of the Euro-E.W.I.N.G group. *Eur J Cancer*, 2016. **61**: 128–136.

92. Mizumoto, M. et al., Proton beam therapy for pediatric malignancies: A retrospective observational multicenter study in Japan. *Cancer Med*, 2016. **5**(7): 1519–1525.

93. Rombi, B. et al., Proton radiotherapy for pediatric Ewing's sarcoma: Initial clinical outcomes. *Int J Radiat Oncol Biol Phys*, 2012. **82**(3): 1142–1148.

CHAPTER 10

Proton Therapy for Paediatric Patients

Masashi Mizumoto, Yoshiko Oshiro and Hideyuki Sakurai

CONTENTS

MOST PAEDIATRIC NON-HEMATOLOGIC MALIGNANCIES have high radiation sensitivity, and thus radiotherapy plays an important role in treatment of these diseases. However, children generally also have higher radiation sensitivity compared to adults, and characteristic radiotoxicities occur in paediatric patients. Proton therapy (PT) may be advantageous for paediatric patients due to its ability to reduce the radiation dose to healthy tissue close to the tumour.

RADIOTHERAPY FOR PAEDIATRIC PATIENTS

Recent progress in multimodal treatment for paediatric malignancies has improved survival, and almost 70% of patients can now be cured [1]. However, this growing population of survivors remains at risk for disease- and treatment-associated late mortality. According to Mertens et al. [2], the all-cause 30-year cumulative mortality is 18.1% (95% CI = 17.3–18.9) for five-year survivors of childhood cancer. The rate of death due to recurrence or progression was highest in the 5- to 10-year period after diagnosis, at 0.99% per year (95% CI: 0.93–1.06), but decreased to 0.10% per year (95% CI: 0.06–0.16) at 25–29 years after diagnosis. The death rate due to second malignancies exceeded the death rate from recurrence at 20–24 years follow-up, and radiotherapy was an independent risk factor for late mortality [2].

Secondary malignancy due to radiotherapy is a severe problem in paediatric patients. There is no threshold dose for avoidance of secondary cancer; therefore, all patients who receive radiotherapy have a risk for secondary cancer. However, the rate of secondary cancer for adults who receive radiotherapy for prostate cancer is only 2.1%, which is only 0.16% higher than patients who do not receive radiotherapy [3]. In contrast, the secondary cancer risk is higher for children than adults because cell division is more active in children. A recent study revealed that repeated computed tomography (CT) in childhood at 50–60 milligray increases brain tumours and leukaemia [4]. Patients receiving radiotherapy of 60 Gray have a 40 times higher risk of osteosarcoma [5], and those receiving chest irradiation have a 16 times higher risk of breast cancer and a >35 times higher risk of thyroid cancer [6]. After cranial irradiation, meningioma and glioma are common as secondary brain tumours, with an increased risk of 3%–20% [7].

Growth hormone deficiency and bone and soft-tissue retardation after radiotherapy can significantly affect growth. The threshold dose for bone retardation is around 20 Gray [8]. Inhomogeneous vertebral body irradiation induces scoliosis or kyphosis, and whole spinal irradiation may cause a short stature. Facial bone irradiation may induce facial deformation, and irradiation may cause premature eruption of teeth [9]. Soft tissues are more sensitive to radiation than bone, and Guyron et al. suggested that the harmful dose for soft tissue was as low as 400 rads [9]. Change in the skin may occur as early as 3–4 years and as late as 9–10 years after radiation [10]. Patients who receive radiotherapy at an early age may not understand the reason for growth retardation, and this may cause mental distress [11]. This may include depression due to facial deformity, and long-term follow-up and support are necessary [11].

In treatment of paediatric brain tumour, radiotherapy plays an important role because obtaining a sufficient surgical margin is difficult. However, the cumulative late mortality of five-year survivors of central nervous system (CNS) malignancies is high, at 25.8% (95% CI = 23.4%–28.3%), and compared with the U.S. population, the risk of death is increased thirteenfold in these survivors [12]. In addition, intelligence retardation is a significant problem after irradiation of a CNS tumour. The decline of intelligence after cranial irradiation is affected by irradiation dose and volume and age at irradiation [13–17], with intelligence quotient (IQ) estimated to decline 3.7–4.3 points per year after cranial irradiation for medulloblastoma [15,16]. In addition, fibrosis and atrophy progress in irradiated tissue, and blood flow abnormalities may occur [18].

PROTON THERAPY

Protons have no radiobiological advantage over photon beam, and the incidence of local control should be equivalent to that of photon beams for the same irradiated dose and time-dose schedule. However, an insufficient dose to the tumour may lead to local recurrence, and PT allows delivery of an increased dose while sparing adjacent tissue. Intensity-modulated radiotherapy (IMRT) has also been developed to improve dose concentration and reduce the dose to normal tissue. IMRT has made it possible to irradiate a complex tumour that is difficult to treat with regular photon radiotherapy. However, the lower dose area is still increased with this technique, which leads to a significant risk of secondary cancer [19]. In contrast, PT has an ideal dose distribution using a few ports, and therefore the low-dose area is significantly smaller than that with photon radiotherapy [20–25]. This is a major advantage for reduction of the risk of secondary malignancy and intelligence retardation.

Many studies comparing the dose distributions in photon radiotherapy and PT have shown that PT reduces the dose to normal organs and lowers the risk of secondary cancer [26–28]. Miralbell et al. found that IMRT for parameningeal rhabdomyosarcma and medulloblastoma has a more than twice and 8–15 times higher risk of secondary cancer, respectively, compared to PT [26]. In craniospinal irradiation (CSI), the risk of secondary cancer of 3%–20% with regular radiotherapy is reduced to 3.4% using PT [7].

Dosimetry comparisons in cranial irradiation have all suggested advantages of PT that lead to preservation of cognitive function. Merchant et al. reported that PT is useful to spare critical normal tissue such as the cochlea and hypothalamus when these areas are not adjacent to the primary tumour, and to reduce the low-to-moderate dose area [29]. In a report of clinical outcomes of cognitive function after PT for 60 paediatric patients with brain tumour, Pulsifer et al. [30] found that Full scale IQ, verbal and nonverbal intelligence, and working memory were stable at a mean of 2.5 years follow-up, whereas progressive cognitive decline was evident at 1–2 years after photon radiotherapy.

From the perspective of tumour control, proton beams can be safely delivered to a tumour that cannot be treated by photon radiotherapy by sparing critical organs due to the high degree of dose conformity [31,32]. Using PT, a dose of 74 Gray was delivered for hypothalamus astrocytoma without severe toxicity, whereas 54 Gray is the usual dose for this tumour using photon radiotherapy. In the 1990s, large unresectable hepatoblastoma was cured by PT at a time when liver transplantation was less common. Patients with large rhabdomyosarcoma undergo palliative photon irradiation because the liver cannot tolerate the curative dose, whereas local control can be achieved using PT.

CLINICAL OUTCOMES OF PROTON THERAPY FOR PAEDIATRIC MALIGNANCIES

Many studies have shown that PT can reduce the dose to normal tissue, compared to photon radiotherapy, but there are few clinical reports of PT for paediatric tumours because this treatment has a relatively short history and there are fewer cases of paediatric tumours.

Rhabdomyosarcoma

There are several clinical reports of PT for rhabdomyosarcoma [31,33–41]. In the first such report in 2000, Hug et al. used proton beams for two cases of orbital rhabdomyosarcoma, and suggested that PT can spare the lens and intra-orbital and ocular normal structures, while maintaining conformal target-dose coverage [33]. The first clinical prospective study of PT for rhabdomyosarcoma was a phase II study by Ladra et al. [34], in which the five-year disease-free and overall survival (OS) rates were 69% and 78%, respectively. These outcomes were similar to previous results with photon beams in the Children's Oncology Group (COG) trial. However, the rates of acute and late toxicities of 13% and 7% were significantly lower than those with 3D photon radiotherapy. In the largest clinical study of PT performed to

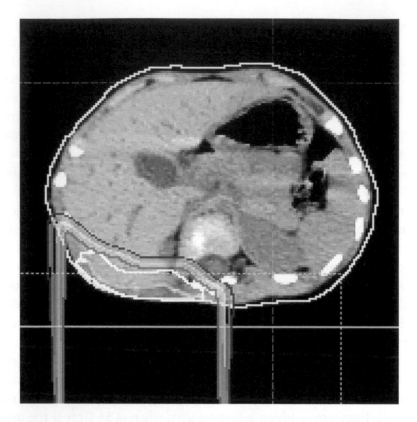

FIGURE 10.1   Dose distribution for dorsal rhabdomyosarcoma (one-year-old). Irradiation of the liver and spinal cord is easily avoided with proton therapy.

date, in 83 patients, Leiser et al. [35] found a five-year local control rate of 78% and an OS rate of 80.6%. The five-year non-ocular late toxicity was only 3.6%, even though more than half of the patients had Stage 3 or 4 disease and 71% had intracranial progression [35]. An irradiation dose of 40–50 Gray is necessary to treat rhabdomyosarcoma, and this dose exceeds the threshold for bone growth retardation. PT can spare normal tissue compared to photon radiotherapy, but not in the target region (Figure 10.1). Thus, growth retardation in the target region is still a concern with the use of PT.

Retinoblastoma

Retinoblastoma has high sensitivity to radiotherapy; therefore, definitive radiotherapy of 40–45 Gray in 20–25 fractions is the first treatment choice. However, secondary cancer and growth retardation are severe problems after radiotherapy. PT can achieve excellent dose distribution with 1–2 ports, and this reduces the doses to the orbital bone and brain. Therefore, PT is well indicated for cases with optic nerve and extraocular invasion for eye preservation (Figure 10.2). Mouw et al. found no secondary cancer or cosmetic problems after PT for 49 patients with retinoblastoma [42]. These findings suggest that PT may also be applicable for other orbital diseases, such as haemangioma.

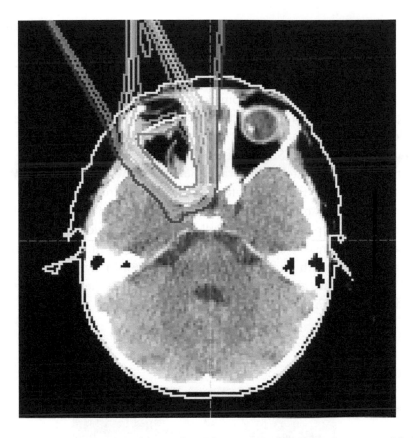

FIGURE 10.2   Postoperative proton therapy for right eye retinoblastoma (two-year-old). The proton beam stops just anterior to the pituitary gland.

Central Nervous System Tumours

1. *Glioma*: Survival for patients with glioma depends on the pathologic grade and degree of surgical resection. The 10-year OS of patients with low-grade glioma is 80%–90% when complete resection is achieved [43], but for paediatric patients with high-grade glioma treated with postoperative photon radiotherapy of 59.4 Gray in 33 fractions, the five-year OS is only 24% [44]. There are few reports of the results of PT for glioma. The treatment dose for low-grade glioma ranges from 48.6 to 54.0 Gray equivalent, similar to photon radiotherapy. In a preliminary study, Hauswald et al. found tumour progression in only one patient of the 19 and no severe acute toxicity for patients with low-grade glioma [45]. Hug et al. reported an OS of 85% for 27 patients with low-grade glioma that was mostly unresectable or residual disease in a follow-up period of 3.3 years [46]. Greenberger et al. found eight-year OS and progression-free survival (PFS) rates of 100% and 82.8%, respectively, over a mean follow-up period of 7.6 years [47] in 32 patients with low-grade glioma, and stabilisation or improvement of visual acuity was achieved in 83% of patients at risk for radiation-induced injury to optic pathways.

2. *Medulloblastoma:* CSI is required in radiotherapy for medulloblastoma. The CSI treatment volume is large, and the risks of late toxicity and secondary cancer are higher than for other irradiation fields. Intracranial toxicities are a significant problem in whole-brain irradiation and posterior fossa boost, with intelligence retardation, hormonal deficiency and ototoxicity being common after irradiation for medulloblastoma [48]. PT is particularly useful for reduction of the risk of these late toxicities. The delivered dose with PT is similar to that in photon radiotherapy: a total dose of 54.0 Gray equivalents, including a CSI dose of 18.0–36.0 Gray equivalents [49–51]. PT achieves similar disease control to photon radiotherapy, but with less toxicity. With PT and photon radiotherapy, Eaton et al. found six-year recurrence-free rates of 78.8% and 76.5%, respectively, and OS of 82.0% and 87.6%, respectively [51]. A phase II study of PT in paediatric patients with medulloblastoma suggested an IQ loss of 1.5 points per year [52], whereas Walter and Ris found that IQ declined by 3.7 and 4.2 points per year after CSI with photon radiotherapy [15,16]. Many dose-volume histogram (DVH) analyses have also suggested that PT can reduce the risks of late toxicity and secondary cancer (Figure 10.3) [53–57].

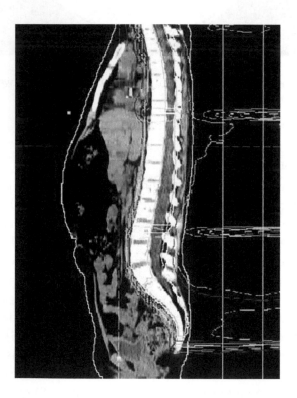

FIGURE 10.3   Dose distribution curve of whole-spine irradiation. All organs anterior to vertebrae were not irradiated.

3. *Ependymoma:* The treatment volume in early studies included prophylactic CSI, but efficacy was not established. Therefore, the clinical target volume (CTV) has generally been defined as the gross target volume (GTV) (remnant tumour or tumour bed) plus a 1.5-centimetre margin. More recently, Merchant et al. showed favourable results with a 1.0-centimetre margin (seven-year local control rate of 83.7% and OS of 85.0% for ependymoma or anaplastic ependymoma) [58], and a reduced treatment field is under discussion (Figure 10.4). Ares et al. used PT with a 0.5–1.0-centimetre margin and obtained five-year OS and local control rates of 84% and 78%, respectively, with side effects of unilateral deafness in two patients (4%) and brain stem necrosis in one (2%) [59].

4. *Germinoma:* Germinoma is the most common type of germ cell tumour (GCT) and has a favourable prognosis, with a 10-year OS of 90% [60,61]. Germinoma has high sensitivity to radiotherapy, and chemoradiotherapy also plays an important role in treatment. However, long-term survival has revealed late-treatment toxicities.

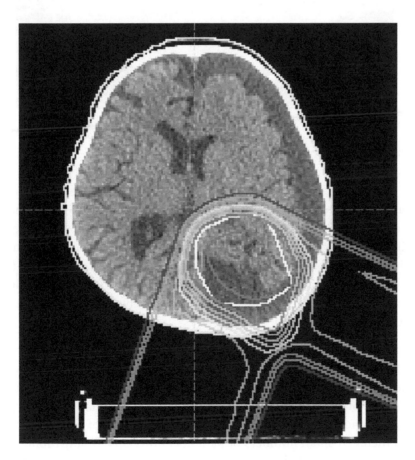

FIGURE 10.4 Postoperative proton therapy for anaplastic ependymoma (one-year-old). Radiation to most of the normal brain is avoided using proton therapy.

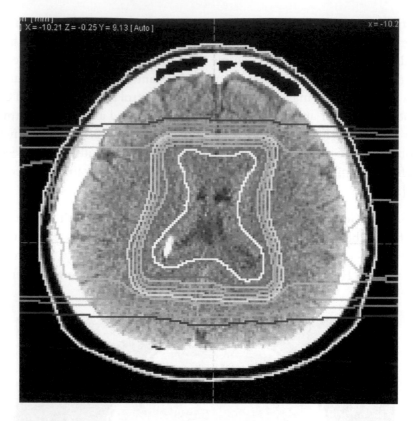

FIGURE 10.5   Dose distribution in whole ventricular irradiation by proton therapy. The dose to the normal brain is decreased by half using proton therapy.

Currently, standard radiotherapy for an intratubular GCT (ICGT) is whole ventricular irradiation (WVI) at a dose of 23.4–30.6 Gray in 13–17 fractions, followed by neoadjuvant chemotherapy [62,63]. PT is also advantageous for WVI, based on comparisons with photon beams for irradiation of intracranial germinoma (Figure 10.5) [24,64]. One early clinical result for 22 patients with ICGT suggested local control, PFS and OS rates of 100, 95% and 100%, respectively, and normal tissue was spared more using PT.

5. *Craniopharyngioma:* A five-year PFS of about 90% has been reported after photon radiotherapy of 50–55 Gray. There are a few reports of clinical outcomes of PT for paediatric craniopharyngioma [65–67]. The delivered dose was 50.4–59.4 Gray and the tumour was well controlled. Luu et al. found that one of 16 patients had local recurrence after PT in a median follow-up period of 60.2 months [65]. These studies included small numbers of patients and relatively short follow-up periods, but the results suggest that PT can achieve similar clinical outcomes to photon radiotherapy.

Other Tumours

There are a few reports on PT for paediatric tumours such as neuroblastoma and Ewing's sarcoma [68–70]. These reports basically show that PT can reduce the dose to normal tissue compared to photon radiotherapy. However, the small number of patients and short follow-up periods make it difficult to evaluate the merits of PT in these studies.

SUMMARY

PT for paediatric patients is only available at a few centres, and patients need to travel long distances to visit these hospitals. Parental support is required, and this can create major physical and financial burdens, for which more assistance is needed. Intensity-modulated PT is more sophisticated than passive scanning, and this technique is likely to be widely applied for paediatric malignancies in the future to reduce the risk of late toxicity and secondary cancer.

REFERENCES

1. A. Jemal, L.X. Clegg, E. Ward, L.A. Ries, X. Wu, P.M. Jamison, P.A. Wingo, H.L. Howe, R.N. Anderson, B.K. Edwards, Annual report to the nation on the status of cancer, 1975–2001, with a special feature regarding survival, *Cancer*, 101 (2004) 3–27.
2. A.C. Mertens, Q. Liu, J.P. Neglia, K. Wasilewski, W. Leisenring, G.T. Armstrong, L.L. Robison, Y. Yasui, Cause-specific late mortality among 5-year survivors of childhood cancer: The childhood cancer survivor study, *J Natl Cancer Inst*, 100 (2008) 1368–1379.
3. M. Abdel-Wahab, I.M. Reis, K. Hamilton, Second primary cancer after radiotherapy for prostate cancer—A seer analysis of brachytherapy versus external beam radiotherapy, *Int J Radiat Oncol Biol Phys*, 72 (2008) 58–68.
4. M.S. Pearce, J.A. Salotti, M.P. Little, K. McHugh, C. Lee, K.P. Kim, N.L. Howe et al., Radiation exposure from CT scans in childhood and subsequent risk of leukaemia and brain tumours: A retrospective cohort study, *Lancet*, 380 (2012) 499–505.
5. M.A. Tucker, G.J. D'Angio, J.D. Boice Jr., L.C. Strong, F.P. Li, M. Stovall, B.J. Stone et al., Bone sarcomas linked to radiotherapy and chemotherapy in children, *N Engl J Med*, 317 (1987) 588–593.
6. F. de Vathaire, C. Hardiman, A. Shamsaldin, S. Campbell, E. Grimaud, M. Hawkins, M. Raquin et al., Thyroid carcinomas after irradiation for a first cancer during childhood, *Arch Intern Med*, 159 (1999) 2713–2719.
7. P.J. Taddei, D. Mirkovic, J.D. Fontenot, A. Giebeler, Y. Zheng, D. Kornguth, R. Mohan, W.D. Newhauser, Stray radiation dose and second cancer risk for a pediatric patient receiving craniospinal irradiation with proton beams, *Phys Med Biol*, 54 (2009) 2259–2275.
8. S.S. Donaldson, Pediatric patients. Tolerance levels and effects of treatment, *Front Radiat Ther Oncol*, 23 (1989) 390–407.
9. B. Guyuron, A.P. Dagys, I.R. Munro, R.B. Ross, Effect of irradiation on facial growth: A 7–25-year follow-up, *Ann Plast Surg*, 11 (1983) 423–427.
10. D.M. Ju, M. Moss, G.F. Crikelair, Effect of radiation on the development of facial structures in retinoblastoma cases, *Am J Surg*, 106 (1963) 807–815.
11. M. Fromm, P. Littman, R.B. Raney, L. Nelson, S. Handler, G. Diamond, C. Stanley, Late effects after treatment of twenty children with soft tissue sarcomas of the head and neck: Experience at a single institution with a review of the literature, *Cancer*, 57 (1986) 2070–2076.

12. G.T. Armstrong, Q. Liu, Y. Yasui, J.P. Neglia, W. Leisenring, L.L. Robison, A.C. Mertens, Late mortality among 5-year survivors of childhood cancer: A summary from the childhood cancer survivor study, *J Clin Oncol*, 27 (2009) 2328–2338.

13. R. Miralbell, A. Lomax, T. Bortfeld, M. Rouzaud, C. Carrie, Potential role of proton therapy in the treatment of pediatric medulloblastoma/primitive neuroectodermal tumors: Reduction of the supratentorial target volume, *Int J Radiat Oncol Biol Phys*, 38 (1997) 477–484.

14. A.T. Meadows, J. Gordon, D.J. Massari, P. Littman, J. Fergusson, K. Moss, Declines in IQ scores and cognitive dysfunctions in children with acute lymphocytic leukaemia treated with cranial irradiation, *Lancet*, 2 (1981) 1015–1018.

15. M.D. Ris, R. Packer, J. Goldwein, D. Jones-Wallace, J.M. Boyett, Intellectual outcome after reduced-dose radiation therapy plus adjuvant chemotherapy for medulloblastoma: A Children's Cancer Group study, *J Clin Oncol*, 19 (2001) 3470–3476.

16. A.W. Walter, R.K. Mulhern, A. Gajjar, R.L. Heideman, D. Reardon, R.A. Sanford, X. Xiong, L.E. Kun, Survival and neurodevelopmental outcome of young children with medulloblastoma at St Jude Children's Research Hospital, *J Clin Oncol*, 17 (1999) 3720–3728.

17. T.E. Merchant, E.N. Kiehna, C. Li, H. Shukla, S. Sengupta, X. Xiong, A. Gajjar, R.K. Mulhern, Modeling radiation dosimetry to predict cognitive outcomes in pediatric patients with CNS embryonal tumors including medulloblastoma, *Int J Radiat Oncol Biol Phys*, 65 (2006) 210–221.

18. Y. Oshiro, T. Okumura, M. Mizumoto, T. Fukushima, H. Ishikawa, T. Hashimoto, K. Tsuboi, M. Kaneko, H. Sakurai, Proton beam therapy for unresectable hepatoblastoma in children: Survival in one case, *Acta Oncol*, 52 (2013) 600–603.

19. E.J. Hall, Intensity-modulated radiation therapy, protons, and the risk of second cancers, *Int J Radiat Oncol Biol Phys*, 65 (2006) 1–7.

20. S.E. Cotter, D.A. Herrup, A. Friedmann, S.M. Macdonald, R.V. Pieretti, G. Robinson, J. Adams, N.J. Tarbell, T.I. Yock, Proton radiotherapy for pediatric bladder/prostate rhabdomyosarcoma: Clinical outcomes and dosimetry compared to intensity-modulated radiation therapy, *Int J Radiat Oncol Biol Phys*, 81 (2011) 1367–1373.

21. E.B. Hug, M. Nevinny-Stickel, M. Fuss, D.W. Miller, R.A. Schaefer, J.D. Slater, Conformal proton radiation treatment for retroperitoneal neuroblastoma: Introduction of a novel technique, *Med Pediatr Oncol*, 37 (2001) 36–41.

22. K.R. Kozak, J. Adams, S.J. Krejcarek, N.J. Tarbell, T.I. Yock, A dosimetric comparison of proton and intensity-modulated photon radiotherapy for pediatric parameningeal rhabdomyosarcomas, *Int J Radiat Oncol Biol Phys*, 74 (2009) 179–186.

23. N.S. Boehling, D.R. Grosshans, J.B. Bluett, M.T. Palmer, X. Song, R.A. Amos, N. Sahoo, J.J. Meyer, A. Mahajan, S.Y. Woo, Dosimetric comparison of three-dimensional conformal proton radiotherapy, intensity-modulated proton therapy, and intensity-modulated radiotherapy for treatment of pediatric craniopharyngiomas, *Int J Radiat Oncol Biol Phys*, 82 (2012) 643–652.

24. S.M. MacDonald, A. Trofimov, S. Safai, J. Adams, B. Fullerton, D. Ebb, N.J. Tarbell, T.I. Yock, Proton radiotherapy for pediatric central nervous system germ cell tumors: early clinical outcomes, *Int J Radiat Oncol Biol Phys*, 79 (2011) 121–129.

25. S.M. MacDonald, S. Safai, A. Trofimov, J. Wolfgang, B. Fullerton, B.Y. Yeap, T. Bortfeld, N.J. Tarbell, T. Yock, Proton radiotherapy for childhood ependymoma: Initial clinical outcomes and dose comparisons, *Int J Radiat Oncol Biol Phys*, 71 (2008) 979–986.

26. R. Miralbell, A. Lomax, L. Cella, U. Schneider, Potential reduction of the incidence of radiation-induced second cancers by using proton beams in the treatment of pediatric tumors, *Int J Radiat Oncol Biol Phys*, 54 (2002) 824–829.

27. M. Hillbrand, D. Georg, H. Gadner, R. Potter, K. Dieckmann, Abdominal cancer during early childhood: A dosimetric comparison of proton beams to standard and advanced photon radiotherapy, *Radiother Oncol*, 89 (2008) 141–149.

28. B.S. Athar, H. Paganetti, Comparison of second cancer risk due to out-of-field doses from 6-MV IMRT and proton therapy based on 6 pediatric patient treatment plans, *Radiother Oncol*, 98 (2011) 87–92.

29. T.E. Merchant, C.H. Hua, H. Shukla, X. Ying, S. Nill, U. Oelfke, Proton versus photon radiotherapy for common pediatric brain tumors: Comparison of models of dose characteristics and their relationship to cognitive function, *Pediatr Blood Cancer*, 51 (2008) 110–117.

30. M.B. Pulsifer, R.V. Sethi, K.A. Kuhlthau, S.M. MacDonald, N.J. Tarbell, T.I. Yock, Early cognitive outcomes following proton radiation in pediatric patients with brain and central nervous system tumors, *Int J Radiat Oncol Biol Phys*, 93 (2015) 400–407.

31. D. Takizawa, Y. Oshiro, M. Mizumoto, H. Fukushima, T. Fukushima, H. Sakurai, Proton beam therapy for a patient with large rhabdomyosarcoma of the body trunk, *Ital J Pediatr*, 41 (2015) 90.

32. M. Fuss, E.B. Hug, R.A. Schaefer, M. Nevinny-Stickel, D.W. Miller, J.M. Slater, J.D. Slater, Proton radiation therapy (PRT) for pediatric optic pathway gliomas: Comparison with 3D planned conventional photons and a standard photon technique, *Int J Radiat Oncol Biol Phys*, 45 (1999) 1117–1126.

33. E.B. Hug, J. Adams, M. Fitzek, A. De Vries, J.E. Munzenrider, Fractionated, three-dimensional, planning-assisted proton-radiation therapy for orbital rhabdomyosarcoma: A novel technique, *Int J Radiat Oncol Biol Phys*, 47 (2000) 979–984.

34. M.M. Ladra, J.D. Szymonifka, A. Mahajan, A.M. Friedmann, B. Yong Yeap, C.P. Goebel, S.M. MacDonald et al., Preliminary results of a phase II trial of proton radiotherapy for pediatric rhabdomyosarcoma, *J Clin Oncol*, 32 (2014) 3762–3770.

35. D. Leiser, G. Calaminus, R. Malyapa, B. Bojaxhiu, F. Albertini, U. Kliebsch, L. Mikroutsikos et al., Tumour control and quality of life in children with rhabdomyosarcoma treated with pencil beam scanning proton therapy, *Radiother Oncol*, 120 (2016) 163–168.

36. T. Yock, R. Schneider, A. Friedmann, J. Adams, B. Fullerton, N. Tarbell, Proton radiotherapy for orbital rhabdomyosarcoma: Clinical outcome and a dosimetric comparison with photons, *Int J Radiat Oncol Biol Phys*, 63 (2005) 1161–1168.

37. D. Forstner, M. Borg, B. Saxon, Orbital rhabdomyosarcoma: Multidisciplinary treatment experience, *Australas Radiol*, 50 (2006) 41–45.

38. B. Timmermann, A. Schuck, F. Niggli, M. Weiss, A. Lomax, G. Goitein, "Spot-scanning" proton therapy for rhabdomyosarcomas of early childhood: First experiences at PSI, *Strahlenther Onkol*, 182 (2006) 653–659.

39. S.K. Childs, K.R. Kozak, A.M. Friedmann, B.Y. Yeap, J. Adams, S.M. MacDonald, N.J. Liebsch, N.J. Tarbell, T.I. Yock, Proton radiotherapy for parameningeal rhabdomyosarcoma: Clinical outcomes and late effects, *Int J Radiat Oncol Biol Phys*, 82 (2012) 635–642.

40. H. Fukushima, T. Fukushima, A. Sakai, R. Suzuki, C. Kobayashi, Y. Oshiro, M. Mizumoto et al., Tailor-made treatment combined with proton beam therapy for children with genitourinary/pelvic rhabdomyosarcoma, *Rep Pract Oncol Radiother*, 20 (2015) 217–222.

41. D.C. Weber, C. Ares, F. Albertini, M. Frei-Welte, F.K. Niggli, R. Schneider, A.J. Lomax, Pencil beam scanning proton therapy for pediatric parameningeal rhabdomyosarcomas: Clinical outcome of patients treated at the Paul Scherrer Institute, *Pediatr Blood Cancer*, 63 (2016) 1731–1736.

42. K.W. Mouw, R.V. Sethi, B.Y. Yeap, S.M. MacDonald, Y.L. Chen, N.J. Tarbell, T.I. Yock et al., Proton radiation therapy for the treatment of retinoblastoma, *Int J Radiat Oncol Biol Phys*, 90 (2014) 863–869.

43. T.E. Merchant, L.E. Kun, S. Wu, X. Xiong, R.A. Sanford, F.A. Boop, Phase II trial of conformal radiation therapy for pediatric low-grade glioma, *J Clin Oncol*, 27 (2009) 3598–3604.

44. T.J. MacDonald, E.B. Arenson, J. Ater, R. Sposto, H.E. Bevan, J. Bruner, M. Deutsch et al., Phase II study of high-dose chemotherapy before radiation in children with newly diagnosed high-grade astrocytoma: Final analysis of Children's Cancer Group study 9933, *Cancer*, 104 (2005) 2862–2871.

45. H. Hauswald, S. Rieken, S. Ecker, K.A. Kessel, K. Herfarth, J. Debus, S.E. Combs, First experiences in treatment of low-grade glioma grade I and II with proton therapy, *Radiat Oncol*, 7 (2012) 189.

46. E.B. Hug, M.W. Muenter, J.O. Archambeau, A. DeVries, B. Liwnicz, L.N. Loredo, R.I. Grove, J.D. Slater, Conformal proton radiation therapy for pediatric low-grade astrocytomas, *Strahlenther Onkol*, 178 (2002) 10–17.

47. B.A. Greenberger, M.B. Pulsifer, D.H. Ebb, S.M. MacDonald, R.M. Jones, W.E. Butler, M.S. Huang et al., Clinical outcomes and late endocrine, neurocognitive, and visual profiles of proton radiation for pediatric low-grade gliomas, *Int J Radiat Oncol Biol Phys*, 89 (2014) 1060–1068.

48. J. Grill, C. Sainte-Rose, A. Jouvet, J.C. Gentet, O. Lejars, D. Frappaz, F. Doz et al., Treatment of medulloblastoma with postoperative chemotherapy alone: An SFOP prospective trial in young children, *Lancet Oncol*, 6 (2005) 573–580.

49. R.V. Sethi, D. Giantsoudi, M. Raiford, I. Malhi, A. Niemierko, O. Rapalino, P. Caruso et al., Patterns of failure after proton therapy in medulloblastomaL: Linear energy transfer distributions and relative biological effectiveness associations for relapses, *Int J Radiat Oncol Biol Phys*, 88 (2014) 655–663.

50. R.B. Jimenez, R. Sethi, N. Depauw, M.B. Pulsifer, J. Adams, S.M. McBride, D. Ebb et al., Proton radiation therapy for pediatric medulloblastoma and supratentorial primitive neuro-ectodermal tumors: Outcomes for very young children treated with upfront chemotherapy, *Int J Radiat Oncol Biol Phys*, 87 (2013) 120–126.

51. B.R. Eaton, N. Esiashvili, S. Kim, E.A. Weyman, L.T. Thornton, C. Mazewski, T. MacDonald et al., Clinical outcomes among children with standard-risk medulloblastoma treated with proton and photon radiation therapy: A comparison of disease control and overall survival, *Int J Radiat Oncol Biol Phys*, 94 (2016) 133–138.

52. T.I. Yock, B.Y. Yeap, D.H. Ebb, E. Weyman, B.R. Eaton, N.A. Sherry, R.M. Jones et al., Long-term toxic effects of proton radiotherapy for paediatric medulloblastoma: A phase 2 single-arm study, *Lancet Oncol*, 17 (2016) 287–298.

53. R. Zhang, R.M. Howell, P.J. Taddei, A. Giebeler, A. Mahajan, W.D. Newhauser, A comparative study on the risks of radiogenic second cancers and cardiac mortality in a set of pediatric medulloblastoma patients treated with photon or proton craniospinal irradiation, *Radiother Oncol*, 113 (2014) 84–88.

54. R.M. Howell, A. Giebeler, W. Koontz-Raisig, A. Mahajan, C.J. Etzel, A.M. D'Amelio, K.L. Homann, W.D. Newhauser, Comparison of therapeutic dosimetric data from passively scattered proton and photon craniospinal irradiations for medulloblastoma, *Radiat Oncol*, 7 (2012) 116.

55. W.H. St. Clair, J.A. Adams, M. Bues, B.C. Fullerton, S. La Shell, H.M. Kooy, J.S. Loeffler, N.J. Tarbell, Advantage of protons compared to conventional X-ray or IMRT in the treatment of a pediatric patient with medulloblastoma, *Int J Radiat Oncol Biol Phys*, 58 (2004) 727–734.

56. M. Yoon, D.H. Shin, J. Kim, J.W. Kim, D.W. Kim, S.Y. Park, S.B. Lee et al., Craniospinal irradiation techniques: A dosimetric comparison of proton beams with standard and advanced photon radiotherapy, *Int J Radiat Oncol Biol Phys*, 81 (2011) 637–646.

57. R. Zhang, R.M. Howell, A. Giebeler, P.J. Taddei, A. Mahajan, W.D. Newhauser, Comparison of risk of radiogenic second cancer following photon and proton craniospinal irradiation for a pediatric medulloblastoma patient, *Phys Med Biol*, 58 (2013) 807–823.

58. T.E. Merchant, C. Li, X. Xiong, L.E. Kun, F.A. Boop, R.A. Sanford, Conformal radiotherapy after surgery for paediatric ependymoma: A prospective study, *Lancet Oncol*, 10 (2009) 258–266.

59. C. Ares, F. Albertini, M. Frei-Welte, A. Bolsi, M.A. Grotzer, G. Goitein, D.C. Weber, Pencil beam scanning proton therapy for pediatric intracranial ependymoma, *J Neurooncol*, 128 (2016) 137–145.

60. J.W. Kim, W.C. Kim, J.H. Cho, D.S. Kim, K.W. Shim, C.J. Lyu, S.C. Won, C.O. Suh, A multi-modal approach including craniospinal irradiation improves the treatment outcome of high-risk intracranial nongerminomatous germ cell tumors, *Int J Radiat Oncol Biol Phys*, 84 (2012) 625–631.

61. M. Matsutani, G. Japanese Pediatric Brain Tumor Study, Combined chemotherapy and radiation therapy for CNS germ cell tumors—The Japanese experience, *J Neurooncol*, 54 (2001) 311–316.

62. G. Calaminus, R. Kortmann, J. Worch, J.C. Nicholson, C. Alapetite, M.L. Garre, C. Patte, U. Ricardi, F. Saran, D. Frappaz, SIOP CNS GCT 96: Final report of outcome of a prospective, multinational nonrandomised trial for children and adults with intracranial germinoma, comparing craniospinal irradiation alone with chemotherapy followed by focal primary site irradiation for patients with localized disease, *Neuro-oncology*, 15 (2013) 788–796.

63. C. Alapetite, H. Brisse, C. Patte, M.A. Raquin, G. Gaboriaud, C. Carrie, J.L. Habrand et al., Pattern of relapse and outcome of non-metastatic germinoma patients treated with chemotherapy and limited field radiation: The SFOP experience, *Neuro-oncology*, 12 (2010) 1318–1325.

64. J. Park, Y. Park, S.U. Lee, T. Kim, Y.K. Choi, J.Y. Kim, Differential dosimetric benefit of proton beam therapy over intensity modulated radiotherapy for a variety of targets in patients with intracranial germ cell tumors, *Radiat Oncol*, 10 (2015) 135.

65. Q.T. Luu, L.N. Loredo, J.O. Archambeau, L.T. Yonemoto, J.M. Slater, J.D. Slater, Fractionated proton radiation treatment for pediatric craniopharyngioma: Preliminary report, *Cancer J*, 12 (2006) 155–159.

66. K.M. Winkfield, C. Linsenmeier, T.I. Yock, P.E. Grant, B.Y. Yeap, W.E. Butler, N.J. Tarbell, Surveillance of craniopharyngioma cyst growth in children treated with proton radiotherapy, *Int J Radiat Oncol Biol Phys*, 73 (2009) 716–721.

67. A.J. Bishop, B. Greenfield, A. Mahajan, A.C. Paulino, M.F. Okcu, P.K. Allen, M. Chintagumpala et al., Proton beam therapy versus conformal photon radiation therapy for childhood craniopharyngioma: Multi-institutional analysis of outcomes, cyst dynamics, and toxicity, *Int J Radiat Oncol Biol Phys*, 90 (2014) 354–361.

68. J.T. Lucas Jr., M.M. Ladra, S.M. MacDonald, P.M. Busse, A.M. Friedmann, D.H. Ebb, K.J. Marcus, N.J. Tarbell, T.I. Yock, Proton therapy for pediatric and adolescent esthesioneuroblastoma, *Pediatr Blood Cancer*, 62 (2015) 1523–1528.

69. Y. Oshiro, M. Mizumoto, T. Okumura, S. Sugahara, T. Fukushima, H. Ishikawa, T. Nakao et al., Clinical results of proton beam therapy for advanced neuroblastoma, *Radiat Oncol*, 8 (2013) 142.

70. B. Rombi, T.F. DeLaney, S.M. MacDonald, M.S. Huang, D.H. Ebb, N.J. Liebsch, K.A. Raskin et al., Proton radiotherapy for pediatric Ewing's sarcoma: Initial clinical outcomes, *Int J Radiat Oncol Biol Phys*, 82 (2012) 1142–1148.

CHAPTER 11

Normal Tissue Complication Probability Reduction in Advanced Head and Neck Cancer Patients Using Proton Therapy

Jacques Balosso, Valentin Calugaru, Abdulhamid Chaikh and Juliette Thariat

CONTENTS

INTRODUCTION

What Is the Benefit of Proton Therapy for Advanced Head and Neck Tumours?

Head and neck squamous cell carcinoma (HNSCC), perhaps more than other cancer types, poses the problem of the benefit-risk balance. Indeed, these tumours are radio-curable, but at very high doses in an extremely functional anatomical environment, where early or late toxicity can have devastating effects on the quality of life (QOL). The need to increase as much as possible the therapeutic index leads to consider particle therapy, which offers greater dosimetric discrimination than photon therapy (Rayon X). Nevertheless, the high incidence of HNSCC (3%–5% of cancers) patients often living in poor conditions and the low availability of proton therapy (PT) centres have meant that, until recently, PT has been applied to treatment of head and neck (H&N) tumours that are not HNSCC. These are adenoid cystic carcinomas, chordomas, or sarcomas of the base of the skull, recurrent undifferentiated carcinoma of nasopharyngeal type (UCNT) or sinonasal tumours for which experience has been gained with PT, often as a boost with RX [1]. This initial experience reveals the possible severity of side effects, ranging from blindness to xerostomia and severe and lasting swallowing disorders. Severe toxicities have also been reported in exclusive PT series due to high-dose treatments of skull base tumours [1].

Thus, PT has been considered as an option for HNSCC only recently. In 2011, PT publications about HNSCC are still rare: in a meta-analysis, Ramaekers et al. [2] reported on 86 suitable studies, only 7 concerning PT, 5 carbontherapy and 74 RX. The variability of the risk estimates of normal tissue complication probability (NTCP) was pointed out. Such variability was probably related to the variety of models used. However, PT was able to reduce toxicity for most of these tumours, and PT improved local control only for sinonasal tumours. In their meta-analysis, clinical studies and *in silico* studies were prefiguring future PT developments such as reduced toxicities of HNSCC treatments, which may further translate into better QOL for patients and reduced costs induced by these toxicities. The delayed nature of certain side effects, dosimetric evidence of the potential benefits of PT, but also the extreme variety of situations make it very difficult, if not impossible, to define homogeneous groups of patients for whom PT would certainly achieve a significant benefit, but this could be demonstrated by randomised prospective trials. The high cost of PT adds to this difficulty. Therefore, the principle of addressing this problem by individual modelling of treatments and their results in terms of NTCP and cost effectiveness has been proposed for several years [3–5].

How to Build a Process for Reducing the Normal Tissue Complication Probability for Head and Neck Squamous Cell Carcinoma by Proton Therapy?

The robust and reliable differential dose distribution between RX and PT is the basic data. In radiation oncology, any dose reduction brings the hope of organic toxicity reduction, in other words, a hope of NTCP reduction. However, an NTCP reduction does not directly translate into a real benefit for the patient. This depends on the true clinical impact of the alleviated toxicity. For example, a simple skin sclerosis is not as important as swallowing disabilities, or neurological impairments. Thus, it is critical to choose relevant and accurate

NTCP models, targeting toxicities that influence the QOL. However, we are not able yet to establish a direct quantitative and continuous relation between NTCP and QOL. Overall survival, weighted by QOL, is known as QALY. The QALY, relative to the supplementary cost, will define the utility or cost-effectiveness of PT and can provide a useful guidance for medical decision for patients' treatment choice.

As of 2013, Ramaekers et al. [3] illustrated this modelling method. On the basis of a Markov model, they showed how the comparison of intensity-modulated proton therapy (IMPT) versus intensity-modulated radiation therapy (IMRT) can be done by differential estimation of NTCP. NTCP, through calibration of a QOL questionnaire (the EQ5D, in this case) for a given type of toxicity, may lead to QALY estimation and consequently to medico-economic calculation of the usefulness of treatment. This demonstration was made with reasonable levels of toxicity, grade ≥2, thus at the level of the current toxicity reduction objectives of treatments. A selection process of patients to undergo PT, based on a given level of estimated individual benefit (10% reduction in serious toxicity, ΔNTCP ≥10%), constitutes a cost-effective approach: 'IMPT, if effective, is the best trade-off at individual level.' A critical element of the process is therefore the ability to precisely predict absolute values of NTCP.

Presently Available Approaches for Normal Tissue Complication Probability Calculation and Comparison

NTCP models are mathematical equations with input values derived from provisional or effective dose distribution in the patient's body; the output is the probability of a given normal tissue complication, specific to an organ with known dose and tissue-related parameters. Such models have been developed since the 1980s and are historically univariate (only taking into account the radiation dose for a single tissue) and have been tuned mainly with 3D-CRT experience and high-grade (>2) toxicities. These classical NTCP models need to be re-tuned according to recent dose calculation algorithms, radiation oncology techniques, radiobiology of organs at risk (OAR), adapted type and level of toxicities (some of the predicted complications are no longer possible thanks to the dose-volume recommendations). The main drawbacks of NTCP models are their numerous varieties, their complexity for some of them and their univariate structure when more and more patients' and treatment characteristics are influencing NTCPs [6]. Those classical NTCP models are often imbedded in the treatment planning system (TPS) but, according to the previous caveat, are of limited usefulness for uninformed users (large uncertainties should limit their use as ratio, presented as ΔNTCP differences, for comparisons of treatments, instead of absolute values).

To overdraw the aforementioned drawbacks, many *ad hoc* NTCP models, based on the multivariate logistic regression core Equation 11.1, have been recently developed with highly valuable clinical data.

$$DMF = \frac{1}{1+e^{-s}} \qquad s = \beta_0 + \sum_{i=1}^{n} \beta_i . x_i \qquad (11.1)$$

where *DMF* stands for dosimetric multifactorial model, and a set of n prognostic variables (x_i), have weighting factors defined as regression coefficients (β_i), $(0 < \beta_i < 1)$, (β_0) being the initial value of (β_i). Factors (β_i) can be estimated from a bootstrap simulation and the Spearman correlation test.

This simple model can calculate accurate absolute values of NTCP for RX and PT using multiple variables such as doses (of different tissues all related to the same function, such as swallowing), age, use of chemotherapy and so on. In fact, since 2005 or so, in-depth prospective works have been able to redefine NTCP predictive models, specific to critical toxicities of HNSCC patients. These toxicities are functionally defined to be able to establish a relation with QOL. These works are described as follows.

Rancati et al. [7] established, for the first time, NTCP parameters for late oedema of the irradiated larynx (considered as an OAR in photon therapy). This prospective study, carried out from 2002 to 2006 on 48 patients, used the Lyman EUD and Logit EUD models for grade 2–3 toxicities with TD_{50} values of 47.3 ± 2.1 Gray and 46.7 ± 2.1 Gray, respectively. The larynx behaves as an organ in parallel with an $\alpha/\beta = 3$.

Houweling et al. [6] produced the largest prospective bi-centric study on the parotid by including 347 patients (Utrecht -Croningen, Netherlands and Michigan, Ann Arbor United States). Patients were treated with 3D-CRT or IMRT. Toxicity was defined, one year after treatment, as binary with <25% of the initial salivary flow corresponding to a grade 3 common terminology criteria for adverse events (CTCAE) versus more. Six different NTCP models were compared. No significant difference was found between commonly used NTCP models. The model indicates a parallel organization of the parotid and a TD_{50} of 39 Gray. The *mean dose* model, the simplest model used, was ranked as the best one. The α/β is probably high as no effect of fractionation has been clearly observed.

Eisbruch et al. [8] established relationships between doses to the constrictor muscles of the pharynx and to the supraglottic regions, and swallowing disorders in 73 patients. Fluoroscopic analysis was the most sensitive method (aspirations not perceived by the patient), then came patients' QOL criteria, while physicians' assessment was uninformative. The NTCP LKB model was used, and the dosimetric criterion that best correlated was the *mean dose* ($\alpha/\beta = 3$ for late effects). The TD_{25} in radioscopic study, which is 56 Gray (EQD2), was recommended as an optimisation objective. In this study, NTCPs were different depending on specific methods used to assess dysphagia. This highlights the need for a standardised method for dysphagia evaluation.

Lindblom et al. [9] studied 124 patients randomised between concomitant boost versus sequential boost for a total of 68 Gray. Grade 3–4 patient-rated trismus was defined as 35 millimetres or less interincisal-opening distance (grade 1–4). Using a 'classical' univariate NTCP approach, the best correlation with grade 3–4 trismus was *mean dose* to the homolateral muscular structures, particularly the masseter. DT_{50} for grade 3–4 in QOL and for the interincisal measurement ($\leq$35 millimetres) were 72.3 and 57.2 Gray, respectively.

Beetz et al. [10] and Christianen et al. [11] used the multi-parametric logistic regression model (1) to study salivary disorders (for 167 patients followed in the Netherlands since 1997 and treated in 3D-CRT for HNSCC) and swallowing disorders according to different evaluations (for 354 multi-centre patients). A search for parameters other than the dose, such as age, medical history, chemotherapy and so on, was made. The results gave highly differentiated specific expressions of s (the s value of Equation 11.1) based on strong multivariate statistical correlations, including *mean doses*.

These models are all based on the dose planned, but none has been adapted to the dose actually delivered and received by OARs. Uncertainties in target contouring, treatment planning and delivery of therapy are not taken into account in these models, other than in the confidence intervals around the estimated values.

Results and conclusions of these works, to date, are restricted to RX and mainly 3D-CRT and rarely IMRT. Thus, a fundamental question arises as to whether existing models are valid for PT. This fundamental issue was successfully addressed by Blanchard et al. [4]. The external validity of photon-derived NTCP models for head and neck cancer patients in a population treated with PT showed a mild drop in model performance, about 10% decrease in the area under curve (AUC) value in the receiver operating characteristic (ROC) curve test. However, most of the models kept an AUC $\geq$0.7, showing robustness and validity for PT. These results are a necessary and valuable first step to apply NTCP to PT for HNSCC patients. Thus, this *allows the model-based approach* advocated by the Dutch groups, provided that continuous refinement of the predictive models is undertaken. This is particularly critical as uncertainties in the delivery of PT and in relative biological effectiveness (RBE) are not taken into account in this study by Blanchard et al.

An Example of *In Silico* Comparison of Two Treatment Plans

Tomotherapy was compared to double-scattering PT for an advanced nasopharyngeal HNSCC case treated at the ICPO (Institut Curie, Orsay, France). Using two different NTCP models, the EUD (ratio = R_{EUD}) and LKB ones [12] (ratio = R_{NTCP}), ΔNTCP was computed for toxicity of different OARs. The dosimetric improvement was obvious with respect to dose-volume histogram (DVH) and low doses on computed tomography (CT) images. However, as seen on the ΔNTCP chart, it did not translate into reduced NTCP for all OARs in this clinical scenario with a very large and central target volume. Such a case would deserve very advanced optimisation to maximise the gain from PT. Involvement of some OARs, such as optical pathways, into the target volume raises the critical question of the effect of RBE on NTCP calculations. RBE has not been taken into consideration into NTCP models to date, but it now deserves attention [13].

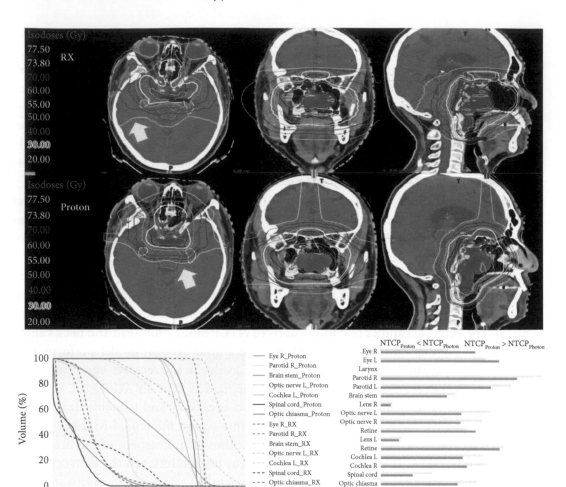

In line with this, Jakobi et al. [14] performed an *in-silico* dose escalation IMRT and IMPT study. They showed that the expected TCP and NTCP changes occurred well, suggesting applicability of NTCP (RX) models to PT: identical improvement for TCP and some increase of toxicities for both, but slightly less for PT. This preliminary study also did not take into account RBE. It did not suggest that dose escalation is less risky by IMPT than by IMRT, nor did it lead to a cost-effectiveness analysis.

In-Silico Testing of Normal Tissue Complication Probability in Head and Neck Squamous Cell Carcinoma Patients

Van der Laan et al. [15] performed *in-silico* models based on valuable works on swallowing dysfunction made for RX. Their work was carried out on the ROCOCO platform: 25 HNSCC IMRT plans optimised for salivary protection were further optimised for swallowing (SW) both with IMRT and IMPT with several levels of technicality (3 and 7 beams). IMPT allowed reduction of mean doses to OARs from −3 to −20 Gray, which translated into a 9% NTCP reduction between standard IMRT and IMRT-SW and by 17% between standard IMRT and very advanced IMPT (7B and SW). These works provide an *a priori* estimate of target NTCP reduction to select patients for PT. The results appear consistent and rather solid. Calibrated QOL and QALY could also be evaluated. This opens the door to the PT guidance using NTCP as a biologically guided radiotherapy (BGRT).

Jakobi et al. [16] compared multiple NTCP end points *in silico* using IMPT as a boost with RX or exclusive IMPT. They used pre-existing RX NTCP models [11,15]. The decision threshold ΔNTCP was 10%. It was reached or exceeded for serious toxicities with the PT boost plan in less than 15% of cases and in about 50% of cases with exclusive IMPT. Interestingly, sprinkling the IMPT as a boost did not bring much reduction of toxicity, but it was less onerous, whereas exclusive IMPT was 'useful' for more than 50% of cases but was probably more onerous. In this case, a medico-economic evaluation devoted to PT would be necessary to establish the relative utility of these two practices as described by Verma et al. [17].

Although the comparative procedure seems to be mastered, the fact remains that the task of individual analysis of each case is time consuming. Jakobi et al. [18] faced this workload and explored possibilities to reduce NTCPs according to tumour site. They showed that oropharyngeal treatments were much more sensitive to SW protection, and selecting patients for PT based on NTCPs seemed useful. However, apart from acute reactions that can evolve without sequelae, tumours of the hypopharynx did not benefit greatly from NTCP comparisons. Thus, comparative analyzes may be limited to oropharyngeal tumours when implementing PT.

Quite different is the Maastricht Group's now well-known approach, which advocates online comparison of RX versus PT to systematically select patients for PT [5]. This group did a real proof of concept for HNSCC patients. They went beyond the preliminary step of ΔNTCP calculations, also achieved by others and well demonstrated. They computed the incremental cost-effectiveness ratio (ICER) at the threshold of 80k€/quality-adjusted life-year (QALY). This retained only 8 cases out of 21 with a multiple cumulative reduction ≥15% at one year, that is, an attrition of 62% compared to solely the BGRT assessment. Finally, only 38% of the cases would be directed towards PT. This exercise used calibrated medical files of the ROCOCO platform and will now have to be adapted to medical records of an ordinary workflow. This raises all the problems of data transfer, limitations of transition from the models NTCP-RX to PT, quality assurance at the various steps and so on.

DISCUSSION OF THRESHOLD DEFINITION

By calculating differential probabilistic complication rates with differently affordable treatments, one introduces necessarily the need for a decision-making process. Thus, a kind of threshold of the magnitude of the expected gain has to be defined to make the decision, to devote more resources to obtain this expected benefit. At the present time, there are no recommendations for this. Up to now, the reported works are *in-silico* studies that do not require a socially accepted decision process, just an hypothesis. This way, the hypothesis, as such, of 10% reduction of toxicity was used in the Langendijk et al. paper of 2013 [19]. However, this hypothesis has been taken as a threshold 'similar to' the previous, by Jakobi et al. in their work of 2015 [18], and finally this threshold, with some refining explanations, is used in the founding paper of Cheng et al. referring to a consensus document of the Dutch health system [5].

If we try to extract general recommendations from the good practices of radiation oncology, for example, the most detailed levels of risks documented by the QUANTEC collaboration [19], it could be stated that three levels of risks are generally accepted in radiation oncology regarding severe late effects: 0% for life-threatening risks as for brainstem and spinal cord; 5% for serious disabling risks as for bowel obstruction, parenchyma destruction; and 10% or even more for less dangerous risks as breath discomfort, intestinal bleeding, skin sclerosis and so on.

Therefore, the Dutch proposal reported by Cheng et al. [5] seems reasonable. It states that a clinical benefit could be assumed if a predicted reduction rate of >10% of a grade ≥ 2 toxicity is predicted. To extend this approach towards a more global assessment, closer to QOL, a concept of *complication profile* is introduced. It is defined as the summation, for a patient, of all the toxicity probability reductions exceeding 5%; in this case, the clinical benefit is set at a total reduction $\geq 15\%$.

Nevertheless, these medical definitions will have to cope with the reality, translate into cost-effectiveness estimates and be backed on acceptable cost per additional QALY. This could need a complementary, and country-dependent, tuning of the considered clinical benefit thresholds [5].

CONCLUSION AND PERSPECTIVES

The works reported in this chapter show that the biologically guided radiation therapy (BGRT) based on NTCP calculations and comparisons is a reality. Classical NTCP models have tended to simplify. *Mean dose* is very often the most relevant quantity among the dosimetric data. For appropriate organs, Dmean may thus be used instead of DVH. On the other hand, new NTCP models can combine several parameters and introduce non-dosimetric aspects such as age, basic functional disorders, comorbidity, associated treatments and so on, all of which could be responsible, to some extent, for toxicities. NTCP models are absolutely critical in this context, which makes their quality and accuracy as necessary as possible.

Considerable effort is being made in Europe to implement online/real-time use of NTCP calculations to produce customised cost-effectiveness estimates for each patient eligible for PT. This still requires a substantial workload and adherence to government health standards, as was done in the Netherlands [20].

NTCP models contribute to a paradigm shift in the practice of evidence-based medicine. While Leeman [21] carried out a thorough, descriptive comparison of PT versus RX in HNSCC with a rich iconography for different tumour sites, he commented on the rapid emergence of prospective clinical trials but omitted NTCP models. He reported 7 trials in progress in 2010, 11 in 2015 and 16 in 2017. Finally, while NTCP modelling is still very specialised and mastered by a few teams, the background work of conventional prospective trials is running. This is absolutely complementary to the model-based approach and rather well advanced in the field of PT applied to HNSCCs. Clinical studies will be useful to demonstrate the accuracy of model predictions and the reality of utility calculations. In the future, they will probably be less and less randomised and will probably be closer to the Phase IV concept of drug marketing. It is reasonable to assume that the application of protons to HNSCC patients will be firmly documented in the coming years by robust and complementary approaches, including NTCP models and clear decision processes.

REFERENCES

1. Chan AW, Liebsch NJ. Proton radiation therapy for head and neck cancer. *J Surg Oncol.* 2008;97(8):697–700.
2. Ramaekers BLT, Pijls-Johannesma M, Joore MA, van den Ende P, Langendijk JA, Lambin P, Kessels AGH, Grutters JPC. Systematic review and meta-analysis of radiotherapy in various head and neck cancers: Comparing photons, carbon-ions and protons. *Cancer Treat Rev.* 2011;37:185–201.
3. Ramaekers BL, Grutters JP, Pijls-Johannesma M, Lambin P, Joore MA, Langendijk JA. Protons in head-and-neck cancer: Bridging the gap of evidence. *Int J Radiat Oncol Biol Phys.* 2013;85(5):1282–1288.
4. Blanchard P, Wong AJ, Gunn GB, Garden AS, Mohamed AS, Rosenthal DI, Crutison J et al. Toward a model-based patient selection strategy for proton therapy: External validation of photon-derived normal tissue complication probability models in a head and neck proton therapy cohort. *Radiother Oncol.* 2016;121(3):381–386.
5. Cheng Q, Roelofs E, Ramaekers BLT, Eekers D, van Soest J, Lustberg T, Hendriks T et al. Development and evaluation of an online three-level proton versus photon decision support prototype for head and neck cancer—Comparison of dose, toxicity and cost-effectiveness. *Radiother Oncol.* 2016. doi:10.1016/j.radonc.2015.12.029.
6. Houweling AC, Philippens ME, Dijkema T, Roesink JM, Terhaard CH, Schilstra C, Ten Haken RK, Eisbruch A, Raaijmakers CP. A comparison of dose-response models for the parotid gland in a large group of head-and-neck cancer patients. *Int J Radiat Oncol Biol Phys.* 2010;76(4):1259–1265.
7. Rancati T, Fiorino C, Sanguineti G. NTCP modeling of subacute/late laryngeal edema scored by fiberoptic examination. *Int J Radiat Oncol Biol Phys.* 2009;75(3):915–923.
8. Eisbruch A, Kim HM, Feng FY, Lyden TH, Haxer MJ, Feng M, Worden FP et al. Chemo-IMRT of oropharyngeal cancer aiming to reduce dysphagia: Swallowing organs late complication probabilities and dosimetric correlates. *Int J Radiat Oncol Biol Phys.* 2011;81(3):e93–e99.
9. Lindblom U, Gärskog O, Kjellén E, Laurell G, Levring Jäghagen E, Wahlberg P, Zackrisson B, Nilsson P. Radiation-induced trismus in the ARTSCAN head and neck trial. *Acta Oncol.* 2014;53(5):620–627.
10. Beetz I, Schilstra C, Burlage FR, Koken PW, Doornaert P, Bijl HP, Chouvalova O et al. Development of NTCP models for head and neck cancer patients treated with three-dimensional conformal radiotherapy for xerostomia and sticky saliva: The role of dosimetric and clinical factors. *Radiother Oncol.* 2012;105:86–93.
11. Christianen ME, Schilstra C, Beetz I, Muijs CT, Chouvalova O, Burlage FR, Doornaert P et al. Predictive modelling for swallowing dysfunction after primary (chemo)radiation: Results of a prospective observational study. *Radiother Oncol.* 2012;105(1):107–114.
12. Burman C, Kutcher GJ, Emami B, Goitein M. Fitting of normal tissue tolerance data to an analytic function. *Int J Radiat Oncol Biol Phys.* 1991;21:123–135.
13. Jones B. Why RBE must be a variable and not a constant in proton therapy. *Br J Radiol.* 2016;89(1063):20160116.
14. Jakobi A, Lühr A, Stützer K, Bandurska-Luque A, Löck S, Krause M, Baumann M, Perrin R, Richter C. Increase in tumor control and normal tissue complication probabilities in advanced head-and-neck cancer for dose-escalated intensity-modulated photon and proton therapy. *Front Oncol.* 2015;5:256.

15. Van der Laan HP, van de Water TA, van Herpt HE, Christianen MEMC, Bijl HP, Korevaar EW, Rasch CR et al. The potential of intensity-modulated proton radiotherapy to reduce swallowing dysfunction in the treatment of head and neck cancer: A planning comparative study. *Acta Oncol.* 2013;52(3):561–569.
16. Jakobi A, Stützer K, Bandurska-Luque A, Löck S, Haase R, Wack LJ, Mönnich D et al. NTCP reduction for advanced head and neck cancer patients using proton therapy for complete or sequential boost treatment versus photon therapy. *Acta Oncol.* 2015;54(9):1658–1664.
17. Verma V, Mishra MV, Mehta MP. A systematic review of the cost and cost-effectiveness studies of proton radiotherapy. *Cancer.* 2016;122(10):1483–1501.
18. Jakobi A, Bandurska-Luque A, Stützer K, Haase R, Löck S, Wack LJ, Mönnich D et al. Identification of patient benefit from proton therapy for advanced head and neck cancer patients based on individual and subgroup normal tissue complication probability analysis. *Int J Radiat Oncol Biol Phys.* 2015;92(5):1165–1174.
19. Marks LB, Ten Haken RK, Martel MK. Guest editor's introduction to QUANTEC: A users guide. *Int J Radiat Oncol Biol Phys.* 2010;76(S3):S1–S160.
20. Langendijk JA, Lambin P, De Ruysscher D, Widder J, Bos M, Verheij M. Selection of patients for radiotherapy with protons aiming at reduction of side effects: The model-based approach. *Radiother Oncol.* 2013;107:267–273.
21. Leeman JE, Romesser PB, Zhou Y, McBride S, Riaz N, Sherman E, Cohen MA, Cahlon O, Lee N. Proton therapy for head and neck cancer: Expanding the therapeutic window. *Lancet Oncol.* 2017. doi:10.1016/S1470-2045(17)30179-1.

CHAPTER 12

The Model-Based Approach to Select Patients for Proton Therapy and to Validate Its Added Value

Johannes A. Langendijk, John H. Maduro,
Anne P.G. Crijns and Christina T. Muijs

CONTENTS

INTRODUCTION

The favourable beam properties of protons over photons can be translated into clinical benefits in two ways. First, protons can be used to escalate the dose to the target with limited or no excess dose deposition to the normal tissues, with the aim to improve loco-regional tumour control without enhancing radiation-induced side effects. However, the effect of target dose escalation on loco-regional control beyond the standard dose is unknown. Furthermore, the effects in terms of acute and late radiation-induced side effects remain to be determined as the dose to the normal tissues in or close to the target may also increase to levels beyond standard. To clinically validate the effect of dose-escalation using protons, randomised controlled trials (RCTs) are still the most optimal design.

Second, protons can be applied primarily to decrease the dose to the normal tissues with an equivalent dose to the target, primarily aiming at prevention of radiation-induced side effects with similar loco-regional control rates. The use of protons for this purpose is heavily based on the large amount of preclinical and clinical data, that higher dose to normal tissues induces more tissue damage and eventually leads to higher rates of clinically apparent radiation-induced side effects. For validating this application of proton therapy, RCTs will be difficult to perform and may yield several methodological problems. However, alternative evidence-based study designs, such as the model-based approach, are also possible and might even be preferred over the classical RCT. It is particularly relevant to validate the advantages of proton therapy to spare normal tissues because, according to the Horizon Scanning Report on Proton Therapy of the Dutch Health Council, proton therapy is expected to be used predominantly to spare normal tissues and thus to prevent radiation-induced side effects and/or secondary tumour induction (85% of the cases).

In this chapter, we will discuss some problems and pitfalls of RCTs when comparing radiation technologies when radiation-induced side effects are used as the primary end point. In addition, the different aspects of the model-based approach, an alternative evidence-based methodology for RCTs, will be discussed.[1]

CONSIDERATIONS FOR RANDOMISED CONTROLLED TRIALS

At present, some RCTs comparing photons with protons are ongoing, but so far, no RCTs have been formally published in peer-reviewed medical journals.[2,3] This has been mainly due to practical problems (such as too limited capacity across countries to perform sufficiently powered RCTs and lack of reimbursement of proton therapy) and to ethical problems (considered lack of equipoise in case of major dose distribution differences).[4,5]

In addition to these problems, some methodological issues should be considered in the design of clinical studies, for example, determining the added value of new radiation technologies primarily to reduce the rate and severity of radiation-induced side effects. A major issue here is that a standard radiation technology, either standard Intensity-Modulated Radio Therapy (IMRT) or standard Intensity-Modulated Proton Therapy (IMPT), does not exist. First, radiation technology gradually evolves due to minor to major adjustments over time. For example, our prospective head and neck cancer data registration program revealed a gradual decrease of the dose to the most critical organs at risk (OAR) among patients treated with IMRT, due to the clinical introduction of several improvements, such as image-guidance, multi-criteria optimisation, NTCP-guided treatment optimisation and most recently adaptive radiotherapy (Figure 12.1).[6-8] Consequently, toxicity rates declined as well, as shown by Christianen et al., who showed that reducing the dose to the swallowing structures, so-called swallowing–sparing IMRT, indeed resulted in lower rates of dysphagia after completion of treatment.[9] Some radiation-induced toxicities develop many years after treatment. Therefore, the time required between study initiation and publication of the final results can take over 10–15 years. Especially, if major improvements have

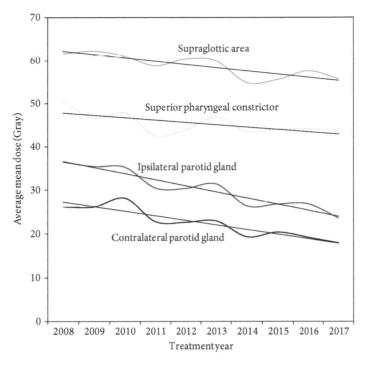

FIGURE 12.1 Average mean dose in four important organs at risk for xerostomia and dysphagia among patients treated for head and neck cancer with intensity modulated radiotherapy (photons). The results show a gradual improvement over time due to multiple subsequent technological improvements.

been made in one or both technologies under study, there is a major risk that these results will not be accepted at the time of publication as the results will be based on radiation technologies considered outdated. Another problem with RCTs is that major differences exist between proton therapy centres as illustrated by studies on proton therapy clinical trial credentialing.[10] Similar differences have been found for IMRT.[11-12] Heterogeneity among centres may result from many factors, such as differences in proton therapy equipment, personnel expertise, delineation of targets and OAR, treatment planning systems, treatment planning techniques, centre-specific workflows, fixation techniques, motion mitigation strategies and machine- and patient-specific quality assurance procedures. In this regard, it is difficult to compare heterogeneous techniques, as neither standard IMRT nor standard IMPT exist. This problem becomes increasingly relevant when results of RCTs are translated into routine clinical practice. The eventual goal is to offer patients the most optimal technique, but how should technology selection in individual patients be done properly when heterogeneity in centre performance both for protons and photons is not considered in decision making? There is even a risk to refer patients for proton therapy in another centre far away from home while photon-based radiation in the referring centre yields better results. Here, the model-based approach may offer a solution, which is beneficial for patients (unnecessary referral and lengthy stays away from home) and better for society (cost-effectiveness).

The model-based approach can be used both for selecting patients for proton therapy as well as for validation of proton therapy when primarily aiming at reduction of side effects with equivalent loco-regional control.[1]

MODEL-BASED SELECTION

The incentive of model-based selection is based on the principle that applying a more complex and expensive technology with limited availability is reserved to patients who are expected to benefit most from protons compared to the current standard technology (photons). When aiming at minimizing the risk of radiation-induced side effects with similar loco-regional control, this principle is met if two general conditions are fulfilled: (1) the dose to the healthy tissues with protons is lower than with photons (Δdose) and (2) this Δdose translates into a clinically relevant reduction of normal tissue complication probabilities (ΔNTCP).[1,13] To test these two main conditions in individual patients, a model-based selection procedure can be used consisting of three steps: (1) selection of NTCP-models suitable for model-based selection, (2) plan comparison (photons versus protons) to assess Δdose and (3) translation of Δdose into ΔNTCP using the selected NTCP-models (Figure 12.2).

Which Normal Tissue Complication Probability Models Can Be Used for Model-Based Selection?

NTCP models describe the relationship between the dose distributions in one or more OAR and the risk of a given radiation-induced side effect. Reliable model-based selection requires NTCP models that enable the most accurate prediction of the risk of side effects and thus should meet several quality criteria. In the Netherlands, the following quality criteria for NTCP models have been defined:

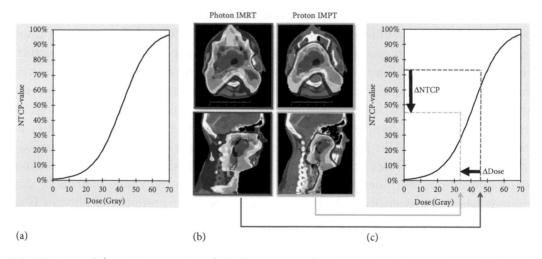

FIGURE 12.2 Schematic overview of the three steps of model-based selection. (a) NTCP model selection, (b) *in-silico* planning comparison between photon IMRT and proton IMPT to assess Δdose, (c) integration of results of planning comparison into the NTCP model to assess ΔNTCP.

- NTCP models are preferably developed in prospective cohort studies as retrospective assessment of many radiation-induced side effects is more likely to result in an underestimation of the incidence and severity of side effects. Other study designs, such as retrospective cohort studies and/or nested case control studies, are acceptable in case prospective assessment is difficult to perform, for example, side effects with very long latency times and/or very low incidences, such as cardiac toxicity after breast cancer radiotherapy and secondary tumour induction.

- NTCP models are preferably multivariable as the risk of many radiation-induced side effects does not only depend on one dose-volume parameter but also on other predictors such as the multiple dose–volume parameters, the addition of chemotherapy or baseline function.[14–18] Moreover, the effect of dose–volume parameters on the risk of side effects may be affected by other factors (confounding or effect modulation), which needs to be corrected for.

- Information on model performance should be available; that is, it should be clear how good the model predicts radiation-induced side effects. The most relevant model performance measures are discrimination and calibration. Discrimination refers to the ability of an NTCP model to distinguish patients who will develop a given side effect versus the ones who will not and is mostly reported as the area under the curve (AUC) or c-statistic.[19] Calibration refers to level of correspondence between predicted risks (NTCP) and observed rates of side effects. Important end points for model calibration are the calibration intercept and slope and the Hosmer–Lemeshow test, testing the hypothesis that the predicted (NTCP values) and observed rates are different.

- To prevent model optimism, that is, overestimating the effect of predictors in the model, internal validation procedures are performed, by proper statistical techniques like bootstrapping and/or cross-validation.

- NTCP models are externally validated in one or more independent patient cohorts to test generalisability in different populations.

- Information on how the NTCP values can be calculated in individual patients is available (e.g. an equation or nomogram).

Based on these quality criteria, the Dutch Proton Therapy Platform (Landelijk Platform Protonentherapie [LPPT]) defined a proposal for levels of evidence for NTCP models (Table 12.1) to facilitate selection of the most appropriate NTCP models for model-based selection.

Plan Comparison

Protons will only result in less radiation-induced side effects if the relevant dose-volume parameters are lower than those obtained with photons. As primary tumour location and local and/or regional extension are different among individual patients, consequential dose distributions in OARs and thus the benefit of protons in terms of Δdose and ΔNTCP may

TABLE 12.1   Levels of Evidence for NTCP Models

Level	Description
Level 1a	External validation of NTCP model in independent dataset in another centre *and* with the new technique (protons)
Level 1b	External validation of NTCP model in independent dataset in another centre with the same technique (photons) *or* in the same centre with the new technique (protons)
Level 1c	NTCP model based on meta-analysis
Level 2a	External validation of NTCP model in independent dataset with *non-random* split in two groups: one for NTCP model development and one for evaluating model performance (external validation)
Level 2b	External validation of NTCP model in independent dataset with *random* split in two groups: one for NTCP model development and one for evaluating model performance (external validation)
Level 3	NTCP model with internal validation only
Level 4a	Multivariable NTCP model without internal and external validation
Level 4b	Univariable NTCP model

still vary considerably, even when the primary tumour site, tumour stage and prescribed target dose are similar. Therefore, an essential step in the proper selection of patients for proton therapy requires a direct comparison of the most optimal photon plan with the most optimal proton plan in every single patient (Figure 12.2). It should be emphasised that a fair comparison between photons and protons can be guaranteed only if clinical target volumes (CTVs), prescribed total dose and dose constraints for the CTV coverage are similar. In addition, constraints for dose–volume parameters of OARs used for plan optimisation should be similar and should follow the equation of the selected NTCP models, also referred to as model-based or NTCP-guided plan optimisation.[6,20] Eventually, the plan comparison provides information on the dose distributions in the targets and OARs for the photon and proton plans and, when subtracted, provides information on Δdose. In this way, a Δdose print can be produced (Figure 12.3).

Assessment of the Reduction in Normal Tissue Complication Probability

The third and final step in model-based selection is to translate Δdose into ΔNTCP by using the dose-volume parameters of the photon and proton plan, together with the other predictors of the selected NTCP models (Figure 12.2). By integrating the results of the individual plan comparison into the selected NTCP models, the corresponding NTCP values can be assessed and the ΔNTCP can be calculated. For tumour types in which multiple toxicity end points are relevant, a ΔNTCP print can be constructed (Figure 12.3), which can then be used for selection of patients for either photons or protons. A ΔNTCP print is a graphical representation of the ΔNTCP values of all toxicities taken into consideration for model-based selection.

It should be noted that NTCP models obtained from patient cohorts treated with photons may be different when used among patients treated with protons. Complication risk estimates based on photon-based NTCP models may lead to either over- or underestimation

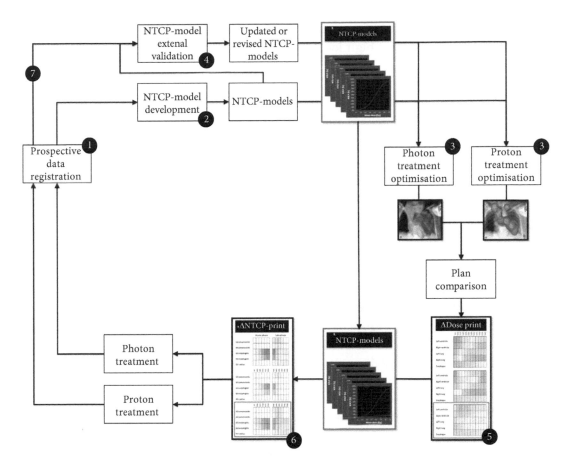

FIGURE 12.3 Rapid learning health care (RLHC) system. The prospective data registration ❶ is used for NTCP model development ❷, which can be used directly for model-based dose optimisation ❸ or after model update or model revision (external validation) ❹. Δdose is assessed by the plan comparison, and if multiple dose parameters are relevant, a Δdose print ❺ is produced. The NTCP models are then used to translate the Δdose print into the ΔNTCP print ❻, which is in fact the biomarker that can be used to decide which patients qualify for proton therapy. In addition, all patients are treated with protons or photons and are included in the same prospective data registry ❶. Validation of the benefit of protons is performed by externally validating the NTCP model among those treated with protons ❼.

of these risks when applied for proton therapy. Overestimation may occur due to the phenomenon of unidentified organs at risk (UFOs). An important problem in the development of NTCP models is multicollinearity, that is, strong correlations between dose-volume parameters of various OAR, resulting in statistical competition between these parameters in the model and resulting in removing dose-volume parameters from the model, while they are actually important. As with IMPT, the problem of redistribution of dose is much lower than with IMRT, and it is more likely that these UFOs are spared, resulting in lower observed complication rates than expected based on the IMRT-based models. Conversely, underestimation of complication risks may occur in case of the presence of a higher relative biological effectiveness (RBE) for normal tissue. This problem may be particularly relevant for complications related to high local dose (e.g. for the central nervous system [CNS]) rather than for parallel organised OAR (e.g. lungs and parotid glands).

Thresholds for Model-Based Selection

The Netherlands Society for Radiotherapy and Oncology (NVRO) established thresholds for ΔNTCP that are considered clinically relevant. It was decided to use the Common Terminology Criteria for Adverse Events (CTCAE) version 4.0 as the basis for the assessment of radiation-induced side effects. The main reasons to use this toxicity classification system are that it is generally accepted as the current standard in the oncology community, it allows for scoring side effects independent of the treatment modality used, and it is commonly applied in clinical trials. In addition, the grading corresponds with the clinical relevance of side effects with higher grading in case of more impact on daily living, hospitalisation and urgency of medical intervention. The decision about which threshold should be used for model-based selection is not straightforward and is arbitrary. For example, the relevance of a grade 2 acute side effect that recovers over time may be different than a grade 2 late side effects that persists until death or progresses over time. However, to enhance clinical utility, it was decided to start with a pragmatic solution, establishing uniform thresholds for ΔNTCP depending on toxicity grading (Table 12.2).

TABLE 12.2   Threshold for ΔNTCP to Quality Patients for Proton Therapy

Grade of Toxicity	Description According to CTCAE v4.0	ΔNTCP Threshold for Model-Based Selection
Grade 1	Mild; asymptomatic or mild symptoms; clinical or diagnostic observations only; intervention not indicated	No indication
Grade 2	Moderate; minimal, local, or non-invasive intervention indicated; limiting age-appropriate instrumental ADLs[a]	ΔNTCP ≥10%
Grade 3	Severe or medically significant but not immediately life-threatening; hospitalisation or prolongation of hospitalisation indicated; disabling; limited self-care ADLs[b]	ΔNTCP ≥5%
Grade 4	Life-threatening consequences; urgent intervention indicated	ΔNTCP ≥2%
Grade 5	Death related to adverse event	

Notes: In case multiple NTCP-models are used for model-based selection, the following criteria are used: in case of multiple grade ≥2 toxicities, the ΣΔNTCP for these toxicities should be ≥15% (with only grade ≥2 toxicities with ΔNTCP >5% taken into account); in case of multiple grade ≥3 toxicities, the ΣΔNTCP for these toxicities should be ≥7.5% (with only grade ≥3 toxicities with ΔNTCP >3.75% taken into account), and in case of multiple grade ≥4 toxicities, the ΣΔNTCP for these toxicities should be ≥3% (given the severity of these side effects, all toxicities count here).

[a] Instrumental ADLs (activities of daily living) refer to preparing meals, shopping for groceries or clothes, using the telephone, managing money and so on.

[b] Self-care ADL refer to bathing, dressing and undressing, feeding self, using the toilet, taking medications and not being bedridden.

Model-Based Selection in Randomised Controlled Trials

As pointed out recently by Widder et al., RCTs investigating the potential benefits of protons against photons when aiming at reduction of side effects with equivalent tumour control should be enriched using model-based selection as well before study entry.[13] In fact, ΔNTCP prints can be considered a biomarker for identifying patients that have the potential to benefit from this technology as patients meet the two previously mentioned principal requirements: that is, protons yield a dose benefit (ΔDose) over photons, and this Δdose is expected to translate into a clinical benefit in terms of reduction of side effects (ΔDose).[1] In this respect, an RCT is not the most suitable study design to answer the question, Who will benefit from proton therapy? But an RCT will yield an all-or-nothing result. Especially if the proportion of patients that will benefit from proton in terms of ΔNTCP is small (e.g. 20%), the overall outcome of an RCT may be negative, while in fact this small subset of patients may yield a major benefit from protons. In contrast, when the proportion of patients with ΔNTCP prints qualifying for protons is relatively large (e.g. 50%), the overall outcome of this RCT may be positive, while in fact the benefit is restricted to only 50% of the patient population included. Model-based selection to enrich populations included in RCT may thus contribute to a more cost-effective introduction of protons into routine clinical practice. In addition, population enrichment implies higher expected differences in primary end points and thus would require a lower number of patients, decreasing the costs and thus increasing the feasibility of successful RCTs.[13]

MODEL-BASED VALIDATION

Model-based selection is increasingly accepted as an appropriate way to select patients for proton therapy when aiming at prevention of radiation-induced side effects with equivalent loco-regional control.[21–25] In addition, clinical validation of the added value of protons in reducing radiation-induced side effects remains crucial, even when RCTs are not considered feasible to perform. As such, a model-based approach can be applied for validation as well. The basic principle of model-based validation is to validate the performance of photon-based NTCP models among patients treated with protons (Figure 12.2).

In a prospective model-based validation study, patients are included only if they meet the predefined criteria for ΔNTCP, as mentioned in the Table 12.1. The main objective in a model-based validation study is to test the hypothesis that proton therapy decreases the complication rates compared to photons and is confirmed if the observed complication rate with protons is significantly lower than the average NTCP values of the photon plans. In such a design, the photon plan of each individual patient that was used to assess ΔNTCP is used as a control. Like in an RCT, a power analysis to calculate the number of required patients based on the expected ΔNTCP of the study cohort is performed prior to the study.

In addition, further testing of NTCP model performance in patients treated with protons, that is, to test the observed complication rate against NTCP based on the proton plans, may reveal model deficiencies and possibly trigger model updates. In general, validation of model performance can be tested using data from a subsequent cohort of patients treated with the same radiation technique later in time (referred to as temporal validation);

using data prospectively collected in different centres (referred to as geographical validation); and/or using data collected in patients treated with another radiation technology, for which we propose the term 'technology validation'.[26,27]

When model performance of photon-based NTCP models is tested in a patient cohort treated with protons, the ideal scenario would be that the original NTCP model fits well and that no adjustments to the NTCP model are needed when applied in patients treated with protons. Alternatively, the risk of radiation-induced side effects using the photon-based NTCP model may be underestimated when applied in patients treated with protons, for example, in case the RBE of protons for that side effect is higher than expected. In contrast, the probability of side effects may also be overestimated, for example, in the case of UFOs. UFOs refer to the phenomenon that not all OARs for a given side effect have been identified in a modelling study, for example, due to the very low power of many studies reporting on NTCP model development or due to multicollinearity problems of highly correlated, competing predictors.

Previous studies indeed showed that model performance of NTCP models may change when tested in patient cohorts treated with other radiation techniques. For example, Beetz et al. found that an NTCP model for moderate to severe xerostomia in head and neck cancer patients treated with 3D-CRT performed less accurately when applied among patients treated with IMRT.[14-16] In this regard, it should be noted that it is unlikely that one single NTCP model will work in all settings due to differences in patient selection, combination with other treatment modalities, radiotherapy treatment planning objectives, image-guidance and adaptive radiotherapy procedures and supportive care management protocols. However, lower performance of NTCP models in different patient cohorts does not necessarily imply that this NTCP model should be completely discarded as it may still contain valuable information on which predictors are considered relevant in relation to the development of a given side effect. In this case, an attractive alternative approach is to perform different levels of model updating.[28] Recently, Vergouwe et al. reported on the so-called closed testing procedure, a strategy for selecting the most appropriate updating method that considers the balance between the amount of evidence for updating an NTCP model and the danger of overfitting. In such an approach, the performance of the original NTCP model, with the same intercept, predictors and corresponding regression coefficients, is tested in the patient cohort treated with protons.[29] The outcome of the closed testing procedure may reveal that no adjustments are needed, or that the model needs to be updated, including recalibration-in-the-large (i.e. re-estimation of the model intercept only), recalibration (i.e. re-estimation of the intercept and slope), or model revision (re-estimation of all regression coefficients of the original model).[30-31] Consequently, if the closed testing procedure indicates that updating the NTCP model is required when used among patients treated with protons, NTCP models for plan optimisation and the equations used to produce NTCP prints may be different for photons than those used for protons. From a methodological point of view, this is not a problem because, for model-based optimisation, selection and technology validation, different models for photons and protons can be used simultaneously.

CLINICAL IMPLEMENTATION AND MAINTENANCE

The model-based approach requires continuous prospective assessment of radiation-induced toxicity over time, in which every patient treated with any form of radiotherapy is included. By linking this data to dose distributions, a comprehensive set of multivariable NTCP models can be developed, updated, revised and/or extended to obtain the most optimal radiotherapy treatment plans for each technology used among patients, which then are evaluated with in the same prospective data registration program. Such a system is also referred to as a rapid learning health care (RLHC) system (Figure 12.3) and provides major opportunities for continuous improvement and subsequent evaluation of the beneficial effects of new radiation technology.[32-34] In the Netherlands, the introduction of proton therapy will be realised within the framework of RLHC systems on a national basis.

CONCLUSION

Model-based selection, using ΔNTCP-prints as biomarkers, is an evidence-based method for identifying patients for proton therapy who are expected to benefit most from this new technology in terms of reduction of radiation-induced side effects with equivalent tumour control. The model-based approach can also be used for validation of new radiation technologies aimed at the reduction of side effects and is based on the external validation of multivariable NTCP models in patients treated with the new technique. When performing RCTs, model-based enrichment is required to ensure a fair generalisability and a proper and cost-effective translation into routine clinical practice. The model-based approach can be used successfully only if it is integrated in an RLHC system.

REFERENCES

1. Langendijk JA, Lambin P, De Ruysscher D, Widder J, Bos M, Verheij M. Selection of patients for radiotherapy with protons aiming at reduction of side effects: The model-based approach. *Radiother Oncol.* 2013;107(3):267–273.
2. Mishra MV, Aggarwal S, Bentzen SM, Knight N, Mehta MP, Regine WF. Establishing evidence-based indications for proton therapy: An overview of current clinical trials. *Int J Radiat Oncol Biol Phys.* 2017;97(2):228–235.
3. Combs SE. Does proton therapy have a future in CNS tumors? *Curr Treat Options Neurol.* 2017;19(3):12. doi:10.1007/s11940-017-0447-4.
4. Glimelius B, Montelius A. Proton beam therapy—Do we need the randomized trials and can we do them? *Radiother Oncol.* 2007;83(2):105–109.
5. Bentzen SM. Randomized controlled trials in health technology assessment: Overkill or overdue? *Radiother Oncol.* 2008;86(2):142–147.
6. Kierkels RG, Visser R, Bijl HP, Langendijk JA, van 't Veld AA, Steenbakkers RJ, Korevaar EW. Multicriteria optimization enables less experienced planners to efficiently produce high quality treatment plans in head and neck cancer radiotherapy. *Radiat Oncol.* 2015;10:87.
7. Kierkels RG, Korevaar EW, Steenbakkers RJ, Janssen T, van't Veld AA, Langendijk JA, Schilstra C, van der Schaaf A. Direct use of multivariable normal tissue complication probability models in treatment plan optimisation for individualised head and neck cancer radiotherapy produces clinically acceptable treatment plans. *Radiother Oncol.* 2014;112(3):430–436.

8. Van der Laan HP, Gawryszuk A, Christianen ME, Steenbakkers RJ, Korevaar EW, Chouvalova O, Wopken K, Bijl HP, Langendijk JA. Swallowing-sparing intensity-modulated radiotherapy for head and neck cancer patients: Treatment planning optimization and clinical introduction. *Radiother Oncol.* 2013;107(3):282–287.
9. Christianen ME, van der Schaaf A, van der Laan HP, Verdonck-de Leeuw IM, Doornaert P, Chouvalova O, Steenbakkers RJ et al. Swallowing sparing intensity modulated radiotherapy (SW-IMRT) in head and neck cancer: Clinical validation according to the model-based approach. *Radiother Oncol.* 2016;118(2):298–303.
10. Taylor PA, Kry SF, Alvarez P, Keith T, Lujano C, Hernandez N, Followill DS. Results from the imaging and radiation oncology core Houston's anthropomorphic phantoms used for proton therapy clinical trial credentialing. *Int J Radiat Oncol Biol Phys.* 2016;95(1):242–248.
11. Das IJ, Andersen A, Chen ZJ, Dimofte A, Glatstein E, Hoisak J, Huang L et al. State of dose prescription and compliance to international standard (ICRU-83) in intensity modulated radiation therapy among academic institutions. *Pract Radiat Oncol.* 2017;7(2):e145–e155.
12. Fairchild A, Langendijk JA, Nuyts S, Scrase C, Tomsej M, Schuring D, Gulyban A, Ghosh S, Weber DC, Budach W. Quality assurance for the EORTC 22071–26071 study: Dummy run prospective analysis. *Radiat Oncol.* 2014;9:248.
13. Widder J, van der Schaaf A, Lambin P, Marijnen CA, Pignol JP, Rasch CR, Slotman BJ, Verheij M, Langendijk JA. The quest for evidence for proton therapy: Model-based approach and precision medicine. *Int J Radiat Oncol Biol Phys.* 2016;95(1):30–36.
14. Beetz I, Schilstra C, van der Schaaf A, van den Heuvel ER, Doornaert P, van Luijk P, Vissink A et al. NTCP models for patient-rated xerostomia and sticky saliva after treatment with intensity modulated radiotherapy for head and neck cancer: The role of dosimetric and clinical factors. *Radiother Oncol.* 2012;105(1):101–106.
15. Beetz I, Schilstra C, van Luijk P, Christianen ME, Doornaert P, Bijl HP, Chouvalova O, van den Heuvel ER, Steenbakkers RJ, Langendijk JA. External validation of three dimensional conformal radiotherapy based NTCP models for patient-rated xerostomia and sticky saliva among patients treated with intensity modulated radiotherapy. *Radiother Oncol.* 2012;105(1):94–100.
16. Beetz I, Schilstra C, Burlage FR, Koken PW, Doornaert P, Bijl HP, Chouvalova O et al. Development of NTCP models for head and neck cancer patients treated with three-dimensional conformal radiotherapy for xerostomia and sticky saliva: The role of dosimetric and clinical factors. *Radiother Oncol.* 2012;105(1):86–93.
17. Christianen ME, Schilstra C, Beetz I, Muijs CT, Chouvalova O, Burlage FR, Doornaert P et al. Predictive modelling for swallowing dysfunction after primary (chemo)radiation: Results of a prospective observational study. *Radiother Oncol.* 2012;105(1):107–114.
18. Wopken K, Bijl HP, van der Schaaf A, van der Laan HP, Chouvalova O, Steenbakkers RJ, Doornaert P et al. Development of a multivariable normal tissue complication probability (NTCP) model for tube feeding dependence after curative radiotherapy/chemo-radiotherapy in head and neck cancer. *Radiother Oncol.* 2014;113(1):95–101.
19. Steyerberg EW, Vedder MM, Leening MJ, Postmus D, D'Agostino RB Sr, Van Calster B, Pencina MJ. Graphical assessment of incremental value of novel markers in prediction models: From statistical to decision analytical perspectives. *Biom J.* 2015;57(4):556–570.
20. Kierkels RG, Wopken K, Visser R, Korevaar EW, van der Schaaf A, Bijl HP, Langendijk JA. Multivariable normal tissue complication probability model-based treatment plan optimization for grade 2–4 dysphagia and tube feeding dependence in head and neck radiotherapy. *Radiother Oncol.* 2016;121(3):374–380.
21. Azria D, Lapierre A, Gourgou S, De Ruysscher D, Colinge J, Lambin P, Brengues M et al. Data-based radiation oncology: Design of clinical trials in the toxicity biomarkers era. *Front Oncol.* 2017;7:83. doi:10.3389/fonc.2017.00083.

22. Stick LB, Yu J, Maraldo MV, Aznar MC, Pedersen AN, Bentzen SM, Vogelius IR. Joint estimation of cardiac toxicity and recurrence risks after comprehensive nodal photon versus proton therapy for breast cancer. *Int J Radiat Oncol Biol Phys*. 2017;97(4):754–761.
23. Blanchard P, Wong AJ, Gunn GB, Garden AS, Mohamed AS, Rosenthal DI, Crutison J et al. Toward a model-based patient selection strategy for proton therapy: External validation of photon-derived normal tissue complication probability models in a head and neck proton therapy cohort. *Radiother Oncol*. 2016;121(3):381–386.
24. Stromberger C, Cozzi L, Budach V, Fogliata A, Ghadjar P, Wlodarczyk W, Jamil B, Raguse JD, Böttcher A, Marnitz S. Unilateral and bilateral neck SIB for head and neck cancer patients: Intensity-modulated proton therapy, tomotherapy, and RapidArc. *Strahlenther Onkol*. 2016;192(4):232–239.
25. Chargari C, Goodman KA, Diallo I, Guy JB, Rancoule C, Cosset JM, Deutsch E, Magne N. Risk of second cancers in the era of modern radiation therapy: Does the risk/benefit analysis overcome theoretical models? *Cancer Metastasis Rev*. 2016;35(2):277–288. doi:10.1007/s10555-016-9616-2.
26. Altman DG, Royston P. What do we mean by validating a prognostic model? *Stat Med*. 2000;19(4):453–473.
27. Siontis GC, Tzoulaki I, Castaldi PJ, Ioannidis JP. External validation of new risk prediction models is infrequent and reveals worse prognostic discrimination. *J Clin Epidemiol*. 2015;68(1):25–34.
28. Toll DB, Janssen KJ, Vergouwe Y, Moons KG. Validation, updating and impact of clinical prediction rules: A review. *J Clin Epidemiol*. 2008;61(11):1085–1094.
29. Vergouwe Y, Nieboer D, Oostenbrink R, Debray TP, Murray GD, Kattan MW, Koffijberg H, Moons KG, Steyerberg EW. A closed testing procedure to select an appropriate method for updating prediction models. *Stat Med*. 2016;36(28):4529–4539.
30. Steyerberg EW, Borsboom GJ, van Houwelingen HC, Eijkemans MJ, Habbema JD. Validation and updating of predictive logistic regression models: A study on sample size and shrinkage. *Stat Med*. 2004;23(16):2567–2586.
31. Janssen KJ, Moons KG, Kalkman CJ, Grobbee DE, Vergouwe Y. Updating methods improved the performance of a clinical prediction model in new patients. *J Clin Epidemiol*. 2008;61(1):76–86.
32. Lambin P, Zindler J, Vanneste B, van de Voorde L, Jacobs M, Eekers D, Peerlings J et al. Modern clinical research: How rapid learning health care and cohort multiple randomised clinical trials complement traditional evidence based medicine. *Acta Oncol*. 2015;54(9):1289–1300.
33. Lambin P, Roelofs E, Reymen B, Velazquez ER, Buijsen J, Zegers CM, Carvalho S et al. Rapid learning health care in oncology—An approach towards decision support systems enabling customised radiotherapy. *Radiother Oncol*. 2013;109(1):159–164.
34. Lustberg T, Bailey M, Thwaites DI, Miller A, Carolan M, Holloway L, Rios Velazquez E, Hoebers F, Dekker A. Implementation of a rapid learning platform: Predicting 2-year survival in laryngeal carcinoma patients in a clinical setting. *Oncotarget*. 2016;7(24):37288–37296.

CHAPTER 13

Overview of Carbon-Ion Radiotherapy in Japan

Hirohiko Tsujii and Tadashi Kamada

CONTENTS

THE PRIMARY PRINCIPLE OF radiotherapy (RT) lies in precise dose localization in the target with minimizing damage to the surrounding normal tissues. This is particularly so when we note the fact that the photon energy on the order of megavoltage (MV) contributed significantly to the improvement of treatment outcome. High-precision RT, such as intensity-modulated photon RT (IMRT) and stereotactic body RT (SBRT), was developed in the late twentieth century. At around the same time, charged particle RT such as proton beam therapy (PBT) and carbon-ion RT (CIRT) progressed rapidly, which consequently improved the efficacy of RT. CIRT is a novel method offering precise dose localization and high radiobiological effectiveness (RBE) to overcome radioresistant, hypoxic tumours that are difficult to control with low linear energy transfer (LET) photon beams [1].

Historically, the Lawrence Berkeley National Laboratory (LBNL) in the United States pioneered the medical application of charged particles. They embarked on PBT in 1954, followed by helium ions in 1957 and neon ions in 1975 [2,3]. However, LBNL terminated clinical programs in 1992 because of funding difficulties. As if the baton were passed across the Pacific, the National Institute of Radiological Sciences (NIRS) in Japan started CIRT in 1994 [4–6], followed by GSI in Germany [7] and several other facilities around the world.

CHARACTERISTICS OF CARBON IONS FOR CLINICAL APPLICATION

Among several types of ion species, carbon-ion beams have attracted strong interest because they exhibit Bragg peak in the body, followed by rapid dose falloff beyond the peak, enabling delivery of sufficient dose to the target volume while minimizing the dose to the surrounding tissues. Furthermore, compared with proton beams, carbon-ion beams cause less scattering and range-struggling in the body, providing smaller lateral penumbra in dose distribution. In the region distal to the rapid falloff, the tail develops from fragmentations due to nuclear interaction with atoms in the body. However, the influence from this tail is clinically negligible because the fragments deliver only low dose with predominantly low LET components.

The difference in dose distribution between carbon ions and protons is observed in their pattern of ionization density, which increases with depth along the beam path in carbon-ion beams but not in proton beams (Figure 13.1). This is indeed a beneficial property for treatment of deep-seated tumours. In fact, the ionization density follows closely the LET that is proportional to RBE within the clinically available range. This means that, as the RBE of carbon-ion beams increases with depth, the tumour is irradiated in the peak region with both physically and biologically higher dose than in the plateau region.

In connection with high RBE, there are other important differences between proton beams and carbon-ion beams. Carbon ions offer potential radiobiological advantages such as: (1) reduced oxygen enhancement ratio (OER), which makes them highly effective for photon-resistant, hypoxic tumours; (2) reduced repair capacity; (3) decreased cell-cycle dependence; and (4) possibly higher immunological responses owing to predominant occurrence of double-strand DNA breaks in the body. It is therefore claimed that CIRT

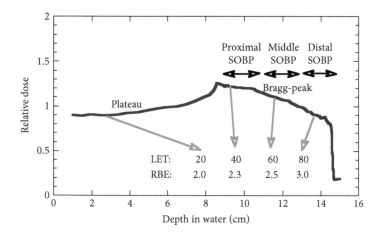

FIGURE 13.1   LET and RBE in the depth-dose profile of carbon-ion beams (290MeV, SOBP=60mm). The LET (hence RBE) increases toward the end of the beam range.

could provide a potential breakthrough for treatment of various cancers that are resistant to photons and protons.

There is a rationale for hypo-fractionated RT in CIRT. Experimental evaluation of therapeutic gain for fractionated CIRT using the tumour growth delay and crypt survival assays has demonstrated that increasing their fraction dose tended to lower the RBE for both the tumour and normal tissues; however, the RBE for the tumour did not decrease as rapidly as the RBE for the normal tissues [8,9]. These results prove that the therapeutic ratio increases rather than decreases, even though the fraction dose is increased.

FACILITIES FOR CARBON-ION RADIOTHERAPY IN JAPAN

Almost 70 facilities for PBT and CIRT are in operation around the world (Table 13.1). In Japan, there are a total of five CIRT facilities in operation, including four facilities for CIRT alone, one facility for both CIRT and PBT, and two new facilities are under construction (Table 13.2). CIRT was started in 1994 at NIRS using the world's first accelerator complex (HIMAC: Heavy Ion Medical Accelerator in Chiba) dedicated to cancer therapy. Following HIMAC, Hyogo Ion Beam Medical Center (HIBMC) was established in 2001, which was the first dual CIRT/PBT centre in the world. Based on technical development at NIRS, a downsized facility was constructed at Gunma (GHMC: Gunma University Heavy Ion Medical Center) and Saga (SAGA HIMAT: Heavy Ion Medical Accelerator in Tosu), which started CIRT in 2010 and 2013, respectively. The fifth facility, started in 2015 in Kanagawa (i-ROCK: Ion-beam Radiation Oncology Center in Kanagawa), is a unique facility for

TABLE 13.1 Particle Therapy Facilities in Operation in the World

		Proton	Carbon	P + C		Total	
West & Central Europe	UK	1			1	16	23.5%
	France	2			2		
	Germany	4		2	6		
	Italy	2		1	3		
	Sweden	2			2		
	Swiss	1			1		
	Austria			1	1		
East Europe	Czech	1			1	2	2.9%
	Poland	1			1		
	Russia	3			3	3	4.4%
Africa	S. Africa	1			1	1	1.5%
Asia	Japan	10	4	1	15	22	32.3%
	China	1	?	1	4		
	Korea	2			2		
	Taiwan	1			1		
North America	Canada	1			1	24	35.4%
	USA	23			23		
Total		56	6	6	68		100%

Source: PTCOG (2016).

TABLE 13.2   Carbon-Ion Radiotherapy Facilities in Japan

Facility	Start	Particle (Max Energy)	Treatment Room	Vendor
HIMAC (Chiba)	1994	C (430)	Carbon Broad beam 3 (H/V, H, V) Scanning 3 (H/V×2, Gantry)	4 Vendors (M+H+T+S)
HIBMC (Hyogo)	2001	C (375) P (230)	Carbon Broad beam 3 (H/V, H, 45°) Proton Broad beam 2 (Gantry×2)	Mitsubishi
GHMC (Gunma)	2010	C (430)	Carbon Broad beam 3 (H/V, H, V)	Mitsubishi
SAGA HIMAT (Saga)	2013	C (400)	Carbon Broad beam 2: H/V, H/45° Scanning 1: H/V	Mitsubishi
iROCK (Kanagawa)	2015	C (430)	Carbon Scanning 4 (H/V×2, H×2)	Toshiba
Yamagata Univ (Yamagata)	Under construction	C (430)	Carbon Scanning 2 (H, Gantry)	Toshiba
Osaka (Osak)	Under construction	C (430)	Carbon Scanning 3 (H, H/V, H/45°)	Hitachi

using only pencil beam scanning method. Two other facilities are under construction at Yamagata and Osaka, and both plan to employ the scanning method from the beginning.

Thus, the number of facilities for PBT and CIRT has rapidly increased in Japan. In 2003 we established a society named 'Japanese Clinical Study Group of Particle Therapy (JCPT)' for exchanging information on clinical aspects of charged particle therapy as well as therapy-related issues. Subsequently, a new study group named 'Japan Carbon-ion Radiation Oncology Study Group (J-CROS)', was organised by the 5 operating facilities. The purpose was to conduct multi-institutional clinical trials and obtain evidence regarding clinical superiority of CIRT over other treatments. Currently, five prospective trials are ongoing for tumours of the lung, liver, pancreas, prostate and rectum (post-operative recurrence).

CLINICAL RESULTS

As of June 2017, more than 10,000 patients with various types of malignancies have been treated at NIRS, and at other facilities the patient number has rapidly increased. Based on clinical outcomes of these patients, a panel of radiation oncologists, radiobiologists and medical physicists from the United States and Europe performed an external peer review for NIRS-CIRT. Detailed comparisons between the results of CIRT and most advanced photon therapy and PBT were then possible [10]. The evaluation also included technical developments such as the respiratory-gated irradiation technique for moving targets with implanting metal markers for tumour orientation (Figure 13.2) [11]. The status of development for pencil beam scanning and superconducting rotating gantry was also included for evaluation. Both of them are currently realised and are used in clinical practice at NIRS.

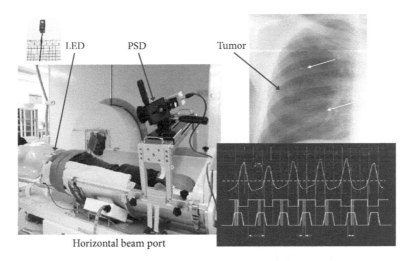

FIGURE 13.2 Respiratory gated irradiation system. Metal markers (arrow) are implanted in the lung for tumor orientation. Position sensitive detector (PSD) receives infrared rays from LED to generate respiratory wave.

Although there have been no randomised trials comparing CIRT with PBT or IMRT, clinical results so far demonstrate a definite advantage of CIRT over other modalities. It is also demonstrated that CIRT allows hypo-fractionated RT (Table 13.3). The average number of fractions is around 12 in three weeks, including a single fraction treatment for early stage lung cancer; two fractions for hepatocellular cancer; 12 fractions for prostate cancer; and 16 fractions for head/neck cancer, bone/soft-tissue cancer and several others.

Bone and Soft-Tissue Sarcoma

CIRT for unresectable bone/soft-tissue sarcomas has been fully covered by national insurance application since March 2017 in Japan. Paediatric cancer was approved for PBT.

Chordoma is a rare type of cancer originating from ectopic remnants of the embryonal notochord, which commonly develop in the sacrum and skull base. They are difficult to treat due to involvement of critical structures such as the brainstem, spinal cord, important nerves and bowels. In about 30%–40% of chordoma patients, the tumour eventually spreads or metastasizes to other parts of the body. Chordoma is diagnosed in about one in one million population per year. This means that about 300 patients are diagnosed with chordoma each year in the United States, about 700 in all of Europe and about 120 in Japan.

According to the materials presented at the committee meeting, Ministry of Health, Labor and Welfare in Japan, for evaluation of charged particle therapy, a total of 50 patients with skull base and upper cervical chordoma were collected by J-CROS. Most patients received 60.8 Gray (RBE) in 16 fractions over four weeks. Overall local control (LC) at 5 and 10 years was 82.4% and 80.0%, respectively, and overall survival (OS) at 5 and 10 years was 90.2% and 68.1%, respectively. The results at 10 years after CIRT were particularly superior to PBT or photon therapy.

TABLE 13.3   Dose Fractionations Used at NIRS

Site	Type of Tumor	Total Dose Gy (RBE)	Fractions	Week	Fraction Size Gy (RBE)
Head & Neck	Adenoca, ACC, MMM	57.6	16	4	3.6
		64.0	16	4	4.0
	Sarcoma	70.4	16	4	4.4
Skull base	Chordoma, Chondrosarcoma	60.8	16	4	3.8
Lung	Peripheral Stage I (T1-2N0M0)	50.0	1	1 day	50.0
		60.0	4	4 days	15.0
	Mediastinal lymph node	48.0	12	3	4.0
	Early central type (superficial)	54.0	9	3	6.0
	Early hilar type (T1-2N0M0)	68.4	12	3	5.7
	Locally advanced (T1-3 N1-2 M0)	72.0	16	4	4.5
Liver	HCC	48.0	2	2 days	24
	Metastasis of colorectal ca	58.0	1	1 day	–
Bone & Soft	Osteosarcoma	70.4	16	4	4.4
	Chordoma, Chondrosarcoma	67.2	16	4	4.2
	Spine, Paraspinal	64.0	16	4	4.0
Prostate	Low / Medium / High risk	57.6	16	4	3.6
		51.6	12	3	4.3
Pancreas	Locally advanced unresectable C-ions + GEM(1000 mg/m2) ×3	55.2	12	3	4.6
	Resectable: Pre-operative CIRT + GEM (1000 mg/m2) ×3	36.8	8	2	4.6
Rectum	Post-ope recurrence	73.6	16	4	4.6
	Re-irradaition of pelvic tumor	70.4	16	4	4.4
Eye	Choroidal melanoma	70.0	5	1	14.0
	Lacrimal galand ACC/Adenoca.	52.8	12	3	4.4
Abd lymph nodes	Lymph node metastasis	52.8	12	3	4.4

Sacral chordoma, which comprises about 50% of chordoma cases, is very difficult to control for a large tumour. PBT has been mostly performed following surgical resection in the United States, whereas CIRT has been used definitively as a sole treatment in Japan. Between 1996 and 2013, a total of 188 patients with unresectable sacral chordoma were treated at NIRS, delivering 67.2–73.6 Gray (RBE) in 16 fractions. The five-year LC, OS and disease-free survival rates were 77.2%, 81.1% and 50.3%, respectively, which is favourably compared with LC of 35%–50% after resection [12].

Osteosarcomas arising from the pelvis and vertebra are rare, representing 6.3% of all osteosarcoma patients. They are usually difficult to operate on and are resistant to X-rays. For this tumour, CIRT combined with chemotherapy offered superior results. The five-year LC and OS rates were 62% and 33%, respectively, which were better than the results after combined surgical resection and PBT [13].

Soft-tissue sarcomas develop from the soft tissues like fat, muscle, nerves, fibrous tissues, blood vessels, or deep skin tissues. They can be found in any part of the body, and most of them develop in the arms or legs. They also develop in the trunk, head and neck and retroperitoneum, which are the major candidates for CIRT. The five-year LC and five-year OS for unresectable retroperitoneal tumours ($n = 24$) was 69% and 50%, respectively [14]. Considering that most patients were not eligible for surgical resection and had high-grade sarcomas, these results are very promising.

Head and Neck Tumour

CIRT has provided favourable results for photon-resistant tumours such as mucosal malignant melanoma (MMM), adenocarcinoma, adenoid cystic carcinoma (ACC), mucoepidermoid carcinoma and other histologies. For treatment of ACC, CIRT as a sole treatment offered the most favourable results. A comparison of LC achieved with a total dose of 57.6 Gray (RBE) or 64.0 Gray (RBE) showed that late local recurrence was significantly lower in the higher-dose group.

The prognosis of MMM is grave. In the Surveillance, Epidemiology and End Results registry, the five-year OS of 815 MMM cases was 25.2%, and even in patients with operable tumours, the five-year OS was limited to 25%–46% [15]. When the patients were treated with CIRT alone, OS was not much improved because of the high rate of metastatic disease. This was successfully improved, however, by combined CIRT and chemotherapy. According to the analysis of 260 patients with unresectable MMM enrolled in J-CROS study, the five-year LC and five-year OS were improved to 72.3% and 44.6%, respectively [16]. Multivariate analysis showed that gross tumour volume and concurrent chemotherapy were significant prognostic factors for OS.

Non-Small-Cell Lung Cancer

For treatment of early-stage non-small cell lung cancer (NSCLC), 10–20 fractions are most commonly used in PBT. In contrast, the ultimate hypo-fractionated RT with only a single-fraction in one day is established based on a dose-escalation study at NIRS [17]. Since 2003 more than 200 patients have been treated with single-fraction irradiation, where a statistically significant difference in LC, in favour of the higher-dose group, is found between the patients receiving 36 Gray (RBE) or more and those receiving less than 36 Gray (RBE). In 20 patients irradiated with 48–50 Gray (RBE), the five-year LC, OS and progression-free survival were 95.0%, 69.2% and 60.0%, respectively. A single fraction dose of 50 Gray (RBE) resulted in excellent LC of nearly 100%.

For inoperable, locally advanced NCSLC, many studies addressed treatment outcome with concurrent chemoradiotherapy (CCRT), including RTOG 0117 trial and JCOG 0301 phase III trial. At NIRS, CIRT have been using 16 fractions, which was associated with manageable toxicity and encouraging LC rates. The two-year LC and OS rates in 62 patients were 93.1% and 51.9%, respectively [18].

Liver Cancer

The prognosis of hepatocellular carcinoma (HCC) is generally poor because the patients commonly have underlying liver disease and only 10%–20% of the cases are surgically resectable. For patients who are not candidates for resection, loco-regional therapies have been employed, such as radiofrequency ablation, transarterial chemoembolisation or SBRT.

Among them, RT has been generally difficult to perform because the liver is relatively radio-sensitive, and better understanding of the dose–response relationship is needed for tumour control and normal tissue toxicities. This has led to the use of charged particle therapy.

In CIRT, a four-fractionated regimen with 52.8 Gray (RBE) given in one week was established at NIRS, resulting in improved LC: the three- and five-year LC were 95.5% and 91.6%, respectively, with acceptable toxicities [19]. Subsequently, the total dose was increased to 60 Gray (RBE) in four fractions, which has become the standard fractionation for all CIRT facilities in Japan. At NIRS, an even more hypo-frationated regimen with two fractions in two days has been realised with scanning beam under respiratory-gated irradiation. This means that, compared with PBT using 10–20 fractions in most facilities, a larger number of HCC patients can be treated with CIRT, which should compensate for high medical cost.

Prostate Cancer

In CIRT for prostate cancer, excellent outcome for biochemical relapse-free survival has been reported for the high-risk group compared with photon or proton series [20]. At NIRS more than 2,000 patients with prostate cancer have been treated with a hypo-fractionated regimen of 12–16 fractions in 3–4 weeks [21]. This is contrary to PBT using longer regimen of 35–40 fractions in 7–8 weeks. Evaluation of PBT for prostate cancer is very controversial, because retrospective analysis of the U.S. Medicare database did not find a reduced toxicity for the patients treated with protons when compared with IMRT. The follow-up data has demonstrated higher survival rate after CIRT than other modalities, particularly in a high-risk prostate cancer. This should be due to high RBE effects of carbon ions.

For the past 10 years, NIRS has been treating prostate cancer, delivering 51.6 Gray (RBE) in 12 fractions over three weeks, by which even a reduction of late toxicities has been achieved. In the near future, a regimen of less than 12 fractions will be possible with a scanning beam because sparing of the urethra and rectum will become easier.

Pancreatic Cancer

The prognosis of pancreatic cancer (PC) is poor because many cases are not detected until the cancer has progressed to the stage where the tumour is unresectable. For those tumours which are advanced but still remain locally confined, IMRT has been employed in combination with chemotherapy, with a five-year survival rate of 10%–30%.

Since 20014, CIRT for such locally advanced unresectable pancreatic cancer (LAUPC) has been conducted at NIRS, based on prospective study with increasing the carbon dose and combined drug dose. Improved results were obtained with a regimen using 55.2 Gray (RBE) in 12 fractions plus concurrent gemcitabine of 1,000 mg/m^2 weekly: two-year OS was 40%–50% [22]. This should be the best ever obtained in LAUPC, when compared to the standard option (IMRT-gemcitabine) showing 10%–20% two-year OS.

It was recommended that NIRS should coordinate a confirmatory, preferentially multi-institutional trial, using the best schedule for LAUPC. We have therefore decided to establish an international collaboration to perform a randomised Phase 3 trial on 'Carbon Ion versus Photon Therapy for Pancreatic Cancer' (CIPHER-PC) trial among 4 institutions including UTSW (United States), NIRS (Japan), GHMC (Japan) and CNAO (Italy).

Locally Recurrent Rectal Cancer

Colorectal cancer is the second most common malignancy in Japan, and rectal cancer constitutes approximately one third of these cases. Although surgical technique has been improved, local recurrence still occurs in 4%–13%. Locally recurrent rectal cancer is generally resistant to conventional RT and is considered a good candidate for CIRT. Using combined IMRT and chemotherapy, five-year LC and OS in inoperable patients are less than 50% and 20%, respectively, while after CIRT, results have been improved: five-year LC and OS for 136 patients treated with 73.6 Gray (RBE) in 16 fractions were 93% and 59%, respectively [23]. There have been no available data for PBT to compare with these results.

Cervical Cancer

CIRT has been targeted primarily at potentially unresectable adenocarcinoma of the uterine cervix at NIRS. Preliminary results demonstrated that LC and OS achieved in 57 patients were superior to conventional photon therapy with or without chemotherapy. It is planned to increase the patient number and follow-up period in a prospective protocol with an upfront hypothesis, and with sufficient power to prove these results in order to provide convincing evidence.

Other Tumours

Breast cancer is the most common invasive cancer in women worldwide. PBT has been employed as post-operative irradiation of left breast cancer in order to reduce the dose to the heart and lung. In contrast, CIRT has been conducted to establish a 'breast-conserving treatment', for which carbon-ion irradiation of breast cancer without lumpectomy surgery has been conducted. The preliminary results demonstrate the benefit of CIRT for local control.

Other tumours that could benefit from CIRT include oesophageal cancer, eye tumours such as malignant melanoma and lachrymal gland tumour, renal cancer and metastatic tumours that are locally confined.

CONCLUSION

Coupled with the opportunity for substantially abbreviated treatment of common cancer, updated experiences have shown that CIRT offered superior results in: (1) bone and soft-tissue sarcomas developing in the skull base, head/neck, pelvis, spine/paraspinal regions and retroperitoneum; (2) histologically non-SCC type of tumours developing in the head/neck and any other sites; (3) locally advanced or early stage tumours of the lung, liver, prostate and pancreas; (4) metastatic or recurrent tumours such as locally recurrent rectal cancer as well as locally confined advanced lesions with non-SCC histology; and (5) those tumours which could benefit from achieving definite local control.

Among them, bone and soft-tissue sarcomas are currently approved by the national health insurance system in Japan. Head and neck cancer is expected to be the next approval. In order to obtain the evidence of clinical efficacy, under the initiative by J-CROS, we are running prospective studies on five types of tumours including stage I NSCLC, HCC, pancreatic cancer (LAUPC), high-risk prostate cancer and pelvic recurrent rectal cancer.

The results of CIRT for LAUPC were excellent and created worldwide interest in performing an international comparative randomised study (CIPHER-PC trial).

In CIRT a significant reduction in overall treatment time and fractions has been obtained with minor toxicities, such as single-fraction RT for early-stage lung cancer, single or two-fraction RT for liver cancer and 12-fraction RT for prostate cancer. Also, for other tumour sites, 16 or even smaller fractions have been sufficient. Therefore, the CIRT facility can be operated more efficiently, permitting treatment of a larger number of patients than is possible with PBT over the same period of time. Further reduction of fraction numbers would increase cost utility and patient throughput, facilitating the use of CIRT for more patients and in more centres.

REFERENCES

1. Chen, G.T.Y., Castro, J.R., Quivey, J.M. 1981. Heavy charged particle radiotherapy. *Ann Rev Biophys Bioeng* 10:499–529.
2. Castro, J.R. 1995. Results of heavy ion radiotherapy. *Radiat Environ Biophys* 34:45–48.
3. Schulz –Ertner, D., Tsujii, H. 2007. Particle radiation therapy using proton and heavier ion beams. *J Clin Oncol* 25:953–964.
4. Tsujii, H., Kamada, T., Baba, M. et al. 2008. Clinical advantages of carbon-ion radiotherapy. *New J Phys* 10:1367–2630.
5. Tsujii, H., Kamada, T. 2012. A review of update clinical results of carbon-ion radiotherapy. *Jpn J Clin Oncol* 42:670–685.
6. Tsujii, H., Kamada, T., Shirai, T. et al. (Eds.). 2014. *Carbon-Ion Radiotherapy: Principles, Practices and Treatment Planning.* Springer, Tokyo, Japan.
7. Kraft, G. 2000. Tumor therapy with heavy charged particles. *Prog Part Nucl Phys* 45:S473–S544.
8. Ando, K., Koike, S., Uzawa, A. et al. 2005. Biological gain of carbon-ion radiotherapy for the early response of tumor growth delay and against early response of skin reaction in mice. *J Radiat Res* 46:51–57. doi:10.1269/jrr.46.51.
9. Yoshida, Y., Ando, K., Murata, K. et al. 2015. Evaluation of therapeutic gain for fractionated carbon-ion radiotherapy using the tumor growth delay and crypt survival assays. *Radiother Oncol* 117:351–357. doi:10.1016/j.radonc.2015.09.027.
10. Kamada, T., Tsujii, H., Debus, J. et al. 2015. Carbon ion radiotherapy in Japan: An assessment of 20 years of clinical experience. *Lancet Oncol* 16(2):e93–e100.
11. Minohara, S., Kanai, T., Endo, M. et al. 2000. Respiratory gated irradiation system for heavy-ion radiotherapy. *Int J Radiat Oncol Biol Phys* 47:1097–1103.
12. Imai, R., Kamada, T., Araki, N. et al. 2016. Carbon ion radiation therapy for unresectable sacral chordoma: An analysis of 188 cases. *Int J Radiat Oncol Biol Phys* 95:322–327.
13. Imai, R., Kamada, T. 2016. Carbon-ion radiotherapy for unresectable osteosarcoma of the trunk. In Ueda, T. and Kawai, A. (Eds.). *Osteosarcoma.* Springer, Tokyo, Japan. doi:10.1007/978-4-431-55696-1_18.
14. Serizawa, I., Kagei, K., Kamada, T. et al. 2009. Carbon ion radiotherapy for unresectable retroperitoneal sarcomas. *Int J Radiat Oncol Biol Phys* 75:1105–1110.
15. Jethanamest, D., Vita, PM., Sikora, A. et al. 2011. Predictors of survival in mucosal melanoma of the head and neck. *Ann Surg Oncol* 18:2748–2756.
16. Koto, M., Demizu, Y., Saitoh, J. et al. 2017. Multicenter study of carbon-ion radiation therapy for mucosal melanoma of the head and neck. *Int J Radiat Oncol Biol Phys* 97:1054–1060.
17. Yamamoto, N., Miyamoto, T., Nakajima, M. et al. 2017. A dose escalation clinical trial of single-fraction carbon ion radiotherapy for peripheral Stage I non-small cell lung cancer. *J Thor Oncol* 12:673–680.

18. Takahashi, W., Nakajima, M., Yamamoto, N. et al. 2015. A prospective nonrandomized phase I/II study of carbon ion radiotherapy in a favorable subset of locally advanced non-small cell lung cancer. *Cancer* 121:1321–1327.
19. Kasuya, G., Yasuda, S., Tsuji, H. et al. 2017. Progressive hypofractionated carbon-ion radiotherapy for hepatocellular carcinoma: Combined analyses of 2 prospective trials. *Cancer.* doi:10.1002/cncr.30816.
20. Shioyama, Y., Tsuji, H, Suefuji, H. et al. 2014. Particle radiotherapy for prostate cancer. *Urol* 22:33–39.
21. Nomiya., Tsuji, H., Maruyama, K. et al. 2014. Phase I/II trial of definitive carbon ion radiotherapy for prostate cancer: Evaluation of shortening of treatment period to 3 weeks. *Br J Cancer* 110:2389–2395.
22. Shinoto, M., Yamada, S., Terashima, K. et al. 2016. Carbon ion radiation therapy with concurrent gemcitabine for patients with locally advanced pancreatic cancer. *Int J Radiat Oncol Biol Phys* 95:498–504.
23. Yamada, S., Kamada, T., Ebner, D. et al. 2016. Carbon-ion radiation therapy for pelvic recurrence of rectal cancer. *Int J Radiat Oncol Biol Phys* 96:93–101.

CHAPTER 14

Particle Therapy Clinical Trials

*Perspectives from Europe, Japan
and the United States*

Cai Grau, Damien Charles Weber, Johannes A. Langendijk,

James D. Cox, Tadashi Kamada and Hirohiko Tsujii

CONTENTS

T HE OBJECTIVE OF THIS chapter is to provide an update on the current status of clinical research with respect to particle therapy trials in Europe, the United States and Japan. This contribution also allows the reader to have a clearer vision of the strategies already developed and those that are being developed in these three regions of the world regarding the design of clinical trials and their main targets.

THE PERSPECTIVE FROM EUROPE

The need to generate clinical evidence for particle therapy is extremely important to the radiation oncology community. Most European countries have a high degree of public coverage of health care, and thus have also a very regulated system for investments in new and costly technology. While most agree that particle therapy offers great opportunities to further improve the therapeutic ratio of radiotherapy, there is widespread discussion regarding lack of evidence for this treatment modality for a wide range of potential indications.

There have been several European networks working in the field of particle therapy, including Union of Light Ion Centres in Europe (ULICE), the European Network for Light Ion Hadron Therapy (ENLIGHT) and European Particle Therapy Network (EPTN) [1]. The two former were funded by EU grants and have contributed significantly to the science and early clinical development of particle therapy in Europe. ULICE was a four-year project set up by 20 leading European research organisations, including two leading European industrial partners, to respond to the need for greater access to hadron-therapy facilities for particle therapy research. The project is built around three pillars: Joint Research Activities, Networking and Transnational access. The ULICE project ended in 2014, leaving a substantial contribution of reports and white papers in the public domain. The ENLIGHT network was established in 2002 to coordinate European efforts in hadron therapy, and today has more than 300 participants from 20 European countries (see Chapter 20). While the funding period of ENLIGHT has ended, the network continues to meet annually for a plenary meeting and educational sessions.

The current European need for clinical collaboration, including clinical trials, is facilitated by the EPTN, which was initiated in 2015. The aim is to promote clinical and research collaboration between the rapidly increasing numbers of European particle therapy centres and to ensure that particle therapy becomes integrated in the overall radiation oncology community. Virtually all major European centres interested in particle therapy, either operational, under construction or planned, as well as two research organisations (EORTC, CERN), are represented in EPTN. In 2017, the EPTN became an official task force of European Society for Radiotherapy & Oncology (ESTRO). There is currently a total of seven thematic groups (work packages):

1. Clinical trials and data registry

2. Dose assessment, quality assurance, dummy runs, technology inventory

3. Education

4. Image guidance in particle therapy

5. TPS in particle therapy

6. Radiobiology, RBE

7. Health economy

The vision and scope of EPTN within clinical trials can be described under the following bullet points:

- Emphasis should be on performing high-quality trials with properly selected candidates and using relevant, validated clinical endpoints.

- A small number of pivotal randomised controlled trials (RCTs) are urgently needed, but most patients will enter other types of controlled trials, and we need to develop/test/validate the methodologies (e.g. 'cohort multiple RCT').

- Model-based selection (as predictive biomarker) is a useful concept for normal tissue complication probability (NTCP) – based studies, and this concept should later be extended to also incorporate tumour control probability (TCP).

- European centres must join forces to create such trials and evidence soon.

- European trials should be open to accredited centres with expertise and relative high numbers who wish to collaborate.

- Prospective collection of high-quality data for proton patients treated outside of clinical trials (using common ontology and data collection forms).

- There is a need to develop a European QA platform for particle therapy trials.

RCTs are considered the gold standard of evidence-based medicine also in European radiotherapy. There are currently three European randomised trials listed at ClinicalTrials.gov, all involving carbon-ion therapy: two studies at Heidelberg Ion-Beam Therapy Center (HIT), Heidelberg randomising between protons and carbon ions in skull base chordoma and chondrosarcoma, respectively, and the ETOILE multicentre trial randomising between carbon ion and IMRT for chordoma, adenoid cystic carcinoma and sarcomas. There is no doubt that an RCT is the most appropriate study design when particle therapy is intended to increase treatment efficacy in terms of local control or survival by target dose escalation beyond the dose considered to be the current standard. However, it is expected that in most cases, particle therapy will be applied to prevent radiation-induced side effects and/or induction of secondary tumours. For the validation of these types of applications, there is a growing awareness that equating evidence-based medicine with RCTs is an undue simplification and other methodologies, such as the model-based approach, are available and need further exploitation. The model-based approach (see Chapter 20) is an alternative evidence-based methodology designed to yield evidence to inform rational selection of patients who would most likely derive clinically relevant benefit from particle therapy in terms of prevention of radiation-induced side effects. The rationale behind the model-based approach is that particle therapy will only lead to improved clinical outcome due to less toxicity in patients, when clinically relevant normal tissue sparing can be expected with particles. Transforming dose into complication risk requires multi-factorial (NTCP) models, including non-dosimetric features (e.g. patients' age and concomitant chemotherapy), and therefore a decrease of dose will not always translate into a relevant decrease of complication risk. The key research agenda for the near future should therefore be to validate this thesis by attempting to falsify the hypothesis that NTCP reduction leads to less clinical toxicity. For this purpose, uniform prospective data registration on a European level or beyond of all patients treated with proton therapy is essential.

EPTN will also work to identify the methodological issues related to phase I and II studies, as well as to RCTs comparing photons with particles, and to define general guidelines for the design of clinical trials to overcome these issues.

PERSPECTIVE FROM THE UNITED STATES

Clinical trials involving particle therapy are complex, not only because of the technology itself but also because of the need for credentialing and quality assurance of participating institutions and investigators, and the need for funding. These elements vary in the different countries that have particle capability. Notably, carbon-ion facilities are yet to be realised in the United States.

The rationale and basic clinical applications for particle therapy are presented well in the preceding chapters. They will be assumed for the following discussion.

It is generally agreed that it would be valuable to have state-of-the-art metabolic and even molecular imaging as a basis for pre-treatment evaluation, target delineation, treatment planning and also for assessing outcomes.

Three categories of clinical trials are generally accepted in the United States. Phase I trials evaluate toxicity; phase II trials assess efficacy; and phase III trials compare new or experimental treatments with standard treatments (which may include altered fractionation, chemotherapy and immunotherapy). Drug trials differ substantially from trials involving advanced technology or new devices in that radiation therapy studies have different endpoints than do drug trials (Table 14.1). Such trials are all predicated on either increasing the dose to the tumour to achieve better control (TCP), or limiting toxicity by avoiding irradiation of organs at risk (NTCP). Although assessing toxic effects on specific organs is extremely valuable, that is often a long-term endpoint. Experience has shown that the only fully reliable endpoint for outcome is survival. Imaging is usually important for assessment of toxicity, as considerable subjectivity often exists in grading particle effects on organs at risk, in particular the lungs.

The gold standard for cancer clinical trials, that which gives rise to 'level I' evidence, is the prospective, comparative randomised clinical trial. These trials are classified as phase III if a new treatment is compared with other widely accepted treatments. However, if two or more new technologies are compared, even if the trials are prospective and randomised, they are usually called phase II because the treatment in neither arm can be considered 'standard'. This is the case for particle therapies such as those comparing treatment with carbon ions and protons or even trials comparing particles with intensity-modulated X-ray

TABLE 14.1   Clinical Trials in Oncology: Phase/Endpoint Designations for Trials of Drugs versus Radiation

Phase	Goal	Drug Trials	Radiation Trials
	Endpoints		
I	Establish maximum tolerated dose	Toxicity (usually acute effects)	Toxicity (usually late effects)
II	Establish activity	Tumour shrinkage (complete or partial response)	Local tumour control
III	Establish quantitative effect in specific tumour types	Disease-free survival, Overall survival	Disease-free survival, Overall survival

Source: Reprinted with permission from Cox, J.D., Design and implementation of clinical trials of ion beam therapy, in Linz U (Ed.), *Ion Beam Therapy: Fundamentals, Technology, Clinical Applications*, Springer, Berlin, Germany, 2012 ed., pp. 311–324, 2011.

radiation therapy (IMRT), because IMRT is still considered experimental by some governmental bodies and insurance companies in the United States.

These trials require advanced imaging capabilities for accurate target delineation and treatment planning. Such imaging is also highly recommended for assessing tumour response.

In a large database of clinical trials maintained by the U.S. federal government (https://clinicaltrials.gov/), 118 studies of proton therapy for cancer in the United States are listed, of which 80 are currently open. Only 20 studies are labelled 'randomised', of which 12 are currently open (3 for cancer of the head and neck; 3 for cancer of the prostate; 2 for cancer of the breast, and 1 each for cancer of the lung, oesophagus, newly diagnosed glioblastoma, and recurrent glioblastoma). As one might imagine, enrolling sufficient numbers of patients with these diseases to arrive at an answer in a randomised trial is a formidable task. Ethical constraints may be involved as well, such as with infants and children for whom randomisation is inappropriate.

Even if sufficient patients were available in principle, the impediments beyond credentialing and quality assurance as noted earlier can produce a biased sample of patients, resulting in a flawed trial. The major distortion in sample is patient preference. Another is payment for each treatment. Payment for the care of patients in the United States is not uniform. Medicare, the payment mechanism for most adults over the age of 65, does cover proton therapy. Younger patients are covered by private insurance companies according to policies that may be quite variable, or they may have to pay for treatment from their personal resources. To demonstrate the value of particle therapy compared with photons from computer images (*in silico*) is rarely considered satisfactory evidence by the gatekeepers at the insurance companies. Thus a biased sample could easily occur in the arms of a prospective trial.

Among the 12 currently open randomised studies, trials of lung cancer are prominent. This is natural, considering that lung cancer is the leading cause of cancer death worldwide. Lung cancer trials are complicated by respiration- and cardiac-induced tumour motion, which can be compensated for by using passive scattering proton techniques, but spot scanning is largely defeated. Prospective trials comparing passive scattered protons to IMRT have thus far revealed little benefit from particles, although the biases noted above clearly taint some studies. The joint Massachusetts General Hospital–MD Anderson Cancer Center study (NCT00915005), the most mature study to date, was designed to increase local control and reduce the risk of radiation pneumonitis. A successor study (NCT01993810) is being conducted by the national cooperative group NRG Oncology. Both of these studies add concurrent chemotherapy with paclitaxel, carboplatin, etoposide or cisplatin to each treatment group.

Cancer of the breast is another common disease being studied. Here, the long-term goal is to reduce the morbidity associated with post-operative irradiation by sparing the skin. One large study of postoperative irradiation comparing photons versus protons (NCT02603341) is specifically termed a comparative effectiveness trial. Other trials for breast cancer involve altered fractionation with both arms receiving scanning beam proton therapy. The problem with such studies is the long period of follow-up required to assess the endpoints, even cutaneous effects.

Similarly, cancer of the prostate is common but less often life-threatening. Because prostate cancer was the one type of cancer treated in large numbers at the Harvard Cyclotron Lab, it has been approved by insurance carriers as part of their standard policies. Nevertheless, because so many men who are covered by Medicare refer themselves for proton treatment, it is quite difficult to mount randomised studies. One trial compares proton therapy with and without androgen suppression for intermediate-risk prostate cancer (NCT01492972). Another compares hypo-fractionated versus standard-fractionation radiation therapy with protons and photons for low-risk prostate cancer (NCT01230866). 'Level I' evidence is virtually impossible because of the required periods of observation.

A major effort is going into use of proton therapy for treating adults with glioblastoma. A large study is under way through NRG Oncology comparing photons with protons, both combined with temozolomide, for newly diagnosed glioblastoma (NCT02179086). Another compares IMRT with intensity-modulated proton therapy (IMPT), with an emphasis on cognitive function (NCT01854554). A trial for patients with recurrent glioblastoma compares IMRT and proton beam therapy, each combined with bevacizumab (NCT01730950).

Proton beam therapy plus the kinase inhibitor sorafenib is being compared with sorafenib alone for patients with hepatocellular carcinoma (NCT01141478).

Finally, an important study of IMRT versus IMPT for carcinoma of the oropharynx (NCT01893307) that evaluates quality-of-life endpoints is actively recruiting.

In each of these studies, investigators are diligently struggling to overcome the hurdles outlined earlier in this discussion.

As mentioned, the United States currently has no facilities for treatment with carbon ions or other heavy particles; no doubt this area will be covered fully by Dr Tsujii in the remaining section of this chapter.

JAPANESE PERSPECTIVE OF PARTICLE THERAPY, WITH SPECIAL EMPHASIS ON CARBON-ION RADIOTHERAPY

Background

In 1975, prior to the start of the carbon-ion radiotherapy (CIRT) project, the National Institute of Radiological Sciences (NIRS) carried out fast neutron (FN) therapy using a cyclotron (30 megavolts deuterons on Be). About 2,200 patients with various clinical settings were treated with FN until 1994 [2]. This experience demonstrated that a high linear energy transfer (LET) beam without conformal dose distribution could never be used in the treatment of major cancers due to severe late damage to normal tissues. As such, based on this FN experience in Japan, the decision was made in 1984 to build the Heavy Ion Medical Accelerator in Chiba (HIMAC) as an integral part of the nation's 'Overall 10-Year Anti-Cancer Strategy'. The accelerator complex took almost a decade to build, and was completed at the end of 1993. A half year later, clinical study with carbon ions for cancer therapy was initiated. Since then more than 10,000 patients have received phase I/II and phase II CIRT at NIRS. Today, five carbon-ion centres, Chiba, Hyogo, Gunma, Saga and Kanagawa, are in operation, and another two carbon facilities are under construction in Japan. These five centres use as standard protocol the treatment regimens established through NIRS's clinical trial work.

CIRT Clinical Trials

Clinical trials on CIRT were initiated at the NIRS with the rationale of combining a high physical selectivity almost similar to that of protons with the potential radiobiological advantages of high-LET for selected tumour types. Clinical applications were started after extensive dosimetric and radiobiological investigation. The clinical indications for carbon ions were to a large extent based on the past FN experience and the Lawrence Berkeley National Laboratory heavy-ion experience, all the while taking advantage of new possibilities offered by the physical selectivity of carbon ions along with the new imaging and radiotherapeutic technologies established in the 1990s.

The 'Heavy Ion Radiotherapy Network Committee' was designated the supreme organisation responsible for clinical trials at NIRS. All clinical trial protocols were first prepared by the planning team, then evaluated by the disease-specific subcommittees, and finally approved by the Network Committee after investigation by the NIRS Ethical Committee. An Evaluation Committee was appointed to deliberate on the validity of whether the individual clinical studies should be continued, and the results of all clinical studies are submitted to the Network Committee, whose sessions were invariably held in public.

When presenting clinical trials, it is recommended to standardise the method of reporting patient data and treatment modalities. This is required to facilitate the exchange of information with other radiation oncology centres, and in particular those centres involved in CIRT. The ICRU and IAEA/ICRU report on ion beams is now in press [3].

In particular, the following information should be reported in a systematic way:

1. Volumes

    a. Gross tumour volume (GTV)

        i. *GTV and the methods used to evaluate the GTV*: Clinical, CT, MRI, PET and so on.

        ii. *Dimensions of the GTV*: All three dimensions and not only their product.

    b. *Clinical target volume (CTV)*: Indicating the margin(s) used when going from GTV to CTV (e.g. 5 or 10 millimetres in all directions, or different margins in different directions)

    c. *Planning target volume (PTV)*: Indicating the margin(s) adopted when going from CTV to PTV (same margin in all directions or not)

2. Prescribed and delivered dose

    a. The (physical) absorbed dose in Gray

    b. The selected 'clinical RBE'

    c. The biologically weighted dose for (mainly) RBE, in Gy(RBE)

3. Where to specify the dose

    a. In relation to the PTV (centre or central region) and/or

    b. In relation to the beam (beam axis, centre of the SOBP) and/or

    c. The minimum dose to the PTV or isodose surface surrounding the PTV.

4. *The time–dose pattern*: Number of fractions, size of the fractions, overall time

Clinical RBE and Gray (RBE)

The RBE and the selection of the appropriate weighting factor to take into account is one of most complex issues when using carbon ions and often a source of confusion. Specific to carbon ions is that the RBE varies with LET and thus with beam depth. In contrast, with FN the radiation quality and RBE remain constant with depth. A series of RBE studies were performed at NIRS on a variety of systems: different cell lines in vitro, and a mouse model. These experiments provided a wide range of RBE values. Currently, a 'clinical RBE' of 3.0 is adopted for carbon-ion treatments at NIRS, with a higher value used when CNS is involved. This value of 3.0 has been verified through the large clinical experience obtained at NIRS with FN. The fact that the product of the (physical) absorbed dose and 3 is expressed in Gray (RBE) should not obscure the fact that the RBE of carbon varies (within a large range) with dose, biological effect and irradiation conditions. The 'equivalence' exists only for one (or a few) conditions. For this point also, it is recommended to be aware of the ongoing evolution of these concepts and definitions as published by international organisations, in particular the ICRU and IAEA/ICRU. The details of the RBE issues and dose prescription are described elsewhere, and are included in the current ICRU report on ion beams.

Conducted Clinical Trials and Their Results at NIRS

From the beginning, all NIRS carbon-ion radiotherapy clinical trials were initiated as prospective dose-searching phase I studies, and then stepped up to phase II fixed-dose studies. These studies were conducted to identify tumour sites suitable for this treatment, including radio-resistant tumours, and to determine optimal dose-fractionation. Hypo-fractionated carbon-ion radiotherapies were additionally conducted so as to improve efficiency while maintaining efficacy and safety in these trials, as the 4 R's of conventional radiotherapy' (repair, reoxygenation, repopulation and redistribution) are less applicable to high-LET radiation [4]. These trials are summarised in Table 14.2.

Comparative Clinical Trials

Randomised controlled clinical trials provide a high level of evidence with regard to beam efficacy, but can be difficult to perform, particularly in the setting of rare diseases limiting available patient recruitment. With regard to common cancers, RCTs are the logical progression following successful Phase II trial results. Based on the results of NIRS's phase I/II trials on locally advanced pancreatic cancer, the US NCI funded a randomised clinical

TABLE 14.2   Conducted Clinical Trials at NIRS

Site	Phase	Stage	Total Dose Gray (RBE)	Number of Fraction	Weeks	References
Head&Neck-1 + 2	I/II	Locally advanced	49 ~ 70	16–18	4–6	[5]
Head&Neck-3	II	Locally advanced	57.6 or 64.0	16	4	[6]
Head&Neck-4	I/II	Sarcoma	70.4	16	4	[7]
Head&Neck-5	II	Malignant melanoma (C-ion +Chemotherapy)	57.6 or 64.0	16	4	[8]
Skull base/cervical spine-1	I/II	Skull base/cervical spine	48.0 ~ 60.8	16	4	[9]
Skull base/cervical spine-2	II		60.8	16	4	
Lung-1	I/II	Stage I (Peripheral type)	59.4 ~ 95.4	18	6	[10]
Lung-2	I/II	Stage I (Peripheral type)	72.0 ~ 79.2	9	3	[10]
Lung-3	II	Stage I (Peripheral type)	72	9	3	[11]
Lung-4	I/II	Stage I (Peripheral type)	52.8 ~ 60.0	4	1	[12]
Lung-5	II	Stage I (Peripheral type)	52.8(T1), 60.0(T2)	4	1	
Lung-6	I/II	Stage I (Peripheral type)	28 ~ 50 (Single irrad)	1	1 (day)	[13]
Lung-7	II	Stage I (Peripheral type)	50 (Single irrad)	1	1 (day)	[14]
Lung-8	I/II	Stage I (Central type)	57.6 ~ 61.2	9	3	
Lung-9	I/II	Locally advanced	68.0 ~ 76.0	16	4	[14]
Lung-10	II	Locally advanced	72	16	4	[15]
Liver-1	I/II	T2 ~ 4 MONO	49.5 ~ 79.5	15	5	[16]
Liver-2	I/II	T2 ~ 4 MONO	48 ~ 70	4 ~ 12	1 ~ 3	[17]
Liver-3	II	T2 ~ 4 MONO	52.8	4	1	[18]
Liver-4	I/II	T2 ~ 4 MONO	32.0 ~ 48.0	2	2 (days)	[19]
Liver-5	II	T2 ~ 4 MONO	48	2	2 (days)	
Pancreas -1	I/II	Pre-op Resectable All	44.8 ~ 48.0	16	4	
Pancreas -2	I/II	Pre-op Resectable All	30.0 ~ 36.8	8	2	[20]
Pancreas -3	I/II	Unresectable (locally advanced)	38.4 ~ 48.0	12	3	
Pancreas -4	I/II	Unresectable + GEM	43.2 ~ 55.2	12	3	[21]
Pancreas -5	II	Unresectable + GEM	55.2	12	3	[22]
Prostate-1	I/II	B2~C	54 ~ 72	20	5	[23]
Prostate-2	I/II	A2~C	63 ~ 66	20	5	[23]
Prostate-3	II	T1C~C	66	20	5	[24]
Prostate-4	II	A2~C	63	20	5	[25]
Prostate-5	II	A2~C	57.6	16	4	[26]
Prostate-6	II	A2~C	51.6	12	3	[27]
Bone/Soft Tissue-1	I/II	Unresectable	52.8 ~ 73.6	16	4	[28]
Bone/Soft Tissue-2	II	Unresectable	70.4	16	4	
Bone/Soft Tissue-3	II	Chordoma (sacrum)	67.2(S1-2),70.4(S3~)	16	4	[29]
Bone/Soft Tissue-4	II	Mobile spine	64	16	4	[30]
Rectum-1	I/II	Post-operative pelvic recurrence	67.2 ~ 73.6	16	4	[31]
Rectum-2	II	Post-operative pelvic recurrence	73.6	16	4	

trial on locally advanced pancreatic cancer in 2015 (https://www.fbo.gov/spg/HHS/NIH/RCB/BAA-N01CM51007-51/listing.html). At present, the NIRS and UTSW are planning to conduct a separate, independently organised RCT on locally advanced pancreatic cancer. Other currently feasible comparative studies to further clarify the efficacy of CIRT for varying indications include those in which the same protocol is applied with differing (neo)adjunct therapies, background-matched subject comparisons, or retrospective evaluation of patients who seek CIRT privately. In these cases, consent from study participants can be easily obtained, the study cost might be low, and an agreement among the participating facilities is relatively easily obtained, with indicated treatment provided to patients by co-operating institutions.

At present, most of the patients receiving CIRT in Japan visit NIRS seeking CIRT in particular. Consequently, it is difficult to convince these patients to participate in randomised trials, in which they may be given a treatment other than CIRT. However, in selected tumours where we believe high-LET radiation may offer benefit, we can participate in randomised studies. Finally, studies aimed at clarifying the usefulness of carbon-ion radiotherapy and elucidating any advantages from hypo-fractionation should be considered. A multi-institutional prospective randomised and/or nonrandomised concurrent phase II clinical trial is one such new approach, and we would like to propose that not only Japanese, but also international centres take part [32].

REFERENCES

1. Weber DC, Abrunhosa-Branquinho A, Bolsi A. Profile of European proton and carbon ion therapy centers assessed by the EORTC facility questionnaire. *Radiother Oncol.* 2017; 124(2):185–189.
2. Tsunemoto H, Yoo SY. Present status of fast neutron therapy in Asian countries. *Bull Cancer Radiother.* 1996; 83(Supple):93s–100s.
3. The ICRU and IAEA/ICRU report on ion beam. 2018, in press.
4. Furusawa Y. Heavy ion radiobiology. In Tsujii H et al. (Eds.), *Carbon Ion Radiotherapy: Principle, Practice, and Treatment Planning*, Springer, Tokyo, Japan, pp. 25–38, 2014.
5. Mizoe JE, Tsujii H, Kamada T et al. Dose escalation study of carbon ion radiotherapy for locally advanced head-and-neck cancer. *Int J Radiat Oncol Biol Phys.* 2004; 60(2):358–364.
6. Mizoe JE, Hasegawa A, Jingu K et al. Results of carbon ion radiotherapy for head and neck cancer. *Radiother Oncol.* 2012; 103(1):32–37.
7. Jingu K, Tsujii H, Mizoe JE et al. Carbon ion radiation therapy improves the prognosis of unresectable adult bone and soft-tissue sarcoma of the head and neck. *Int J Radiat Oncol Biol Phys.* 2012; 82(5):2125–2131.
8. Koto M, Demizu Y, Saitoh JI et al. Japan carbon-ion radiation oncology study group. Multicenter study of carbon-ion radiation therapy for mucosal melanoma of the head and neck: Subanalysis of the Japan Carbon-Ion Radiation Oncology Study group (J-CROS) study (1402 HN). *Int J Radiat Oncol Biol Phys.* 2017; 97(5):1054–1060.
9. Mizoe JE, Hasegawa A, Takagi R et al. Carbon ion radiotherapy for skull base chordoma. *Skull Base.* 2009; 19(3):219–224.
10. Miyamoto T, Yamamoto N, Nishimura H et al. Carbon ion radiotherapy for stage I non-small cell lung cancer. *Radiother Oncol.* 2003; 66:127–140.

11. Miyamoto T, Baba M, Yamamoto N et al. Curative treatment of stage I non-small-cell lung cancer with carbon ion beams using a hypo-fractionated regimen. *Int J Radiat Oncol Biol Phys*. 2007; 67:750–758.
12. Miyamoto T, Baba M, Sugane T et al. Carbon ion radiotherapy for stage I non-small cell lung cancer using a regimen of four fractions during 1 week. *J Thorac Oncol*. 2007; 2:916–926.
13. Yamamoto N, Miyamoto T, Nakajima M et al. A dose escalation clinical trial of single-fraction carbon-ion radiotherapy for peripheral stage I non-small-cell lung cancer. *J Thorac Oncol*. 2017; 12:673–681.
14. Takahashi W, Nakajima M, Yamamoto N et al. A prospective nonrandomized phase I/II study of carbon ion radiotherapy in a favorable subset of locally advanced non-small cell lung cancer. *Cancer*. 2015; 121:1321–1327.
15. Hayashi K, Yamamoto N, Karube M et al. Prognostic analysis of radiation pneumonitis: Carbon-ion radiotherapy in patients with locally advanced lung cancer. *Radiat Oncol*. 2017; 12:91.
16. Kato H, Tsujii H, Miyamoto T et al. Results of the first prospective study of carbon ion radiotherapy for hepatocellular carcinoma with liver cirrhosis. *Int J Radiat Oncol Biol Phys*. 2004; 59:1468–1476.
17. Imada H, Kato H, Yasuda S et al. Comparison of efficacy and toxicity of short-course carbon ion radiotherapy for hepatocellular carcinoma depending on their proximity to the porta hepatis. *Radiother Oncol*. 2010; 96:231–235.
18. Kasuya G, Kato H, Yasuda S et al. Progressive hypofractionated carbon-ion radiotherapy for hepatocellular carcinoma: Combined analyses of two prospective trials. *Cancer*. 2017; 123(20):3955–3965.
19. Yasuda S. Hepatocellular carcinoma. In Tsujii H et al. (Eds.), *Carbon Ion Radiotherapy: Principle, Practice, and Treatment Planning*, Springer, Tokyo, Japan, pp. 213–218, 2014.
20. Shinoto M, Yamada S, Yasuda S et al. Phase 1 trial of preoperative, short-course carbon-ion radiotherapy for patients with resectable pancreatic cancer. *Cancer*. 2013; 119:45–51.
21. Shinoto M, Yamada S, Terashima K et al. Carbon ion radiation therapy with concurrent gemcitabine for patients with locally advanced pancreatic cancer. *Int J Radiat Oncol Biol Phys*. 2016; 95:498–504.
22. Kawashiro S, Yamada S, Okamoto M et al. Multi-institutional study of carbon ion radiation therapy for locally advanced pancreatic cancer: Japan Carbon Ion Radiation Oncology Study group (J-CROS) study 1403. *Int J Radiat Oncol Biol Phys*. 2016; 96:S140–S141.
23. Akakura K, Tsujii H, Morita S et al. Phase I/II clinical trials of carbon ion therapy for prostate cancer. *Prostate*. 2004; 58:252–258.
24. Ishikawa H, Tsuji H, Kamada T et al. Carbon ion radiation therapy for prostate cancer: Results of a prospective phase II study. *Radiother Oncol*. 2006; 81:57–64.
25. Okada T, Tsuji H, Kamada T et al. Carbon ion radiotherapy in advanced hypofractionated regimens for prostate cancer: From 20 to 16 fractions. *Int J Radiat Oncol Biol Phys*. 2012; 84:968–972.
26. Ishikawa H, Tsuji H, Kamada T et al. Carbon-ion radiation therapy for prostate cancer. *Int J Urol*. 2012; 19:296–305.
27. Nomiya T, Tsuji H, Maruyama K et al. Phase I/II trial of definitive carbon ion radiotherapy for prostate cancer: Evaluation of shortening of treatment period to 3 weeks. *Br J Cancer*. 2014; 110(10):2389–2395.
28. Kamada T, Tsujii H, Tsuji H et al. Efficacy and safety of carbon ion radiotherapy in bone and soft tissue sarcomas. *J Clin Oncol*. 2002; 20:4466–4471.

29. Imai R, Kamada T, Araki N et al. Carbon-ion radiation therapy for unresectable sacral chordoma. An analysis of 188 cases. *Int J Radiat Oncol Biol Phys.* 2016; 95:322–327.
30. Matsumoto K, Imai R, Kamada T et al. Working group for bone and soft tissue sarcomas. Impact of carbon ion radiotherapy for primary spinal sarcoma. *Cancer.* 2013; 119:3496–3503.
31. Yamada, S Kamada T, Ebner DK et al. Carbon-ion radiation therapy for pelvic recurrence of rectal cancer. *Int J Radiat Oncol Biol Phys.* 2016; 96:93–101.
32. Kamada, TY. Evaluation of treatment outcomes using the heavy ion medical accelerator in Chiba (HIMAC). In Tsujii H et al. (Eds.), *Carbon Ion Radiotherapy: Principle, Practice, and Treatment Planning,* Springer, Tokyo, Japan, pp. 121–126, 2014.
33. Cox, JD. Design and implementation of clinical trials of ion beam therapy. In Linz U (Ed.), *Ion Beam Therapy: Fundamentals, Technology, Clinical Applications,* Springer, Berlin, Germany, 2012 ed., pp. 311–324, 2011.

CHAPTER 15

Role of Particle Therapy in Future Radiotherapy Practice

Roberto Orecchia

CONTENTS

INTRODUCTION

The number of newly diagnosed cancer patients reported in 2012 was more than 14 million cases worldwide, and this number has been projected to reach 24.6 million by 2030, with an increase of more than 70% [1]. In Europe, about 4.0 million new cancer patients are expected in 2025. This represents an increase of 16% in the absolute number of cases, when compared to the 3.4 million diagnosed in 2012.

Cancer is the leading cause of early death. The number of cancer deaths recorded in 2012 was 8.2 million worldwide, with a projection to rise up to 13.0 million by 2030 [1]. Five sites (lung, breast, colorectal, prostate and stomach) comprised almost half the total number of cancer cases, and caused 53% of all cancer deaths. Around 56% (about 8.0 million) of new cases occurred in higher developed countries, which account for less than a third of the global population. The incidence and mortality are substantially different in medium- and low-developed countries, but the cancer picture is not static or uniform. When the level of human and socioeconomic development increases in these countries, cancer emerges as a major source of morbidity and mortality, creating a new consistent health problem.

Radiotherapy (RT) is a fundamental component of treating most of these cancers, alone or mostly as part of a course on an evidenced-based integrated approach with surgery and/or drugs. When the disease is still localised, RT has been used to treat most common and many rare cancers, but also, in the case of locally advanced or incurable cancer, RT is useful to control gross tumour masses or to palliate symptoms. Estimation of the exact proportion of cases that will need RT is quite complex, due to different patterns of cancer presentation and limited information on the number of patients effectively receiving this treatment [2]. The optimal rate, based on indications of RT, population studies and epidemiologic data, should be more than 50% of all patients with cancer, which means millions of cases every year, but the actual rates are lower in most countries, including Australia, the United States, Great Britain, Canada and Spain [3]. The difference between the expected and current number of cases for RT has been explained by changes in recommendations and changes in the proportion of stages of the disease in each country, the presence of co-morbidity and advanced age of the population, and includes the suboptimal availability of facilities and resources. In Europe, the estimated number of patients having an indication to RT was about 1.7 million in 2012 (50% of the total number of new cancers), with a majority of them having breast, lung, prostate and head and neck cancers. This need is expected to increase up to 2.0 million in 2025, with wide variations in the magnitude (from 0% to 35%) of the demand among the countries [4]. Of these patients, more than 20% will require a re-treatment for a recurrent cancer. All these projections are very useful to assess the expected cancer burden and the related impact for planning the activity of RT centres, including the implementation of new radiation techniques.

TECHNOLOGICAL TOOLS OF HIGH-PRECISION RADIOTHERAPY

Technological innovation has led to remarkable improvements in every phase of RT, related to treatment, from simulation to planning, and to delivery, with the aim to tailor treatment to the specific anatomy of individual patients. The current clinical goal is to minimise healthy organs' toxicity and improve loco-regional control.

The fundamental step towards high-precision RT was the advance in diagnostic imaging, allowing the passage from a two-dimensional to a three-dimensional (3D) perspective. The high conformability of 3D conformal RT, supported by the implementation of advanced treatment planning systems equipped with virtual reconstruction and beam-eye-views, and new devices such as multileaf collimators (MLC), has improved dosimetric accuracy. The optimisation of dose distribution limits hot spots and reduces exposure to the organs at risk (OAR), paving the way to safe dose escalation in the attempt to further improve loco-regional control rates, with the subsequent possible positive impact of increasing overall survival.

Nowadays the technological evolution has taken a step forward, as intensity-modulated radiotherapy (IMRT), based on different types of delivery (step & shoot, sliding windows, Tomotherapy® and volumetric arc therapy), and stereotactic radiotherapy (SRT), which can be undertaken using gamma units or modified linear accelerators, has entered into routine clinical practice.

IMRT consists of the delivery of small beams of X-rays with different intensities. The control of the fluence of each beam improves dose distribution, enabling a real customised design of the target volumes and permitting further shaping of the high-dose volume ('dose-painting'). IMRT plan is more time-consuming and requires special calculation algorithms able to guarantee computer-aided optimisation methods. The IMRT technique also allows the simultaneous delivery of different dose levels to different target volumes within a single treatment fraction: this approach has defined as the 'simultaneous integrated boost' (SIB) technique. The SIB technique is of particular interest because it can be used to yield higher doses to the critical area (boost volume) without increasing the overall treatment time.

SRT allows delivery of a high or very high focal dose (up to 20 Gray and more) per fraction, with significant reduction of the dose-to-no-target tissue, able to get very good local control rates when used in the treatment of small-volume and well-defined primary tumours, but also to provide excellent results in relieving painful metastases. SRT exploits the excellent steep dose gradient to improve tolerance and delivers the treatment in few fractions for an optimal quality of life.

High-precision RT demands accurate identification of target and OARs. Adequate contouring is the fundamental prerequisite for an effective and safe treatment plan. However, contouring is a process prone to errors and inter- and intra-operator variability. The integration of radiological and metabolic imaging such as magnetic resonance (MR) and positron emission tomography (PET) with CT simulation can provide useful information for high-precision definition of all the volumes of interest. An additional but essential aspect of the advances in technology has been expressed in the quality of treatment execution.

High conformability means high sensitivity to any changes occurring in the patient during the course of RT. Displacement of the target due to organ motion, anatomical changes in the patient's body or inaccurate setup affect dose distribution, leading to inadequate target coverage or excessive irradiation of the OARs. Strategies to verify target shape, volume and position, and to correct the topographical inconsistencies with the original plan, are part of image-guided radiotherapy (IGRT), which uses various devices, such as electronic portal imaging, cone beam CT, megavoltage CT, ultrasound, optical imaging and fiducial markers.

In the case of substantial deviations from the original plan, it becomes necessary to replan and re-optimise the dose distribution. This is the concept of 'adaptive radiotherapy'. Besides inter-fraction motion, intra-fraction motion has become relevant because delivery time is longer with modern RT techniques. In order to manage these uncontrolled physiological changes, modern treatment rooms have been equipped with movement tracking systems to monitor and compensate for target motion during irradiation. The new hybrid MR linear accelerators, recently available on the market, may represent a further step towards a fully adaptive intra-fraction planning system [5]. More precise real-time tracking of target and OARs, thanks to advanced soft-tissue visualization, should lead to narrower safety margins around the target and to smaller treated volumes, making it possible to escalate the dose safely.

The combination of dosimetric data and data regarding clinical toxicity makes it possible to chart complex dose–response relationships and define specific tolerance doses for OARs. A comprehensive review of this dose–effect correlations of the most commonly irradiated organs was organised in the Quantitative Analysis of Normal Tissue Effect in the Clinic (QUANTEC) project, published in 2010 [6]. Models linking dose with toxicity and tumour control, based on radiobiological and mathematical principles, have been used to predict the normal tissue complication probability (NTCP) and the tumour control probability (TCP), to evaluate the potential treatment outcome [7].

All these impressive developments in RT have applied extensively in the current clinical practice, and have stimulated a renewed interest in the use of particle beam therapy (protons and carbon ions, in principle), resulting in new design for treatment units and the construction of new facilities. In fact, the superior physical selectively of charged particle beams, even when compared to high-precision X-ray RT, could not be fully exploited without the improved volumes of interest definition, setup and treatment planning accuracy and delivery precision, developed for high-precision RT. The introduction of this new modality of radiation treatment into the clinical routine remains difficult, especially if it is more expensive than the conventional one. The clinical benefit has to be proven in relation to the increased cost, taking into account the socioeconomic and health environment.

PERSPECTIVE AND NEED FOR PARTICLE THERAPY IN FUTURE CLINICAL PRACTICE

To date, the number of patients treated with X-ray RT vastly outnumbers those treated with particle beams. Almost all patients currently treated by RT receive X-ray techniques, and only about 0.1% receive radiation from charged particles. This is due to the historically low number of centres able to offer particle therapy to patients, but now this availability is increasing quickly. The Particle Therapy Co-Operative Group (PTCOG) constantly updates the statistics on patients and centres, and, as reported in July 2017 in the last update of the web page www.ptcog.ch, more than 70 centres are now treating patients, with the great majority, about 60, furnished with proton beams. These centres are not uniformly distributed, having the greatest concentration in the United States, Europe and Japan. An additional 40 centres are planned to be operative in the next five years. At the end of 2015, and from the beginning of particle therapy activities, more than 154,000 patients worldwide received charged particles, mainly by protons (more than 131,000). In the last period, the number of patients yearly treated was strongly increasing (up to 15,000): twofold, when compared with the number treated 10 years before.

There is substantial evidence to continue to support the clinical use and studies on particle therapy being modern RT devoted to realise very selective treatments. The success of RT relies on increasing the therapeutic window between TCP and NTCP. Using particle beams, a given tumour radiation dose, which determines the TCP, can be achieved at a lower integral dose to the OARs, thus reducing the NTCP. Alternatively, at the same NTCP, a higher TCP can be achieved.

Children are a good example of this possible advantage, due to their particular susceptibility to late adverse effects of radiation, including secondary cancers, cardiac and endocrine diseases, cognitive dysfunctions and others [8]. Given the reductions in OARs exposure using particle therapy and, therefore, the potential for reduction in adverse effects, this treatment is widely accepted for childhood cancers. A recent review on 15 different tumour sites suggests that disease control and survival rates seen with particle therapy are comparable to those seen with X-rays, although a definitive evaluation of the cost-effectiveness ratio is still lacking [9]. Some data also suggest a potential reduction in the incidence of secondary malignancies in this population [10]. Participation in large multi-institutional registry trials, common in paediatric oncology, will serve to document the effective reduction in late adverse effects, in the absence of randomised clinical trials that in paediatric patients would be very difficult to conduct, given the ethical as well as practical issues. It could be very challenging to accrue in a randomised trial between X-rays and particles when families have been informed of the potential benefit in reducing the integral dose in the group of patients who will receive particle therapy [11]. A recent analysis from US Cancer Data Base has shown that 8% of the entire cohort of paediatric patients treated by RT between 2004 and 2013, more than 12,000 in total, received proton therapy, with an increased proportion between 2004 (only 1.7%) and 2013 (17.5%). Socioeconomic factors seem to affect the use of particles in children, with more likely referral influenced by private managed care rather than Medicaid or no insurance, higher household income and educational attainment [12].

In particle therapy, some additional biological advantages have also been considered. Treatment with high-LET particles, such as heavy ions, more effectively induces cell death than X-ray RT, and, for this reason, carbon ions are considered to be the most advanced radiation tool for the treatment of radio-resistant tumours [13]. The excellent results obtained in Japan on several types of cancers have encouraged the interest of the oncological community [14]. As an example, at NIRS in Chiba a two-year overall survival rate of 50% was reached in locally advanced and inoperable pancreatic cancer using doses of carbon ions below 53 Gray equivalent, combined with concurrent gemcitabine [15]. This improvement is remarkable, being that the current rate with standard chemo-radiation is lower, no more than 20%.

Particle therapy has solidified its indication for eye melanoma and other ocular tumours [16]. Worldwide, up to the end of 2014, the total number of eye patients treated with protons was about 29,000 in 10 centres from seven countries [17]. A meta-analysis described a five-year local control rate for uveal melanomas of 95% or greater, maintained at 15-year follow-up at about 95%. These results were superior to those obtained by brachytherapy plaques. Enucleation rate at five years was approximately 10%, with a small increase at 15 years at approximately 15% [18]. Large studies have demonstrated high local control and acceptable or low toxicity rate in chordomas and chondrosarcomas of the skull base and spine treated by protons [19,20]. These tumours are also currently treated by carbon ion in Japan, Germany and Italy, but comparative data are not still available [21].

Due to these long-term and well-established results on both eye melanoma and chordomas and chondrosarcomas, in 2012 ASTRO's Emerging Technology Committee Base concluded that there was a consistent evidence of the benefit of treating these tumours by particle therapy [22]. Besides, in paediatric CNS malignancies there is 'a suggestion' that proton therapy is superior to X-ray therapy, but there is still insufficient supporting data. The report also concluded that the current data did not provide sufficient evidence of benefit for the most frequent cancers, and sites have strongly recommended clinical trials for these and other diseases. In 2014, ASTRO Model Policy stated proton therapy 'is considered reasonable in instances where sparing the surrounding normal tissue cannot be adequately achieved with photon-based RT and is of added clinical benefit to the patient', opening for wider indications, but again pointing out the need for the generation of clinical evidence in controlled clinical trials or in multi-institutional patient registries [23].

Many factors should be considered in designing these new clinical protocols [24]. Treatment planning and plan evaluation of particles require special considerations compared to the processes used for X-rays, including the approximations and assumptions of models for computing dose distributions, and the greater vulnerability of protons to uncertainties, especially from anatomy variations and inter- and intra-fractional changes. Besides, the relative biological effectiveness (RBE) of protons has simplistically assumed to have a constant value of 1.1. In reality, the RBE is variable along the Bragg curve and depends on a complex function of the energy of protons, dose per fraction, tissue and cell type, endpoint and other factors. This issue is more complex when high-LET particles, such as carbon ions, are used. In Japan, RBE-weighted dose calculation is based on the model proposed by Kanai and colleagues [25], while in all European centres, a different model is applied, the LEM-1-based approach [26]. The difference of the radiobiological model applied on the actual physical dose has a strong impact, and the deviations in the dose delivered in different centres entail potentially different clinical outcomes, when the same fractionation scheme has been adopted. To overcome this problem, it is necessary to develop and validate in clinical settings the models for RBE-weighted prescription doses that take into account conversion factors according to different target size and shape, and beam configuration [27,28].

The therapeutic effect of particles versus X-ray therapy, or protons versus carbon ions, owing to a high number of parameters to be compared, including fractionation, overall treatment duration, RBE and physical dose distribution still needs more clinical trials [29]. In addition, the implementation of these trials has been hampered by many other problems, such as patient preference and insurance coverage. Many patients in the United States and most European countries who meet trial eligibility requirements have restrictive insurance policies that deny coverage for participation in clinical trials.

In most currently ongoing clinical studies, toxicity is one of the primary endpoints, and the superiority of particles was demonstrated only if a statistically significant rate of complications occurred in the standard arm. Luckily, the rate of adverse events is now decreasing at state-of-the-art facilities, and this strategy is not necessarily the best approach to demonstrating the potential benefits of particles [30]. The results of trials in which the endpoint is the evaluation of the overall survival of patients with highly prevalent tumours (for example, lung, prostate, breast, liver or pancreatic cancer) might provide more definitive answers.

Patient selection is the other major hurdle in trial design. Selection based only on tumour location can also affect the results, because the use of particle therapy should be associated with important biological advantages that have been exploited in clinical trials. Therefore, in principle the best patient-stratification strategy should consider the potential biological advantages derived from physical dose distributions, or be based on the biological response.

EXPECTED NUMBER OF PATIENTS SUITED FOR PARTICLE THERAPY

The construction of many new particle therapy centres raises the question of how many patients have potential benefit, and, consequently, how many centres are needed to treat these patients. The well-established and accepted indications for charged particles are very limited to some rare tumours, and this small number cannot justify so important an investment as opening new centres. For the more frequent types of cancer, the cost–benefit ratio remains highly controversial, even though several recent general reviews of the literature continue to confirm very good results of particle therapy [13,14,22,24,29,31–37]. These results will have interpreted not only in terms of comparative effectiveness but also of incremental effectiveness and cost [33,36]. The initial goal of estimating eligibility for particle therapy for a significant fraction, up to 15%–20%, of all the patients referred for RT was probably too optimistic [38]. Currently, a more realistic view should approximate to 2% and 5% at 5 and 10 years respectively, mainly confined to the geographic areas in which particle therapy is or will be available in the upcoming years. In spite of this apparently low percentage of suitable patients, only in Europe will the potential need be from 40,000 to 100,000 new patients per year, many times more than allowed by the current or future availability of existing and planned centres. Given the lack of recognised levels of evidence, today the clinical benefit derived by patients should be evaluated on an individual basis, which is an achievable task, or by applying novel models for properly selecting them. A Dutch group proposed a model-based approach consisting of two phases: in the first phase, the selection of the patient populations most likely to benefit from particle using NTCP value, and in the second, such results were clinically validated using sequential prospective observational cohort studies [39,40]. Other recommended approaches are to use biomarkers (including hypoxia) and dosimetry, and compare the results with a control cohort of patients.

These patient selection models should be encouraged and used when determining the number of facilities delivering particle therapy that are required in a specific geographical area. The profile of the European proton and carbon ion centres has recently been assessed by EORTC by a facility questionnaire [41]. Results are that the average number of patients treated per year and per particle centre was quite low, 221, with a range from 40 to 557. The majority of these centres are treating chordomas and chondrosarcomas, paediatric cancers, brain tumours and sarcomas and some of them (27%) only eye tumours. The critical reasons for this still-limited activity are, in addition to the availability, the high cost, the non-uniform method of reimbursement, the lack of measurable clinical benefit, and the limited applicability out of some selected tumour sites. Without solving these issues, particle therapy will grow in future clinical practice but not at the level that patients and oncologists expected. Technological advances of RT seem likely to reach a plateau in the near future and, if there is to be a continuation towards ever more cost-effective treatments, we need to accompany this progress by a better understanding of biology, adapting the best tool to the tumour molecular profile.

REFERENCES

1. Ferlay J, Soerjomataram I, Dikshit R et al. 2015. Cancer incidence and mortality worldwide: Sources, methods and major pattern in GLOBOCAN 2012. *Int J Cancer* 136:E359–E386.
2. Atun R, Jaffray DA, Barton M et al. 2015. Expanding global access to radiotherapy. *Lancet Oncol* 16:1153–1186.
3. Barton MB, Jacon S, Shafir J et al. 2014. Estimating the demand for radiotherapy from the evidence. A review of changes from 2003 to 2012. *Radiother Oncol* 112:140–144.
4. Borras JM, Lievens Y, Barton M et al. 2016. How many new cancer patient in Europe will require radiotherapy by 2025? An ESTRO-HERO analysis. *Radiother Oncol* 119:5–11.

5. Lagendijk JJ, Raaymakers BW, Raaijmakers AJ et al. 2008. MRI/linac integration. *Radiother Oncol* 86:25–29.

6. Jackson A, Marks LB, Bentzen SM et al. 2010. The lessons of QUANTEC: Recommendations for reporting and gathering data on dose-volume dependences of treatment outcome. *Int J Radiat Oncol Biol Phys* 76:S155–S160.

7. Marks LB, Yorke ED, Jackson A et al. 2010. Use of normal tissue complication probability models in the clinic. *Int J Radiat Oncol Biol Phys* 76:S10–S19.

8. Rowe LS, Krauze AV, Ning H et al. 2017. Optimizing the benefit of CNS radiation therapy in the pediatric population—PART 2: Novel methods of radiation delivery. *Oncology* (Williston Park) 31:224–226.

9. Leroy R, Benahmed N, Hulstaert F et al. 2016. Proton therapy in children: A systematic review of clinical effectiveness in 15 pediatric cancers. *Int J Radiat Oncol Biol Phys* 95:267–278.

10. Mizumoto M, Murayama S, Akimoto T et al. 2017. Long-term follow-up after proton beam therapy for pediatric tumors: A Japanese national survey. *Cancer Sci* 108:444–447.

11. Chapman TR, Ermoian RP. 2017. Proton therapy for pediatric cancer: Are we ready for prime time? *Future Oncol* 13:5–8.

12. Shen CJ, Hu C, Ladra MM et al. 2017. Socioeconomic factors affect the selection of proton radiation therapy for children. *Cancer.* doi:10.1002/cncr.30849.

13. Pompos A, Durante M, Choy H. 2016. Heavy ions in cancer therapy. *JAMA Oncol* 2:1539–1540.

14. Kamada T, Tsujii H, Blakely EA et al. 2015 Carbon ion radiotherapy in Japan: An assessment of 20 years of clinical experience. *Lancet Oncol* 16:e93–e100.

15. Shinoto M, Yamada S, Terashima K et al. 2016. Carbon ion radiation therapy with concurrent gemcitabine for patients with locally advanced pancreatic cancer. *Int J Radiat Oncol Biol Phys* 95:498–504.

16. Stannard C, Sauerwein W, Maree G et al. 2013. Radiotherapy for ocular tumours. *Eye (Lond)* 27:119–127.

17. Hrbacek J, Mishra KK, Kacperek A et al. 2016. Practice patterns analysis of ocular proton therapy centers: The international OPTIC survey. *Int J Radiat Oncol Biol Phys* 95:336–343.

18. Wang Z, Nabhan M, Schild SE et al. 2013. Charged particle radiation therapy for uveal melanoma: A systematic review and meta-analysis. *Int J Radiat Oncol Biol Phys* 86:18–26.

19. Fossati P, Vavassori A, Deantonio L et al. 2016. Review of photon and proton radiotherapy for skull base tumours. *Rep Pract Oncol Radiother* 21(4):336–355.

20. De Amorim Bernstein K, De Laney T. 2016. Chordomas and chondrosarcomas. The role of radiation therapy. *J Surg Oncol* 114:564–569.

21. Mizoe JE. 2016. Review of carbon ion radiotherapy for skull base tumors (especially chordomas). *Rep Pract Oncol Radiother* 21:356–360.

22. Allen AM, Pawlicki T, Dong L et al. 2012. An evidence based review of proton beam therapy: The report of ASTRO's emerging technology committee. *Radiother Oncol* 103:8–11.

23. Proton beam therapy, ASTRO model policy, ASTRO 2014. www.astro.org/uploadedFiles/Main_Site/Practice_Management/Reimbursement/ASTRO PBT Model Policy FINAL.pdf

24. Mohan R, Grosshans D. 2017. Proton therapy—Present and future. *Adv Drug Deliv Rev* 109:26–44.

25. Kanai T, Matsufuji N, Miyamoto T et al. 2006. Examination of GyE system for HIMAC carbon therapy. *Int J Radiat Oncol Biol Phys* 64:650–656.

26. Krämer M, Scholz M. 2000. Treatment planning for heavy-ion radiotherapy: Calculation and optimization of biologically effective dose. *Phys Med Biol* 45:3319–3330.

27. Fossati P, Molinelli S, Matsufuji N et al. 2012. Dose prescription in carbon ion radiotherapy: A planning study to compare NIRS and LEM approaches with a clinically oriented strategy. *Phys Med Biol* 57:7543–7554.

28. Molinelli S, Magro G, Mairani A et al. 2016. Dose prescription in carbon ion radiotherapy: How to compare two different RBE-weighted dose calculation systems. *Radiother Oncol* 120:307–312.
29. Durante M, Orecchia R, Loeffler JS. 2017. Charged-particle therapy in cancer: Clinical uses and future perspectives. *Nat Rev Clin Oncol* 14:483–495.
30. Cox JD. 2016. Impediments to comparative clinical trials with proton therapy. *Int J Radiat Oncol Biol Phys* 95:4–8.
31. Uhl M, Herfarth K, Debus J. 2014. Comparing the use of protons and carbon ions for treatment. *Cancer J* 20:433–439.
32. Doyen J, Falk AT, Floquet V et al. 2016. Proton beams in cancer treatments: Clinical outcomes and dosimetric comparisons with photon therapy. *Cancer Treat Rev* 43:104–112.
33. Verma V, Mishra MV, Mehta MP. 2016. A systematic review of the cost and cost-effectiveness studies of proton radiotherapy. *Cancer* 122:1483–1501.
34. Ottawa (ON): Canadian Agency for Drugs and Technologies in Health. 2016. Proton beam therapy versus photon radiotherapy for adult and pediatric oncology patients: A review of the clinical and cost-effectiveness [Internet]. CADTH rapid response reports.
35. Sakurai H, Ishikawa H, Okumura T. 2016. Proton beam therapy in Japan: Current and future status. *Jpn J Clin Oncol* 46:885–892.
36. Mishra MV, Aggarwal S, Bentzen SM et al. 2017. Establishing evidence-based indications for proton therapy: An overview of current clinical trials. *Int J Radiat Oncol Biol Phys* 97:228–235.
37. Mohamad O, Sishc BJ, Saha J et al. 2017. Carbon ion radiotherapy: A review of clinical experiences and preclinical research, with an emphasis on DNA damage/repair. *Cancers (Basel)* 9(6):E66.
38. Glimelius B, Ask A, Bjelkengren G et al. 2005. Number of patients potentially eligible for proton therapy. *Acta Oncol* 44:836–849.
39. Langendijk JA, Lambin P, De Ruysscher D et al. 2013. Selection of patients for radiotherapy with protons aiming at reduction of side effects: The model-based approach. *Radiother Oncol* 107:267–273.
40. Widder J, van der Schaaf A, Lambin P et al. 2016. The quest for evidence for proton therapy: Model-based approach and precision medicine. *Int J Radiat Oncol Biol Phys* 95:30–36.
41. Weber DC, Abrunhosa-Branquinho A, Bolsi A et al. 2017. Profile of European proton and carbon ion therapy centers assessed by the EORTC facility questionnaire. *Radiother Oncol* 124:185–189.

CHAPTER 16

Health Economics

Cost, Cost-Effectiveness and Budget Impact of Particle Therapy

Yolande Lievens and Madelon Johannesma

CONTENTS

INTRODUCTION

Health care has become a main economic expenditure in developed countries. More specifically, health expenditures on cancer are a major challenge, with expenses in the European Union, for example, rising from €35.7 billion in 1995 to €83.2 billion in 2014 (Jönsson 2016). Although the ever-growing budget for cancer drugs is an important component in this evolution, the rapid diffusion of new technologies has also been an important driver of increasing health care costs (Bodenheimer 2005, Sorenson 2013).

As a consequence, economic evaluations have gained importance and general acceptance with a view to keeping the health system sustainable. Before introduction and implementation in clinical care, new technologies not only have to prove their clinical superiority but also whether they provide value for money, and hence are worth the investment. Particle therapy is a typical example of an expensive technology that has been heavily debated in this regard. Because of its higher cost in comparison to photon radiotherapy, it is crucial to define whether the investment is acceptable from a societal point of view, whether and for which patients it is cost-effective, and what its expected impact on the health care budget will be. In this chapter, we will discuss the costs, cost-effectiveness and budget impact of particle therapy.

THE COSTS

Resource costs are an often neglected component in the economic assessment of new radiotherapy treatments and technologies. Yet they are an indispensable first step in making investment cases, performing economic evaluations and budget impact analyses (BIA), and, in fine, supporting reimbursement setting.

Ideally, resource costs should reflect the quantity and quality of resources consumed, that is, the actual costs made to deliver state-of-the-art health care interventions. Next, reimbursement tariffs should align with these costs. However, this is rarely the case in radiotherapy, as it is not easy to develop a reimbursement system that accurately mirrors the variability in resource costs by the type of treatment and, moreover, is flexible enough to adjust to the rapid and incremental technological evolution (Kesteloot 1996). As is also the case in other health care domains, radiotherapy reimbursement is typically based on negotiations rather than on accurate cost data derived from consistent costing studies. This is not trivial, as the failure to capture accurate resource costs has been suggested as one of the reasons underlying the actual health care crisis (Kaplan 2011).

The Need for a Sound Costing Methodology

Unfortunately, accurate cost data in radiotherapy remain scarce. A systematic literature review performed on the publication period 1981–2015, focusing on cost calculation studies for external beam photon therapy, revealed only 52 primary publications (Defourny 2016). In addition to the limited availability of solid radiotherapy cost data published over such a long-time horizon, the large heterogeneity in scope of the analysis (in terms of resource inputs included and in cost outputs analysed) hampered comparisons across the studies and made it hard to derive firm conclusions. In order to eliminate many of these discrepancies and to provide sound cost data to support decision-making, it has been strongly advocated to implement conventional cost accounting methods in the health care sector (Kaplan 2011, Defourny 2016). It is striking that in the published literature on external beam radiotherapy costing only 40% of the studies employed validated cost accounting methods, either micro-costing or activity-based costing (ABC) (Defourny 2016). Both methods provide superior insight into resource use and yield improved precision of the cost calculation of new technologies (Barnett 2009).

Microcosting combines detailed data on the resources utilised, obtained through direct measurement (e.g. through time and motion studies), and on their unit costs. This bottom–up analytical approach is typically reserved to provide insight into the cost of novel and very specific interventions or process steps, such as the cost of image-guided radiotherapy for prostate cancer (Perrier 2013). Due to its focus and precision, it is, however, unsuitable and too labour-intensive for calculating costs with a wider scope, such as the cost of entire radiotherapy departments.

ABC, conversely, assigns resource costs via the activities performed at each treatment process step to compute the actual treatment cost. It was originally developed for the manufacturing industry to get a better understanding of the cost of production processes characterised by large product diversity and/or a significant amount of indirect resource costs. Because the large amount of cost-drivers made ABC complex to develop and maintain, it has more recently evolved into time-driven activity-based costing (TD–ABC). This improved version of ABC requires providers to estimate only two parameters at each process step: (1) the cost of each of the resources used and (2) the quantity of time the patient spends with each resource.

Because radiotherapy is a well-structured treatment process, yet with a large range of treatments varying in fractionation and complexity, and with capital and human resource costs that cannot be directly traced to one single treatment, TD–ABC is well suited for costing in the radiotherapy setting. TD–ABC studies have been performed to compute the real-life cost of external beam photon radiotherapy for specific indications and for entire departments (Perez 1993, Lievens 2003, Poon 2004, van de Werf 2012, Yong 2012, Hulstaert 2013). It has also been used to estimate the cost of particle therapy, in particular to evaluate the impact of different technical scenarios' (proton vs. carbon ion [C-ion]) indications and operational models (Perrier 2007, Vanderstraeten 2014).

The Cost of Particle Therapy

Table 16.1 gives an overview of resource cost calculation studies devoted to particle therapy. Cost estimates based on charges or reimbursements were not considered in this overview.

Particle therapy has typically been estimated to cost 2–3 times more than external beam photon therapy. Some 20 years ago, this ratio was calculated by the French National Agency for Medical Evaluation, and others drew similar conclusions in 2003 (Fleurette 1996, Goitein 2003). The latter cost calculation consisted of a theoretical model, assuming fully operational reference departments for photons and protons, with specified construction and operational parameters and costs. The primary output of the model was the average cost per proton fraction delivered and estimated at 2.4 times the fraction cost of photon therapy, yet, with the assumption that this ratio would come down to 2.1 or even 1.7 over the next decade, following improvements in efficiency. In a more recent study comparing the cost of combined particle, proton-only and photon facilities, however, the cost ratio at fraction level for protons compared to photons still amounted to 3.2; and, considering C-ion therapy, even to 4.8 (Peeters 2010).

It is clear that due to its relation to the use of resources, the number of fractions has a considerable impact on the treatment cost. As an example, whereas in the calculation of Peeters the capital investment and annual operational costs were, not surprisingly, estimated about 50% higher for combined C-ion/proton therapy facilities compared to proton-only centres, the treatment cost for early-stage lung cancer was calculated at €10,030 for C-ions and €12,380 for protons, based on the assumption that the former would allow more hypo-fractionated schedules (4 vs. 10 fractions) (Peeters 2010). The average cost per fraction is nevertheless a simplified measure that does not necessarily reflect the reality of the actual treatment costs: it assumes that the entire cost of a treatment scales linearly with the

TABLE 16.1   Cost Calculation Studies on Particle Therapy

First Author	Publication Year	Context	Analysis	Type of Cost Accounting Model	Average Treatment Cost[a]	Main Conclusions
Goitein	2003	Perform a theoretical analysis of proton therapy costs	Relative cost of protons versus photons (IMRT)	Not specified (computer spread sheet model)	Cost per fraction: Proton: €1,025 IMRT: €425	The cost ratio of proton compared to IMRT is about 2.4. Cost differential may reduce due to efficiency improvements.
Huybrechts	2007	Analyse whether particle therapy can be covered within the public health care budget (Belgium)	Review of clinical and economic evidence on particle therapy, investment and cost analysis	Business model	Cost per treatment[b]: Combined C-ion/proton centre: €24,700 Proton only centre: €19,100	Based on the limited available evidence of clinical effectiveness, it is not justified to invest in a particle therapy centre using the public health care budget.
Perrier	2007	Develop a cost simulation model for carbon ion therapy, applied to five European countries	Cost analysis of carbon ion therapy	Activity-based costing	Cost per treatment: C-ion: €19,008	Operational factors influence the cost. The operational costing tool provides input to define reimbursement.
Peeters	2010	Perform an estimation and comparison of the costs of carbon ion, proton and photon therapy	Cost analysis of photon, proton and combined particle facilities	Not specified (Excel analytical framework)	Cost per fraction: Combined C-ion/proton centre: €1,128 Proton only centre: €743 Photon centre: €233	Investment costs are highest for combined particle facility, lowest for photon facility. Cost differences are smaller for total costs/year and for specific treatment costs.

(Continued)

TABLE 16.1 (*Continued*) Cost Calculation Studies on Particle Therapy

First Author	Publication Year	Context	Analysis	Type of Cost Accounting Model	Average Treatment Cost[a]	Main Conclusions
Vanderstraeten	2014	Determine the treatment cost and required reimbursement for a new particle therapy facility in Belgium, define the impact of type of financing	Cost analysis and required reimbursement for carbon ion only, proton only and combined particle facilities	Time-driven activity-based costing vs. business model	Cost per treatment TD–ABC: *Privately financed*: C-ion only: €29,450; Proton only: €46,342; Combined C-ion/proton: €46,443 *Publicly sponsored*: C-ion only: €16,059; Proton only: €28,296; Combined C-ion/proton: €23,956 Cost per treatment business model[b]: *Privately financed*: C-ion only: €32,400; Proton only: €51,200; Combined C-ion/proton: €51,150 *Publicly sponsored*: C-ion only: €18,400; Proton only: €32,300; Combined C-ion/proton: €27,550	The cost ratio of proton compared to photon is 3.2; that of C-ion 4.8. Shorter fractionation schedules of particles might reduce its costs. Publicly sponsored C-ion only facilities are the most attractive from a financial perspective. Reimbursement for privately financed centres is very sensitive to delays in commissioning and to the interest rate. Higher throughput and hypo-fractionation have a positive impact on the treatment costs.

[a] Costs are presented as in publication, without attempt to correct for the year of analysis.
[b] Reimbursement needed per treatment.
C-ion: Carbon ion therapy.
IMRT: Intensity-modulated radiotherapy.
TD–ABC: Time-driven activity-based costing.

number of fractions, whereas it is known that the relative impact of other costs, such as the cost of treatment preparation or quality assurance, typically increases with decreasing fraction number. TD–ABC is a useful method to correctly calculate the treatment costs accounting for such factors. Using this methodology, Vanderstraeten compared proton-only, carbon-only and combined centres. Even with this more refined costing method, the treatment costs were lower in carbon-only centres. Again, this was primarily driven by the broad acceptance of hypo-fractionation for C-ions as opposed to the assumption of conventional fractionation schedules for protons, as evidence on shortened fractionation schedules was still lacking at the time of analysis (Vanderstraeten 2014). It is obvious that if it were not for the assumed higher patient throughput in C-ion centres – ensuing from hypo-fractionation in combination with a typically adult patient population, often requiring shorter time slots than children – the high investment cost of C-ion facilities would translate into higher treatment costs.

The high capital investments required for equipment and buildings – often amounting to more than €100 million – dominate the cost picture of particle therapy, in contrast to what is the case for photon therapy in high-income countries, where the cost of human resources typically accounts for 50%–60% of the costs (Perez 1993, Lievens 2003, Ploquin 2008, van de Werf 2012). This has several consequences. First, operational factors allowing for more efficient use of the equipment, even if at the detriment of higher salary costs related to longer operating hours, have a positive impact on the treatment cost (Goitein 2003, Perrier 2007, Peeters 2010, Vanderstraeten 2014). In addition, economies of scale decrease the overall treatment cost, amongst others due to the cost of the accelerator being spread over different treatment rooms. But the same goes for dedicated proton or C-ion centres as opposed to combined centres, as a second treatment room similar to the first typically requires lower incremental investment costs (Peeters 2010, Vanderstraeten 2014). Lastly, business costs (more specifically the repayment of loans to cover the investment) can become substantial, as calculated at 42% of the total proton centre costs by Goitein (2003). Hence, higher interest rates and delays in the commissioning and ramp-up phase are factors that have been shown to negatively impact the ultimate treatment cost in privately run centres (Peeters 2010, Vanderstraeten 2014), and lower reimbursement rates may as such threaten for-profit, debt-laden centres (Johnstone 2016).

From the above, it is clear that a broad range of factors related to the type of equipment, the size of the centre, the patient population and treatment indications, and the operational scenarios, all play a role in the ultimate treatment cost. In addition, the incremental nature of technology evolution makes it difficult to conclusively define the costs, and calls for continuous reassessment (Lievens 2015, Lievens 2017). This, however, complicates the issue of determining the adequate reimbursement, which, as stated previously, should ideally be aligned with resource costs. Although all described cost analyses add to this knowledge, and hence support decision-making on the appropriate financing levels, two studies have specifically used a business analysis approach to define the amount of reimbursement needed to make a potential particle centre sustainable (Huybrechts 2007, Vanderstraeten 2014). But prior to financing, policy-makers will also require evidence on the cost-effectiveness and on the budget impact of such novel interventions. These aspects are discussed hereafter.

COST-EFFECTIVENESS

Cost effectiveness analysis (CEA) is one of the methods used to conduct economic evaluations: depending on whether both costs and effects are examined and whether two alternatives are compared, different types of economic evaluations are defined (Table 16.2) (Pijls-Johannesma 2008). CEA and cost-utility analysis (CUA) are the most frequent forms of a full economic evaluation, which compares the relative costs and outcomes (effects) of different courses of action. CEA weighs the extra cost of a new intervention to its incremental gain in clinical outcome in terms of a natural unit such as survival, local control, number of cancers averted and so on. Survival, expressed in number of life years gained (LYG), is the favoured outcome measure as it allows broad comparison amongst different diseases. The result of such an evaluation is presented in a ratio of cost per outcome, for example, cost per LYG (€/LYG) called the incremental cost-effectiveness ratio, or ICER.

In CUA it is acknowledged that both the quantity and the quality of outcome may be important to a patient. CUA therefore introduces quality of life into the equation by adjusting the gain in life years with a 'utility factor', typically ranging from 0 (death) to 1 (perfect health). For example, applying a utility or preference of 0.5 to a survival of one year will result in a quality-adjusted survival of six months. The results of a CUA are expressed in cost per quality adjusted life years or €/QALY (Pijls-Johannesma 2008).

Differences in effects between particle therapy and the conventional alternative treatment, typically some form of photon therapy, are expected, and especially improved local control, a gain in long-term (side) effects or a positive impact on quality of life.

Challenges in Determining Cost-Effectiveness of Technologies

Gathering adequate data is often a problem in performing economic evaluations for technologies, as it is for particle therapy. Different published cost-effectiveness studies, mainly on proton therapy, used different methodologics as well different approaches to retrieve data. Some used Markov models while others used estimations from Medicare/ (Surveillance, Epidemiology and End Results program (SEER) data. Centers for Medicare & Medicaid Services (CMS) is the largest public payer in the United States that publishes rates of reimbursement of hospitals and other healthcare providers. In combination with SEER data (from the U.S. National Cancer Institute, under National Institutes of Health), estimates on (cost-) effectiveness can be made. However, as mentioned, these reimbursement

TABLE 16.2   Types of Economic Evaluation

		Are Both Costs (Input) and Effects (Output) Examined?		
		No		Yes
		Only effects	Only costs	
Is there a comparison	No	Outcome description	Cost description	Cost-outcome description
between two alternatives?	Yes	Efficacy or Effectiveness	Cost analysis	Full economic evaluation[a]

[a] Cost-benefit analysis, cost-utility analysis or cost-effectiveness analysis.

rates may be quite different from the cost actually incurred by providers and, depending on the perspective of the analyses, these estimates may need to be adjusted to reflect this. On the other hand, comparing results of various Markov models should also be done with caution. Markov models are decision-analytic models based on a series of 'health states' that a patient can occupy at a given point in time. While patients evolve through these health states during the course of their disease, they accumulate survival and consume costs (Briggs 1998). The impact of variables included in a model – patient and tumour characteristics, assumed impact of the treatment on outcome, treatment costs – is essential for the final result of the analysis (Lievens 2013a). Therefore, comparing results of Markov models where different variables are used risks drawing wrong conclusions.

High-quality economic evaluations for new technologies are scarce. This is in part related to the fact that guidelines on economic evaluations have primarily been written for pharmaceuticals. But the main reason for lack of evidence on the economic aspects of new technologies such as radiotherapy – in contrast to the abundant literature available for oncology drugs – involves the different path and timeline to bring drugs and devices to the market. As neither comparative effectiveness data nor economic evidence is mandatory before market launch, responsibility for evidence generation, if any, is typically shifted to health care providers after investments have been made (Lievens 2015, Lievens 2017).

There are also aspects related to the technology itself that complicate the evaluation of its value. Unlike drugs, health technologies often have multiple applications: when they reach the market, their use is not strictly defined as it is for drugs. Also, important improvements of a new technology may take place over time (Taylor 2009). In contrast with drugs, the technical performance of a medical device involves an interaction between the device and the clinician, which develops incrementally by refining the clinical procedure and with the increasing skill and expertise of the operator. While drugs are a classic case of 'embodied technology', medical technology requires training before its implementation in routine clinical practice. Last but not least, specifically for radiotherapy, it is well known that the clinical benefits typically take years to mature. For all of these reasons, accurate and stable data on the effects and costs of new radiotherapy technologies are hard to obtain (Lievens 2017).

Past, Current and Future Evidence of Cost-Effectiveness

In the past decade, a limited number of systematic reviews of effectiveness and cost-effectiveness studies on particle therapy have been performed (Brada 2007, Lodge 2007, Pijls-Johannesma 2008, Brada 2009, Lievens 2013, Verma 2016). Disappointedly, conclusions on (cost-)effectiveness remain the same over that 10-year period: for most indications, no firm conclusions on (cost-)effectiveness could be drawn due to a lack of evidence, mainly related to inadequate data collection, and as a consequence, limited data to perform good analysis. The basis of this evidence gap has been described earlier.

The last cost-effectiveness review in this list, published in 2016, again stated that it is highly unlikely that protons will be the most economic option for all cancers or even for all patients with a given type of cancer (Verma 2016). It was concluded that with greatly limited amounts of data, proton beam therapy (PBT) offers promising cost-effectiveness for treating paediatric brain tumours, well-selected breast cancers, locally advanced non-small cell lung cancer (NSCLC) and high-risk head and neck cancers. It has not been demonstrated that PBT is cost-effective for prostate cancer or early-stage NSCLC. Included studies used different methods to determine cost-effectiveness.

Below more detailed findings from this review are described for prostate cancer, breast cancer, non-small cell lung cancer (NSCLC), head and neck cancer, paediatrics and uveal melanomas. Where available, results of additional CEA publications on PBT will be mentioned. Table 16.3 shows the CEAs published since 2012.

Prostate Cancer

Although difficult to estimate, literature from 2003 indicated that worldwide about 26% of all cases treated with protons were prostate cancer cases (Sisterson 2005). Despite the relatively long history of treating prostate cancer with protons and the large number of patients, no clear results on (cost-)effectiveness are available (Martin 2014). This is also the overall conclusion of the review by Verma. One study, for example, used Medicare reimbursement data and identified 21,647 patients who received prostate radiotherapy, of which 553 were treated with protons and the remainder with intensity-modulated radiotherapy (IMRT). The median Medicare reimbursement was U.S.$32,428 for PBT and U.S.$18,575 for IMRT. The authors concluded that PBT was not sufficiently cost-effective at similar toxicity levels experienced by both groups (Yu 2013). An additional Markov model was published comparing protons with photon therapy, that is, IMRT and stereotactic body radiotherapy (SBRT) in 2012. It concluded that SBRT is the most cost-effective, compared to IMRT and PBT, from societal and payers' perspectives (Parthan 2012). Based on the results of the review, Verma stated that it is important to consider this lack of proven cost-effectiveness for prostate cancer given the relative ubiquity of patients receiving prostate PBT treatments worldwide (Verma 2016).

Breast Cancer

Verma summarised CEAs that were also incorporated in earlier reviews with no clear conclusions due to the lack of data and the given uncertainty (Brada 2007, Lodge 2007, Pijls-Johannesma 2008, Brada 2009). More recently, an interesting study was published about the impact of patient selection on the cost-effectiveness of PBT for breast cancer (Mailhot Vega 2016). The authors found that at a willingness to pay (WTP) of U.S.$50,000/QALY, proton therapy was cost-effective for women with one cardiac risk factor beginning at mean heart doses of 9 and 10 Gray for women aged 50 and 60 years, respectively. Hence, although not cost-effective for all, their data suggest that proton therapy may be cost-effective for specific patients, that is, those with one cardiac risk factor in cases for which photon plans are unable to achieve a mean heart doses below five Gray.

Non-Small Cell Lung Cancer

Two CEA studies using Markov modelling were included in the review (Grutters 2010, Lievens 2013). Grutters performed a CEA for inoperable stage I NSCLC, comparing four treatment options: PBT, C-ion therapy, three-dimensional conformal radiation therapy (3D-CRT) and SBRT (Grutters 2010). For a five-year time horizon, the costs were

TABLE 16.3 Cost-Effectiveness Analyses of Proton Therapy on Different Tumour Sites Published Since 2012

First Author	Publication Year	Institute	Sites	Design	Subgroup Analysis	Comparator(s)	Cost/QALY	Outcomes	Main Conclusions
Leeung	2017	An Nan Hospital, Taiwan	Inoperable advanced HCC	Markov	Yes	Proton versus SBRT	NT$ 557,907 per QALY gained (ICER: NT$ 213,354)	Time to progression and survival	Protons are cost-effective compared to SBRT. The incremental cost-effectiveness ratio could be considerably lower if patients at higher risk of severe toxicity were chosen for the treatment. The cost-effectiveness is thus highly dependent on the possibility of selecting appropriate risk patients for the therapy.
Mailhot Vega	2016	MGH, U.S.	Breast	Markov, model-based (NTCP)	Yes	proton vs. photon	At a WTP of $50,000/QALY, PBT was cost-effective for women with one CRF beginning at MHDs of 9 and 10 Gray for women aged 50 years and 60 years, respectively.	Presence or lack of cardiac risk factors (CRFs)	Referral for PBT may be cost-effective for patients with one CRF in cases for which photon plans are unable to achieve an MHD <5 Gray.
Moriartry	2015	Mayo Clinic, U.S.	Uveal melanoma	Markov		Proton versus plaque brachytherapy versus enucleation	Compared with enucleation, ICERs for plaque brachytherapy and PBT were $77,500/QALY and $106,100/QALY, respectively.		At a WTP of $50,000/QALY, neither PBT nor plaque brachytherapy were cost-effective
Herano	2014	Shizuoka Cancer Center, Japan	Paediatrics medullobalstoma	Markov	No	Proton versus IMRT	Model obtains population- and patient-specific costs and utilities associated with therapy.	Risks of hearing loss[a] HRQOL (EQ5D, SF6D, HU13)	99% probability of PBT being cost effective at a societal WTP value.

(Continued)

TABLE 16.3 (*Continued*)   Cost-Effectiveness Analyses of Proton Therapy on Different Tumour Sites Published Since 2012

First Author	Publication Year	Institute	Sites	Design	Subgroup Analysis	Comparator(s)	Cost/QALY	Outcomes	Main Conclusions
Mailhot Vega	2013	MGH, U.S.	Paediatrics, medulloblastoma	Monte Carlo	No	Proton versus IMRT	IMRT cost was estimated at $112,790, PBT cost at $80,211, and total QALYs gained favoured the PBT treated patients (17.37 vs. 13.91, respectively): <$5,000 per QALY	Risk of developing 10 adverse events: growth hormone deficiency, coronary artery disease, ototoxicity, secondary malignant neoplasm and death	PBT is cost-effective.
Parthan	2012	Optum, US	Prostate	Markov	No	Proton versus photon (IMRT and SBRT)	Payers' perspective: SBRT: $24,873/8.11 IMRT: $33,068/8.05 PBT: $69,412/8.06	Treatment-related mortality and long-term toxicity: • gastrointestinal • genitourinary • sexual dysfunction (CTCAE/RTOG scale)	SBRT is cost-effective compared to IMRT and PBT from societal and payers' perspectives.
Ramaekers	2012	Maastro Clinic, The Netherlands	Head and neck	Markov, model-based (NTCP)	Yes	IMRT all versus IMPT all versus IMPT if effective	IMPT if efficient as opposed to IMRT for all patients resulted in an ICER of €60,278 per QALY gained. IMPT for all patients compared with IMPT if efficient resulted in an ICER of €127,946 per QALY gained.	Xerostomia and dysphagia	PBT is not cost-effective for all head and neck cancer patients but it is for a subgroup of patients.

WTP = willingness to pay; CRF = cardiac risk factor; ICER = incremental cost-effectiveness ratio; NTCP = normal tissue complication probability; PBT = proton beam therapy; IMPT = intensity-modulated proton therapy.

a Calculated on cochlear dose for each treatment.

respectively €27,567, €19,215, €22,696 and €13,871; the corresponding QALYs gained 2.33, 2.67, 1.98 and 2.59. This translated into SBRT and C-ion therapy being most cost-effective for inoperable stage I NSCLC patients, while SBRT remained most cost-effective for operable patients. Lievens performed a Markov analysis of concurrent chemoradiotherapy for locally advanced NSCLC using PBT, IMRT and 3D-CRT. Based on the then available effectiveness data, PBT increased the QALYs gained by 0.549 and 0.452 compared with 3D-CRT and IMRT, respectively, with resulting ICERs of €34,396 and €31,541 for PBT compared to 3D-CRT and IMRT, respectively (Lievens 2013).

Though limited, these data suggest that protons and C-ions, depending of the indication and patient population considered, might be cost-effective as compared to other radiotherapeutic options, which is in line with published *in-silico* trials (Roelofs 2012). However, since multiple factors are of significant influence to the outcome in lung cancer treatment, multifactor models combining patient, clinical and treatment variables, will be useful to determine the best possible treatment options for individuals or subgroups of NSCLC patients. Work in this context is ongoing. Richard, for example, used a Bayesian network to develop a multifactor model predicting pneumonitis, oesophagitis and patient-specific costs associated with therapy for stage III NSCLC; in the same patient population Oberije developed a nomogram using stratified Cox regression to predict survival (Richard 2014, Oberije 2015).

Head and Neck Cancer

Two Markov model studies evaluate the cost-effectiveness of particle therapy in head and neck cancer. Back in 2005, the Swedish study of Lundkvist (2005) showed that the cost of PBT was only €3,887 higher than conventional photon radiotherapy, yet resulted in a substantial gain in QALYs (1.02). From these results the cost-effectiveness of PBT for head and neck cancer seems obvious, but due to a lack of firm outcome and cost data at the time of analysis, the results were uncertain.

Another Markov modelling study by Ramaekers included patients with stage III and IV oral cavity, laryngeal and pharyngeal cancers. Three groups were compared: IMRT delivered to all patients versus intensity-modulated proton therapy (IMPT) delivered to all patients versus applying IMPT only to those patients where dosimetric data combined with costs suggested efficiency. The results showed that 'IMPT if efficient' as opposed to 'IMRT for all' patients resulted in an acceptable ICER of €60,278 per QALY gained. 'IMPT for all' patients compared with 'IMPT if efficient' was, however, not cost-effective with an ICER of €127,946 per QALY gained. These results showed that, assuming equal survival for IMPT and IMRT, IMPT could still be cost-effective for individually selected patients based on predicted toxicity profile. In practice this means that one should make a trade-off between expected costs and benefits for each individual patient.

In line with the above-mentioned multifactor model for NSCLC, such a model would be valuable for head and neck cancer as well. As a matter of fact, Cheng developed an online three-level proton versus photon decision support prototype for head and neck cancer that allows comparison of the dose, toxicity and cost-effectiveness. Using this model and given nationally accepted guidelines for a 15% reduction of complications, including swallowing dysfunction and xerostomia, all patients would clinically benefit from proton therapy after six months and 91% after 12 months, while only 35% would be considered cost-effective at a threshold of 80,000€ per QALY gained (Cheng 2016).

Paediatrics

The review by Verma included five childhood studies on PBT, all but one of which concerned medulloblastoma (Lundkvist 2005a, 2005b, Mailhot Vega 2013, Hirano 2014, Mailhot Vega 2015). The Swedish group found that the factors most contributing to the costs of adverse events were decrease in IQ, hearing loss and growth hormone deficiency. Their Markov model simulation showed that due to a substantial decrease in these side effects in the lifetime of these children, PBT dominated the cost analyses (€23,647 per patient saved with PBT) along with substantially improving their quality of life (0.683 QALYs gained from PBT). Based on these data, PBT is the dominant situation from a health economic perspective, supporting it as a standard indication. Again, it should be mentioned that there remained a high level of uncertainty around the inputs used in the model (Lundkvist 2005a, 2005b). More recently, a similar CEA was performed by Mailhot Vega (2013). Here IMRT cost was estimated at $112,790, while the PBT cost was estimated at $80,211 and total QALYs gained favoured the PBT-treated patients (17.37 vs. 13.91, respectively). Lastly, a Japanese CEA on medulloblastoma also used a Markov model approach to compare PBT with photons, based on the difference of delivered cochlear dose for each treatment and the subsequent risk of hearing loss (Hirano 2014). Three types of health-related quality of life (HRQOL) measurements were used (EQ-5D, HUI3 and SF-6D) to estimate the QALYs. The analysis showed that computed ICER values differed between EQ-5D, HUI3 and SF-6D, being $21,716/QALY, $11,773/QALY and $20,150/QALY, respectively. Yet, irrespective of the utility index used, the ICERs were lower than $46,729/QALY (JPY 5 million/QALY), which is considered the WTP in Japan.

A last study modelled the effects of hypothalamic radiotherapy dose and the ensuing risk of growth hormone deficiency in a population of children with brain tumours, aged 4–12 years at the time of diagnosis. Various ICERs were computed, depending on age, hypothalamic dose and growth hormone deficiency (Mailhot Vega 2015).

Whereas dosimetric superiority and/or reduced toxicities with PBT have been demonstrated for several other paediatric tumours (such as retinoblastoma, ependymoma, craniopharyngioma, rhabdomyosarcoma and esthesioneuroblastoma), CEAs have not yet been performed for these indications (Verma 2016).

UVEAL MELANOMA

Verma summarised a Markov study by the Mayo Clinic on uveal melanomas (Moriarty 2015). PBT, enucleation and plaque brachytherapy were compared. They reported similar costs of $22,772 for enucleation, $24,894 for PBT and $28,662 for plaque brachytherapy. The QALYs gained were nearly identical at 2.918, 2.938 and 2.994, respectively. These results indicated that PBT was not cost-effective. However, firm conclusions can't be made due to criticism of the methodology and model assumptions. For example, no stratification on tumour size or localisation was performed, whereas evidence on clinical efficacy of PBT seems obvious in large ocular tumours and in some specific anatomical localisations in the eye, while the superiority of PBT in smaller lesions (<4 millimetres) was less clear (Lodge 2007).

Taking the above-mentioned discussion into consideration, it seems hard, or even impossible, to estimate the cost-effectiveness of particle therapy based on the published literature and the currently used methodology of evidence synthesis by clinical indication or by tumour type. It becomes more and more evident that particles will not be cost-effective for unselected patient groups, hence that the choice for particle therapy versus more standard radiotherapy approaches will have to be individualised for each patient, taking into account specific risk factors and tumour anatomy. This calls for a new methodological approach to evidence generation and economic evaluation (Lievens 2017).

A model-based approach could be the solution to assess the cost-effectiveness of particle therapy for specific subgroups or for individual patients. As discussed before, dose response and normal tissue complication probability models can be used to predict the incremental clinical benefit of particle therapy at the patient level, and weight this to its incremental cost to define cost-effectiveness. This approach may also allow for selecting enriched cohorts of patients who are likely to benefit from protons and could be enrolled in more formal clinical trials. The use of decision support systems could be of importance for the practical implementation of such an approach in the clinical setting.

Several of the previously discussed studies indeed show that subgroup analysis (by age, by co-morbidity, by toxicity risk) enhances the cost-effectiveness of proton therapy. The studies on hepatocellular carcinoma (Leung 2017), breast cancer by Mailhot Vega and head and neck cancer by Ramaekers moreover demonstrated that using a model-based approach could successfully be applied to CFA (Ramaekers 2013, Mailhot Vega 2016).

BUDGET IMPACT

In addition to defining the acceptability of care through health economic evaluation, it is imperative to ascertain its affordability. As a consequence, there is growing recognition that a comprehensive economic assessment of new health care interventions requires both cost-effectiveness and BIA. Whereas an economic evaluation is performed to define the monetary value of a new intervention as compared to the actual standard, the aim of a BIA is to estimate the impact of a (possible) policy decision concerning a new intervention on the planned resource use and the health care budget consumed over a period of time (Mauskopf 2007). One major difference between CEA and BIA is the target population: although in both cases consistent with the reimbursement request, it is closed in CEA and open in BIA. This means that in addition to the actual size of the target population conforming to the inclusion criteria of the trial generating the evidence, one should also forecast its evolution over time. However, in a context of residual uncertainty about the clinical outcome, it may be difficult to define this target population, hence the foreseeable costs.

Nevertheless, country-level estimates of the patient population that would benefit from particle therapy have been performed in view of planning and budgeting for particle therapy. Such analyses are typically based on available statistics of tumour incidence, number of patients potentially eligible for radiotherapy and scientific support from clinical trials on the added benefit of particle therapy in terms of improved tumour control or decreased toxicity. In addition, model dose-planning studies and knowledge of the dose-response relations of different tumours and normal tissues can help to further fine-tune the number of cases that would benefit from particle therapy at country level. In Sweden, it was estimated that between 2,200 and 2,500 patients annually – 14%–15% of all irradiated patients – would be eligible for PBT, and that for these patients the potential therapeutic benefit is so great as to justify the additional expense of proton therapy (Glimelius 2005). Of course, the estimated numbers are highly dependent on the strictness of the definition of clinical benefit and are expected to evolve over time due to a changing cancer population mix, clinical evidence and technical capabilities. In Belgium, for example, a first estimate by the Belgian Health Care Knowledge Centre in 2007 judged that on annual basis not more than 100 patients would benefit from particle therapy, while a more recent feasibility study of a hadron therapy centre in Belgium in 2013 estimated the number of eligible patients at 233 for standard indications (such as paediatric tumours) along with 1,820 for seven model indications (a.o. pancreatic and locally advanced NSCLC) (Huybrechts 2007, De Croock 2013). Apart from defining the potential patient population, the latter study also performed economic analyses regarding the investment, operational and treatment costs as well as on required reimbursement considering different technical and infrastructural scenarios (Vanderstraeten 2014).

In addition to the country perspective, important for health care payers and decision makers, such analysis is also important to estimate the clinical and financial impacts at departmental level. To evaluate how PBT utilisation might fit into the existing radiation oncology armamentarium of an academic hospital, all patients treated within the time frame of one year were evaluated for potential PBT treatment using various adoption scenarios. Not unexpectedly, it was found that the degree of PBT utilisation would depend on the strictness of selection criteria, which in the specific centre would range between 6% and 25%, with concomitant cost increases from minimal to 40% (Dvorak 2013).

Although it can be expected that before investing in particle therapy, countries and departments do assess the financial investment and operational impact of such a decision, literature on this aspect is scarce. The publication of formal BIAs would, however, be welcomed, as it would add to our understanding of the economic consequences of implementing such costly technology.

REFERENCES

Barnett PG. 2009. An improved set of standards for finding cost for cost-effectiveness analysis. *Med Care* 47:S82–S88.

Bodenheimer T. 2005. High and rising health care costs. Part 2: Technologic evolution. *Ann Intern Med.* 142:932–937.

Brada M. 2007. Proton therapy in clinical practice: Current clinical evidence. *J Clin Oncol.* 25(8):965–970.

Brada M. 2009. Current clinical evidence for proton therapy. *Cancer J.* 15(4):319–324.

Briggs A. 1998. An introduction for Markov modeling for economic evaluation. *Pharmacoeconomics* 13(4):397–409.

Cheng Q. 2016. Development and evaluation of an online three-level proton vs photon decision support prototype for head and neck cancer—Comparison of dose, toxicity and cost-effectiveness. *Radiother Oncol.* 118(2):281–285.

De Croock R. 2013. Feasibility study of a Hadron Therapy Centre in Belgium. Financed by the Federal Public Service of Health, Food Chain Safety and Environment. *Cancer Plan Action* 30, May 10.

Defourny N. 2016. Cost evaluations of radiotherapy: What do we know? An ESTRO-HERO analysis. *Radiother Oncol.* 121(3):468–474.

Dvorak T. 2013. Utilization of proton therapy: Evidence-based, market-driven, or something in-between? *Am J Clin Oncol.* 36(2):192–196.

Fleurette F. 1996. Proton and neutron radiation in cancer treatment: Clinical and economic outcomes. *Bull Cancer Radiother.* 83:S223–S227.

Glimelius B. 2005. Number of patients potentially eligible for proton therapy. *Acta Oncol.* 44(8):836–849.

Goitein M. 2003. The relative costs of proton and X-ray radiation therapy. *Clin Oncol (R Coll Radiol).* 15(1):S37–S50.

Grutters JP. 2010. The cost-effectiveness of particle therapy in non-small cell lung cancer: Exploring decision uncertainty and areas for future research. *Cancer Treat Rev.* 36:468–476.

Hirano E. 2014. Cost-effectiveness analysis of cochlear dose reduction by proton beam therapy for medulloblastoma in childhood. *J Radiat Res.* 55(2):320–327.

Hulstaert F. 2013. Innovative radiotherapy techniques: A multicentre time-driven activity-based costing study. Health Technology Assessment (HTA). Brussels, Belgium: Belgian Health Care Knowledge Centre (KCE); KCE Reports 198C. D/2013/10.273/9.

Huybrechts M. 2007. Hadrontherapy. Health Technology Assessment (HTA). Brussels, Belgium: Federaal Kenniscentrum voor de Gezondheidszorg (KCE). KCE reports 67A.

Johnstone P. 2016. Reconciling reimbursement for proton therapy. *Int J Radiat Oncol Biol Phys.* 95:9–10.

Jönsson B. 2016. The cost and burden of cancer in the European Union 1995–2014. *Eur J Cancer.* 66:162–170.

Kaplan R. 2011. How to solve the cost crisis in health care? *Harvard Bus Rev.* 9:47–64.

Kesteloot K. 1996. Reimbursement for radiotherapy treatment in the EU countries: How to encourage efficiency, quality and access? *Radiother Oncol.* 38:187–194.

Leung HWC. 2017. Cost-utility of stereotactic radiation therapy versus proton beam therapy for inoperable advanced hepatocellular carcinoma. *Oncotarget.* 8:75568–75576.

Lievens Y. 2003. Activity-Based costing: A practical model for cost calculation in radiotherapy. *Int J Radiat Oncol Biol Phys.* 57:522–535.

Lievens Y. 2013a. Health economic controversy and cost-effectiveness of proton therapy. *Semin Radiat Oncol.* 23(2):134–141.

Lievens Y. 2013b. Proton radiotherapy for locally-advanced non-small cell lung cancer, a cost-effective alternative to photon radiotherapy in Belgium? *J Thorac Oncol.* 8(suppl 2):S839–S840.

Lievens Y. 2015. Cost calculation: A necessary step towards widespread adoption of advanced radiotherapy technology. *Acta Oncol.* 54(9):1275–1281.

Lievens Y. 2017. Access to innovative radiotherapy: How to make it happen from an economic perspective? *Acta Oncol.* doi:10.1080/0284186X.2017.1348622.

Lodge M. 2007. A systematic literature review of the clinical and cost-effectiveness of hadron therapy in cancer. *Radiother Oncol.* 83(2):110–122.

Lundkvist J. 2005a. Cost-effectiveness of proton radiation in the treatment of childhood medulloblastoma. *Cancer* 103:793–801.

Lundkvist J. 2005b. Proton therapy of cancer: Potential clinical advantages and cost-effectiveness. *Acta Oncol.* 44(8):850–861.

Mailhot Vega R. 2015. Cost effectiveness of proton versus photon radiation therapy with respect to the risk of growth hormone deficiency in children. *Cancer* 121:1694–1702.

Mailhot Vega RB. 2013. Cost effectiveness of proton therapy compared with photon therapy in the management of pediatric medulloblastoma. *Cancer* 119:4299–4307.

Mailhot Vega RB. 2016. Establishing cost-effective allocation of proton therapy for breast irradiation. *Int J Radiat Oncol Biol Phys.* 95(1):11–18.

Martin NE. 2014. Progress and controversies: Radiation therapy for prostate cancer. *CA Cancer J Clin.* 64:389–407.

Mauskopf JA. 2007. Principles of good practice for budget impact analysis: Report of the ISPOR Task Force on good research practices-budget impact analysis. *Value Heal.* 10(5):336–347.

Moriarty JP. 2015. Cost-effectiveness of proton beam therapy for intraocular melanoma (serial online). *PLoS One* 10:e0127814.

Oberije C. 2015. A validated prediction model for overall survival from stage III non-small cell lung cancer: Toward survival prediction for individual patients. *Int J Radiat Oncol Biol Phys.* 92(4):935–944.

Parthan A. 2012. Comparative cost-effectiveness of stereotactic body radiation therapy versus intensity-modulated and proton radiation therapy for localized prostate cancer. *Front Oncol.* 2:81.

Peeters A. 2010. How costly is particle therapy? Cost analysis of external beam radiotherapy with carbon-ions, protons and photons. *Radiother Oncol.* 95(1):45–53.

Perez CA. 1993. Cost accounting in radiation oncology: A computer-based model for reimbursement. *Int J Radiat Oncol Biol Phys.* 25:895–906.

Perrier L. 2007. A decision-making tool for a costly innovative technology. The case for carbon ion radiotherapy. *J Econ Méd.* 25(7–8):367–380.

Perrier L. 2013. Cost of prostate image-guided radiation therapy: Results of randomized trial. *Radiother Oncol.* 106:50–58.

Pijls-Johannesma M. 2008. Cost-effectiveness of particle therapy: Current evidence and future needs. *Radiother Oncol.* 89(2):127–134.

Ploquin NP. 2008. The cost of radiation therapy. *Radiother Oncol.* 86:217–223.

Poon I. 2004. The changing costs of radiation treatment for early prostate cancer in Ontario: A comparison between conventional and conformal external beam radiotherapy. *Can J Urol.* 11:2125–2132.

Ramaekers BL. 2013. Protons in head-and-neck cancer: Bridging the gap of evidence. *Int J Radiat Oncol Biol Phys.* 85(5):1282–1288.

Richard P. 2014. Development of a multi parametric cost-effectiveness model for comparison of therapeutic modalities in definitive radiation therapy for stage III non-small cell lung cancer (NSCLC). *Int J Radiat Oncol Biol Phys.* 90:S589.

Roelofs E. 2012. Results of a multicentric in silico clinical trial (ROCOCO): Comparing radiotherapy with photons and protons for non-small cell lung cancer. *J Thorac Oncol.* 7(1):165–176.

Sisterson J. 2005. Ion beam therapy in 2004. *Nucl Inst Meth Phys Res.* 241:713–716.

Sorenson C. 2013. Medical technology as a key driver of rising health expenditure: Disentangling the relationship. *ClinicoEcon Outcomes Res.* 5:223–234.

Taylor RS. 2009. Assessing the clinical and cost-effectiveness of medical devices and drugs: Are they that different? *Value Heal.* 12(4):404–406.

van de Werf E. 2012. The cost of radiotherapy in a decade of technology evolution. *Radiother Oncol.* 102:148–153.

Vanderstraeten B. 2014. In search of the economic sustainability of Hadron therapy: The real cost of setting up and operating a Hadron facility. *Int J Radiat Oncol Biol Phys.* 89(1):152–160.

Verma V. 2016. A systematic review of the cost and cost-effectiveness studies of proton radiotherapy. *Cancer* 122(10):1483–1501.

Yong JH. 2012. Cost-effectiveness of intensity-modulated radiotherapy in oropharyngeal cancer. *Clin Oncol.* 30;24:532–538.

Yu JB. 2013. Proton versus intensity-modulated radiotherapy for prostate cancer: Patterns of care and early toxicity. *J Natl Cancer Inst.* 105:25–32.

CHAPTER 17

Big Data-Based Decision Support Systems for Hadron Therapy

Yvonka van Wijk, Cary Oberije, Erik Roelofs
and Philippe Lambin

CONTENTS

INTRODUCTION

In the last decade, major advances have been made in the field of radiation oncology, bringing new diagnostic techniques and expanding the number of treatment modalities [1]. With a larger number of treatment options comes a heightened potential for personalised medicine, however, this has its challenges. Traditional evidence-based medicine uses randomised trials that are designed to represent homogenous populations of patients, and are not based upon patient, disease and treatment parameters [2].

The challenge in clinical decision-making is that physicians must take into account a large number of characteristics, such as patient biology, pathology, medical images, blood test results, given medication, dose in the organs at risk, dose in the target volume, fractionation and genomic data relevant to radiotherapy. The decisions should take into account the tumour control rate, the survival rate and the radiotoxicity. The human cognitive capacity is limited, however, making predictive modelling and big data in radiation oncology an increasingly essential tool in decision-making [3].

Particle therapy modalities, such as hadron therapy, are expected to be less toxic and more effective than the more conventional photon therapy, due to their favourable dose distributions [4,5]. However, planning studies show that not all patients would benefit from this more expensive treatment [6,7], making decision support systems (DSS) an important tool to justify patient stratification for particle therapy.

This chapter will discuss the process of gathering data, training models and developing DSS using rapid learning health care (RLHC).

RAPID LEARNING HEALTH CARE

During clinical trials, certain patient groups are under-represented, causing the data to be biased [8,9]. RLHC uses and reuses medical data from both clinical trials and from standard clinical practice in novel studies or for decision support in new patients [10,11]. RLHC consists of four sequential phases: (1) the data, (2) knowledge, (3) application and (4) evaluation phases [3]. Figure 17.1 shows an overview of these phases. The first phase includes the obtaining and mining of prior data regarding the patients, disease, treatment and outcome. The second phase analyses this data using methods such as machine learning to

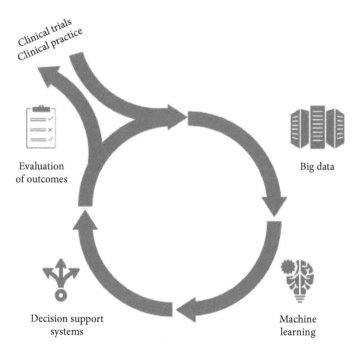

FIGURE 17.1 The phases of rapid learning health care (RLHC): Data collection from clinical practice and clinical trials, development of knowledge on the obtained data, application of the models via decision support systems (DSS), evaluation of the DSS and the application of DSS in clinical practice. (Courtesy of T. van Wijk.)

obtain knowledge from the data. The application phase improves clinical practice using this knowledge in the form of DSS. The final phase evaluates the DSS with regard to outcome, and uses this assessment as input for the first phase.

BIG DATA

Ideally, the data obtained in the first phase of RLHC meets the classification demands of 'Big Data': the data should have *volume* so that enhanced knowledge can be obtained with large amounts of variables; the data should have *velocity* so that the gained knowledge remains practicable; the data should have *variety* in order to find which treatment best fits an individual patient; and the data should have *veracity* so that trust can be placed in the acquired knowledge [12]. The primary challenge in this phase of RLHC is obtaining data that meets these demands. CancerLinQ (http://cancerlinq.org/) is an initiative of the American Society of Clinical Oncology (ASCO) to meet this challenge [13]. Figure 17.2 shows an overview of the CancerLinQ system. The system collects and analyses all patient data obtained from the complete electronic health records (EHRs), and offers the opportunity to learn from previous patients, improving the quality of care. The system is built to transfer data, including patient and care provider demographics, appointments and visit details, medical and prescription history, surgery reports, pathology and lab data and family history. Using language recognition methods, the software will even be able to extract information from clinician notes.

Centralisation of data faces some difficult challenges, such as differences between institutes including language and data storage, and the reluctance to share data due to its value, privacy issues and lack of resources [14]. The first steps to overcoming these issues are

Data from practice
to CancerLinQ

CancerLinQ processes
Data on patient level

De-identification

Data analysis
parameterised reporting

FIGURE 17.2 CancerLinQ is an initiative of the American Society of Clinical Oncology (ASCO) and works in four steps: (1) data is uploaded to CancerLinQ from clinical practice, (2) the data is processed on patient level, (3) patient data is de-identified, (4) the data is analysed and reported and made available to clinical practice. (Courtesy of T. van Wijk.)

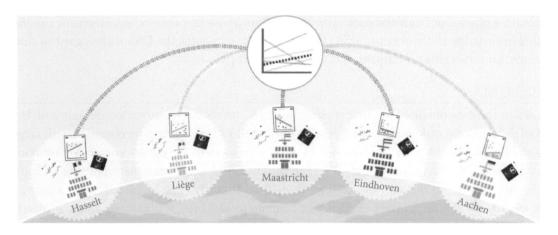

FIGURE 17.3 The graphical abstract for "Infrastructure and distributed learning methodology for privacy-preserving multi-centric rapid learning health care: EuroCAT." (From Deist, T.M. et al., *Clin. Trans. Radiat. Oncol.*, 4, 24–31, 2017.)

being made through cooperation among radiotherapy institutes all over the world (the Netherlands, Germany, Belgium, the United Kingdom, the United States, Italy, Denmark, Australia, India and China, with South Africa, Ireland and Canada as prospective partners), which applies RLHC using the EuroCAT project (www.eurocat.info) [15]. Figure 17.3 shows the infrastructure of the EuroCAT project. This project overcomes some of the classical barriers to data sharing by using 'distributed learning' to train RLHC models on data locally without the data needing to leave its relative institute (please see the animation: https://www.youtube.com/watch?v=nQpqMIuHyOk).

For this project to work, data within the institutes needs to be translated to general, well-defined terms. Machine-readable data needs to be developed, where local terms can be matched to the general ones, for example the NCI Thesaurus. These general terms serve as an interface for each institute, enabling information gathering and interpretation by a semantic gateway to the data. This also stimulates the harmonisation between institutes on which data should be collected and how (disease-specific 'umbrella' protocols) [16].

MACHINE LEARNING

Extracting knowledge from data is the third step in RLHC, and a common method used for this is machine learning [17]. Machine learning trains models on past data and uses it to make predictions regarding outcomes, such as tumour control and complication risks on data obtained from a new patient. Before models obtained with machine learning can be applied, they need to be validated, for which the TRIPOD (transparent reporting of a multivariable prediction model for individual prognosis or diagnosis) statement is a popular template [18]. The TRIPOD statement allows the reporting of the development, validation or updating of prediction tools. Proper validation should be done on external datasets, that is, from a dataset comparable to the training set but from a different institute. There are a number of radiotherapy models available for different types of diseases, for example, http://predictcancer.org, https://mskcc.org/nomograms/prostate, and http://research.nki.nl/ibr/.

DECISION SUPPORT SYSTEMS

The third phase of RLHC is the application of gained knowledge, which can be applied using DSS. DSS are typically software applications such as recursive partitioning analyses models, nomograms or websites such as http://www.predictcancer.org. DSS are used to support the groups responsible for deciding on patient treatment, such as the physician, the tumour board and the patient, in making knowledgeable decisions regarding treatment options. Within the field of radiotherapy, it is acceptable to use physics- and radiobiology-based models to predict complication risk and tumour control using tumour control probability (TCP) models [20] and normal tissue complication probability (NTCP) models [21]. An example of a DSS is the software "PRODECIS", which compares a proton treatment plan to a photon treatment plan for head and neck cancer (Figure 17.4). DSS represent an extension to this practice, integrating validated physics- and radiobiology-based models with patient and disease parameters into a multifactorial structure [22].

The goal of DSS is to use a combination of features (clinical, treatment, imaging, bio-markers, pathology, genetics etc.) to make robust predictions on patient outcome [23], as shown in Figure 17.5. The outcome predicted by DSS can include tumour control, including local recurrence or rate of metastases, complication probability, overall survival or even cost-effectiveness (CE).

Clinical Features

In several studies, clinical features have been found to have a large influence on patient outcome [24,25], and for DSS to make individual predictions for patients, these features should be integrated into the system. Examples of clinical features are disease progression, organ function, age or sex of the patient and so on. The measurement of some of these features, such as performance status, is straightforward and can be done using methods such as questionnaires [26]. However, in order to reach a level of trustworthiness similar to

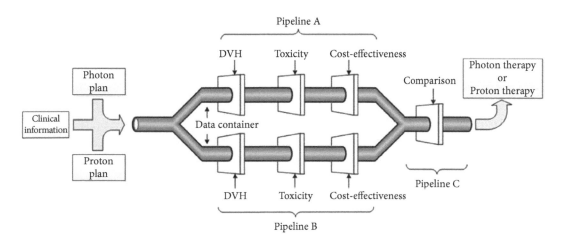

FIGURE 17.4 An example of a multifactorial Decision Support System that compares a photon plan to a proton plan and gives a patient specific recommendation. (From Cheng, Q. et al., *Radiother Oncol*, 118(2), 281–285, 2016.)

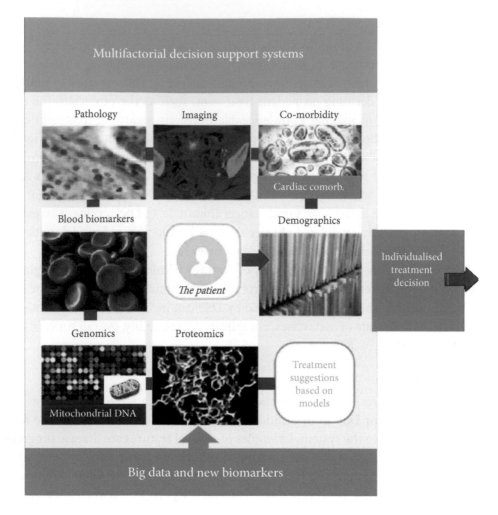

FIGURE 17.5 A multi-factorial decision support system uses existing research evidence and resources to make a customised recommendation for patient treatment.

that of randomised trials, the clinical parameters must be recorded in a meticulous manner. Standardised protocols should be used to this end, for example, 'Umbrella' Protocol NCT01855191 (https://www.cancerdata.org/protocols/eurocat-umbrella-protocol-nsclc). It is crucial that this is done after establishing a standardised protocol to ensure that the results are comparable between different centres and time points [27]. Additionally, prerequisites for the measurements of certain features should be recorded, for example, features evaluated only on patients showing certain symptoms to prevent bias when interpreting the measurements.

Similar to the measurement of features, toxicity should be measured and graded using validated systems, for example, the common terminology criteria for adverse events (CTCAE) [28], which allows events to be reported by either patient or physician. A protocol must be established to report which system was used for observation and how events resulting from the treatment were managed. The reporting of the toxicity should be conducted in line with the Strengthening The Reporting of Observational Studies in Epidemiology (STROBE) statement for observational studies and genetic-association studies [29].

This statement is a protocol which lists all items that should be addressed during reporting to ensure standardised interpretation for these types of studies.

Treatment Features

Radiotherapy treatment features include both dose features and additional, non-radiation treatments given to the patient, such as chemotherapy or hormonal therapy, which have been shown to influence outcomes. Modern technologies such as intensity-modulated radiotherapy (IMRT), brachytherapy (BT), volumetric arc radiotherapy (VMAT) or particle-beam therapy, such as hadron therapy, allow for localised dose delivery around the target volume with maximum sparing of the organs at risk with very high dosimetric accuracy [30]. These techniques, in combination with increasing knowledge of normal tissue response to radiotherapy [31], are used to maintain the delicate balance between TCP and NTCP. Figure 17.6 shows the dose distribution resulting from IMRT, VMAT, intensity-modulated proton therapy (IMPT) and BT treatment planning on the prostate.

To model the dose response in DSS, the relevant features are obtained from the planned dose distributions. To ensure that these features are realistic, measures should be taken to deliver the prescribed dose as planned and these measures should be described [32]. One of the modelling challenges in this field is how best to combine the dose distribution for every sub-volume of a structure into features. Parameters such as mean dose, maximum dose,

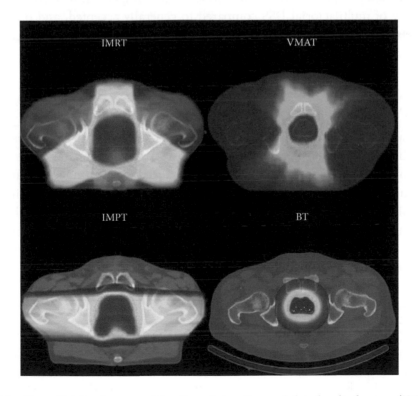

FIGURE 17.6 Dose distributions resulting from intensity-modulated radiotherapy (IMRT), volumetric arc radiotherapy (VMAT), intensity-modulated proton therapy (IMPT) and brachytherapy (BT) treatment planning on a patient with prostate cancer.

volume receiving more than a set amount of dose or equivalent uniform dose (EUD) are easily quantified and are often used, which is acceptable for a large range of applications. However, to improve personalised decision-making, spatial characteristics should be analysed as well to differentiate in radio-sensitivity in different regions of organ or tumour [33].

Cost-Effectiveness

In the past, the endpoints of most DSS have been TCP and/or NTCP. However, due to the limited resources available in health care and considering the increasing costs of newer treatments, CE comparisons are becoming increasingly relevant [34]. This is especially important when considering treatments that are not significantly beneficial to all patients, but are notably more expensive or invasive than traditional treatments, such as hadron therapy [35]. A CE analyses is a good way to weigh the TCP, the NTCP and the treatment costs against each other. Additionally, a CE analysis could improve the insurance perspective regarding more expensive treatments, and could allow reimbursement for patients for whom such a treatment proves cost-effective [36].

EVALUATION

The goal of RLHC is to improve predictability of treatment outcomes to increase effectiveness and efficiency of patient treatment. However, it is important to verify the improved predictability of multifactorial DSS compared to generally accepted, and often less complex, models [37]. Accepted RLHC models should repeatedly be reevaluated using robust data to test if the outcome is as expected and how this relates to evidence-based guideline knowledge.

SUMMARY

Modern health care aspires to optimise personalised cancer therapy and faces many challenges in this respect. The increase in available treatment options and the diversity in patients prove incredibly problematic for individualised decision making. However, DSS developed using RLHC have the potential to bring us one step closer to realising that goal. An essential step that needs to be taken is the standardisation of data acquisition, including data concerning treatment, clinical features, imaging, genetics and outcome. Also, the clinical assessment of developed DSS is critical, as well as standardising the development of robust prediction models.

REFERENCES

1. Vogelzang, N.J., S.I. Benowitz, S. Adams, C. Aghajanian, S.M. Chang, Z.E. Dreyer et al., Clinical cancer advances 2011: Annual report on progress against cancer from the American Society of Clinical Oncology. *J Clin Oncol*, 2012. **30**(1): 88–109.
2. Maitland, M.L., R.L. Schilsky, Clinical trials in the era of personalized oncology. *CA Cancer J Clin*, 2011. **61**(6): 365–381.
3. Abernethy, A.P., L.M. Etheredge, P.A. Ganz, P. Wallace, R.R. German, C. Neti et al., Rapid-learning system for cancer care. *J Clin Oncol*, 2010. **28**(27): 4268–4274.

4. Grutters, J.P., K.R. Abrams, D. de Ruysscher, M. Pijls-Johannesma, H.J. Peters, E. Beutner et al., When to wait for more evidence? Real options analysis in proton therapy. *Oncologist*, 2011. **16**(12): 1752–1761.

5. Pijls-Johannesma, M., J.P. Grutters, P. Lambin, D.D. Ruysscher, Particle therapy in lung cancer: Where do we stand? *Cancer Treat Rev*, 2008. **34**(3): 259–267.

6. Roelofs, E., L. Persoon, S. Qamhiyeh, F. Verhaegen, D. De Ruysscher, M. Scholz et al., Design of and technical challenges involved in a framework for multicentric radiotherapy treatment planning studies. *Radiother Oncol*, 2010. **97**(3): 567–571.

7. van der Laan, H.P., T.A. van de Water, H.E. van Herpt, M.E. Christianen, H.P. Bijl, E.W. Korevaar et al., The potential of intensity-modulated proton radiotherapy to reduce swallowing dysfunction in the treatment of head and neck cancer: A planning comparative study. *Acta Oncol*, 2013. **52**(3): 561–569.

8. Grand, M.M., P.C. O'Brien, Obstacles to participation in randomised cancer clinical trials: A systematic review of the literature. *J Med Imaging Radiat Oncol*, 2012. **56**(1): 31–39.

9. Murthy, V.H., H.M. Krumholz, C.P. Gross, Participation in cancer clinical trials: Race-, sex-, and age-based disparities.*JAMA*, 2004. **291**(22): 2720–2726.

10. Dehing-Oberije, C., S. Yu, D. De Ruysscher, S. Meersschout, K. Van Beek, Y. Lievens et al., Development and external validation of prognostic model for 2-year survival of non-small-cell lung cancer patients treated with chemoradiotherapy. *Int J Radiat Oncol Biol Phys*, 2009. **74**(2): 355–362.

11. Roelofs, E., M. Engelsman, C. Rasch, L. Persoon, S. Qamhiyeh, D. de Ruysscher et al., Results of a multicentric in silico clinical trial (ROCOCO): Comparing radiotherapy with photons and protons for non-small cell lung cancer. *J Thorac Oncol*, 2012. **7**(1): 165–176.

12. Lustberg, T., J. van Soest, A. Jochems, T. Deist, Y. van Wijk, S. Walsh et al., Big Data in radiation therapy: Challenges and opportunities. *Br J Radiol*, 2017. **90**(1069): 20160689.

13. Schilsky, R.L., D.L. Michels, A.H. Kearbey, P.P. Yu, C.A. Hudis, Building a rapid learning health care system for oncology: The regulatory framework of CancerLinQ. *J Clin Oncol*, 2014. **32**(22): 2373–2379.

14. Budin-Ljosne, I., P. Burton, J. Isaeva, A. Gaye, A. Turner, M.J. Murtagh et al., DataSHIELD: An ethically robust solution to multiple-site individual-level data analysis. *Public Health Genomics*, 2015. **18**(2): 87–96.

15. Deist, T.M., A. Jochems, J. van Soest, G. Nalbantov, C. Oberije, S. Walsh et al., Infrastructure and distributed learning methodology for privacy-preserving multi-centric rapid learning health care: EuroCAT. *Clin Transl Radiat Oncol*, 2017. **4**: 24–31.

16. Meldolesi, E., J. van Soest, N. Dinapoli, A. Dekker, A. Damiani, M.A. Gambacorta et al., An umbrella protocol for standardized data collection (SDC) in rectal cancer: a prospective uniform naming and procedure convention to support personalized medicine. *Radiother Oncol*, 2014. **112**(1): 59–62.

17. Kourou, K., T.P. Exarchos, K.P. Exarchos, M.V. Karamouzis, D.I. Fotiadis, Machine learning applications in cancer prognosis and prediction. *Comput Struct Biotechnol J*, 2015. **13**: 8–17.

18. Collins, G.S., J.B. Reitsma, D.G. Altman, K.G. Moons, Transparent reporting of a multivariable prediction model for Individual Prognosis or Diagnosis (TRIPOD): The TRIPOD statement. *Ann Intern Med*, 2015. **162**(1): 55–63.

19. Cheng, Q., E. Roelsof, B.L.T. Ramaekers, D. Eekers, J. van Soest, T. Lustberg, T. Hendriks, F. Hoebers, et al., Development and evaluation of an online three-level proton vs photon decision support prototype for head and neck cancer: Comparison of dose, toxicity and cost-effectiveness. *Radiother Oncol*, 2016. **118**(2): 281–285.

20. Walsh, S., W. van der Putten, A TCP model for external beam treatment of intermediate-risk prostate cancer. *Med Phys*, 2013. **40**(3): 031709.

21. Michalski, J.M., H. Gay, A. Jackson, S.L. Tucker, J.O. Deasy, Radiation dose-volume effects in radiation-induced rectal injury. *Int J Radiat Oncol Biol Phys*, 2010. **76**(3 Suppl): S123–S129.

22. Dekker, A., S. Vinod, L. Holloway, C. Oberije, A. George, G. Goozee et al., Rapid learning in practice: A lung cancer survival decision support system in routine patient care data. *Radiother Oncol*, 2014. **113**(1): 47–53.

23. Bright, T.J., A. Wong, R. Dhurjati, E. Bristow, L. Bastian, R.R. Coeytaux et al., Effect of clinical decision-support systems: A systematic review. *Ann Intern Med*, 2012. **157**(1): 29–43.

24. Klopp, A.H., P.J. Eifel, Biological predictors of cervical cancer response to radiation therapy. *Semin Radiat Oncol*, 2012. **22**(2): 143–150.

25. Kristiansen, G., Diagnostic and prognostic molecular biomarkers for prostate cancer. *Histopathology*, 2012. **60**(1): 125–141.

26. Schmidt, M.E., K. Steindorf, Statistical methods for the validation of questionnaires—discrepancy between theory and practice. *Methods Inf Med*, 2006. **45**(4): 409–413.

27. Garrido-Laguna, I., F. Janku, C. Vaklavas, G.S. Falchook, S. Fu, D.S. Hong et al., Validation of the Royal Marsden Hospital prognostic score in patients treated in the Phase I Clinical Trials Program at the MD Anderson Cancer Center. *Cancer*, 2012. **118**(5): 1422–1428.

28. National Cancer Institute (U.S.), Common terminology criteria for adverse events (CTCAE). Rev. ed. NIH publication. 2009, Bethesda, MD: U.S. Department of Health and Human Services, National Institutes of Health, National Cancer Institute, 194 p.

29. von Elm, E., D.G. Altman, M. Egger, S.J. Pocock, P.C. Gotzsche, J.P. Vandenbroucke et al., The Strengthening the Reporting of Observational Studies in Epidemiology (STROBE) Statement: Guidelines for reporting observational studies. *Int J Surg*, 2014. **12**(12): 1495–1499.

30. Jaffray, D.A., Image-guided radiotherapy: From current concept to future perspectives. *Nat Rev Clin Oncol*, 2012. **9**(12): 688–699.

31. Jaffray, D.A., P.E. Lindsay, K.K. Brock, J.O. Deasy, W.A. Tome, Accurate accumulation of dose for improved understanding of radiation effects in normal tissue. *Int J Radiat Oncol Biol Phys*, 2010. **76**(3 Suppl): S135–S139.

32. Hermans, B.C., L.C. Persoon, M. Podesta, F.J. Hoebers, F. Verhaegen, E.G. Troost, Weekly kilovoltage cone-beam computed tomography for detection of dose discrepancies during (chemo)radiotherapy for head and neck cancer. *Acta Oncol*, 2015. **54**(9): 1483–1489.

33. Bentzen, S.M., Theragnostic imaging for radiation oncology: Dose-painting by numbers. *Lancet Oncol*, 2005. **6**(2): 112–117.

34. Amin, N.P., D.J. Sher, A.A. Konski, Systematic review of the cost effectiveness of radiation therapy for prostate cancer from 2003 to 2013. *Appl Health Econ Health Policy*, 2014. **12**(4): 391–408.

35. Ramaekers, B.L., J.P. Grutters, M. Pijls-Johannesma, P. Lambin, M.A. Joore, J.A. Langendijk, Protons in head-and-neck cancer: bridging the gap of evidence. *Int J Radiat Oncol Biol Phys*, 2013. **85**(5): 1282–1288.

36. Sensible and sustainable care W.a. S. the Council for Public Health and Health Care to the Minister of Health, Editor. 2006: Zoetermeer.

37. Steyerberg, E.W., A.J. Vickers, N.R. Cook, T. Gerds, M. Gonen, N. Obuchowski et al., Assessing the performance of prediction models: A framework for traditional and novel measures. *Epidemiology*, 2010. **21**(1): 128–138.

CERN Collaborations in Hadron Therapy and Future Accelerators

Ugo Amaldi and Adriano Garonna

CONTENTS

STUDIES IN HADRON THERAPY AT CERN: EULIMA AND PIMMS

Introduction

An appropriate introduction to this subject can be found in the report published in 2014 by the Organisation for Economic Co-operation and Development (OECD) for the Global Science Forum, a document in which four case studies of CERN spin-offs are presented[1] (OECD, 2014). The part devoted to tumour therapy states:

> *The hadron therapy story demonstrates the important role of CERN's technical experts, especially the accelerator physicists and engineers (for example, Philip Bryant, Giorgio Brianti and Pierre Lefèvre). To the outside world, they are not as prominently visible as the physicists who are identified with the great scientific discoveries, but they are valued and influential within the councils of the laboratory.[2] For the EULIMA and PIMMS projects, CERN management had the ability to assign some staff members, part-time and even full-time, despite the enormous demands on time and resources in connection with LEP and LHC. In addition, the laboratory's open and dynamic working atmosphere made it possible for many experts to contribute ideas in an ad-hoc manner, based only on personal interest and a desire to help [...]. CERN's major contribution to ion therapy was not mandated "top down" from the highest governance levels. Rather, it grew and evolved gradually as a small fraction of the laboratory's work programme was shifted towards a new goal, based on the creativity, determination and idealism of a small number of staff members. Interestingly, these motivated persons occupied very different positions in the nominal institutional hierarchy. Thus, individuals who played essential roles in the hadron therapy story include (and this enumeration is not meant to be exhaustive) a Director-General (Carlo Rubbia), a senior administrator (Kurt Hübner), two senior physicists (Ugo Amaldi and Meinhard Regler) and an accelerator engineer (Pierre Mandrillon). [...] The laboratory's ability to make a significant contribution to cancer therapy is, to some extent, based on a coincidence of the right topic, the right scale of the undertaking and, above all, the right people.*

The European Light Ion Medical Accelerator – EULIMA

In 1986 Pierre Mandrillon – a French scientist working on proton and neutron therapy in the Centre Antoine Lacassagne hospital in Nice, who had designed and built a 62 mega-electron volts cyclotron for the Laboratoire du Cyclotron – organised at CERN a meeting of European scientists interested in the design of an accelerator for carbon ion therapy.

---

[1] The four subjects of the *Report on the Impacts of Large Research Infrastructures on Economic Innovation and on Society* are (1) the Large Hadron Collider superconducting magnets, (2) the developments in tumour hadron therapy, (3) the commercial utilisation of software packages, (4) the effects of the CERN educational and outreach programme.

[2] The importance of these physicists and engineers in the developments of sub-nuclear physics is underlined in Amaldi (2015).

Mandrillon had worked at CERN and knew the laboratory well. Several European laboratories joined and in 1989, European funds were obtained for a two-year project dubbed EULIMA (Farley and Carli, 1991). A formal agreement was signed to reimburse CERN for the use of various resources.

Mandrillon's initial idea was to design a 400 MeV/u superconducting cyclotron for light ion therapy and he was well advanced with such a design. However, because the design work was at CERN, which was the home of the Low Energy Antiproton Ring (LEAR), a synchrotron storage ring of similar energy, it soon became clear that a synchrotron would have been a cheaper and more straightforward solution than a superconducting cyclotron. This was corroborated by the fact that in Japan two synchrotrons were at the heart of the ion therapy centre HIMAC, which was then under construction. Under Pierre Lefèvre, who had been the designer of LEAR together with Dieter Möhl, a dedicated synchrotron and a proton gantry were designed and, eventually, the supervising committee deliberated that this was the solution to be pursued. The OECD report states: "The final report came squarely down on the side of the synchrotron, a result that was not greeted with universal satisfaction, especially among some of the originators of the project. EULIMA was terminated rather abruptly in 1991 with the end of the European funding."

The Proton and Ion Medical Machine Study – PIMMS

In spring 1991 Ugo Amaldi of CERN and Meinhard Regler of the Austrian Academy of Sciences launched two independent projects: (1) the design and (2) construction of an Italian National Centre for Light Ion Therapy and the creation of a central European neutron spallation source called AUSTRON. They are described in the next two subsections.

By the fall of 1995, the AUSTRON design study, led by Philip Bryant of CERN, was completed and a party was organised for the presentation of the final report. This was the occasion during which Amaldi and Regler decided to join forces to convince CERN to initiate a new project focused on the design of an optimised synchrotron for ion therapy. They involved Kurt Hübner, Director of Accelerators at CERN, who had supported the AUSTRON project from the beginning, and, quite naturally, they asked Phil Bryant to be the leader of this initiative. It was he who introduced the acronym PIMMS.

Before the end of 1995, the TERA Foundation, created in 1992, moved some of its collaborators from Italy to CERN, under the leadership of Sandro Rossi, and activity started before the end of the year while the internal CERN discussions lasted some months, with a very positive contribution from Horst Wenniger, who was at that time Research-Technical Director. The approval of the CERN Directorate arrived only in April 1996, as described in a short memorandum written by Hübner to Amaldi and Bryant.

A Project Advisory Committee was created by TERA with Giorgio Brianti – who had been the CERN Technical Director for about 10 years – as chairman. It is estimated that CERN had invested 10 person-years in PIMMS, TERA 25 person-years and MedAustron 10 person-years (OECD, 2014). Later, the Czech Foundation Oncology 2000 joined the project with three person-years.

The study was joined by GSI where a design effort in favour of hadron therapy was already building up, driven by Dieter Böhne and Gerhard Kraft. However, the collaboration with GSI came to an end when GSI eventually decided to adopt a different design and sought protection by patents. The initiative of GSI led to the Heidelberg Ion-Beam Therapy Center (HIT), funded by the University of Heidelberg and the German federal government and based on the pioneering work at GSI. HIT has been in operation since 2009 and features two proton treatment rooms and a 700 ton carbon ion gantry.

The PIMMS mandate was to design a light ion hadron therapy centre comprising a combination of systems, optimised for medical use – that is, capable of sub-millimetre accuracy for the conformal treatment of complex-shaped tumours by active scanning – without any financial and/or space limitations. Its output was intended to be used as a tool kit for future designs within other study groups. The study started with the design of the synchrotron, optimised to provide ideal conditions for the slow extraction system (Benedikt, 1997), accompanied by a thorough theoretical examination of slow extraction and the techniques to produce a smooth beam-spill (Pullia, 1999). It closed in late 2000 with the publication of two comprehensive reports (Badano et al., 1999). The PIMMS layout is shown in Figure 18.1.

The developments that led to PIMMS and, eventually, to the construction of MedAustron have been described in Benedikt and Bryant (2011). As pointed out, CERN was particularly well positioned to study the synchrotron design in view of the extensive R&D that teams had previously carried out on slow-extraction schemes in LEAR, the Proton Synchrotron (PS) and the Super Proton Synchrotron (SPS), and in view of the work done for EULIMA. These studies ranged from classical quadrupole-driven configurations to elaborate ultra-slow extraction using stochastic noise.

A list of the special features of the PIMMS design includes the following:

- Injection from the inside of the ring

- Slow extraction based on smooth acceleration by a 'betatron core' (Badano and Rossi, 1997), where all the currents in the other machine components are kept constant and, thus, the lattice optics remain unchanged and continue to satisfy the 'Hardt condition' (Hardt, 1981)

- An 'empty' bucket that increases the velocity of the particles entering the extraction resonance, thus reducing the intensity fluctuations of the extracted beam (Crescenti, 1998)

- Orthogonal modules for the high-energy beam lines that provide the various functions such as varying the horizontal and vertical dimensions of the therapy beam and rotating the gantry completely independent (Benedikt et al., 1999a)

- A mobile cabin gantry (named Riesenrad gantry) for carbon ions (Benedikt et al., 1999b; Reimoser et al., 2000)

- A 'rotator module' to make the beam optics of the gantry independent of the gantry rotation angle (Benedikt and Carli, 1996)

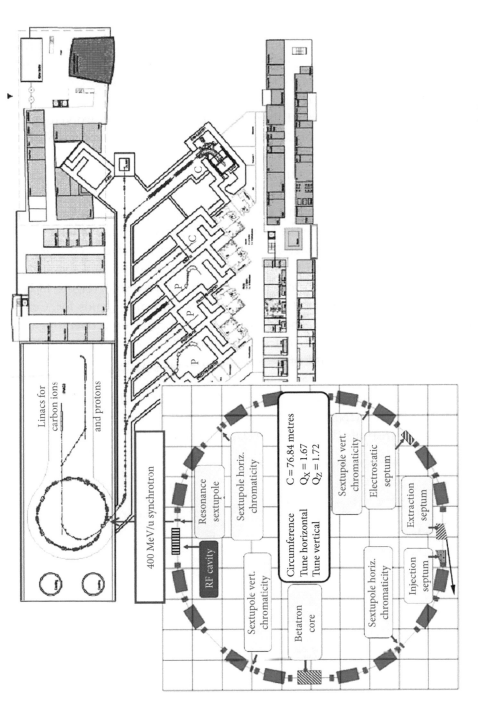

FIGURE 18.1 The PIMMS layout included two injection linacs (one for protons and the other for carbon ions), two different proton gantries and a carbon ion gantry of the type called initially 'mobile cabin' and later 'Riesenrad gantry'. (From Badano, L. et al., Proton–ion medical machine study (PIMMS)—Part I, CERN/PS 99-010 DI, Geneva, Switzerland, March 1999; Borri, G. et al., Part II, CERN/PS 00-007 DR, Geneva, Switzerland, July 2000.)

CNAO – The Centre for Light Ion Therapy in Pavia

As already mentioned, the two roots of the PIMMS were both planted in spring 1991. The Italian story of this endeavour (Amaldi and Magrin, 2005) started in May 1991, when Ugo Amaldi and Giampiero Tosi wrote a report proposing the design and construction of a national particle therapy centre with light ions (Amaldi and Tosi, 1991). In September 1991, INFN financed the 'ATER experiment', initially a collaboration between the Milan Section of INFN and CERN. In the following years, this 'experiment' became a large collaboration involving the INFN branches in Genoa, Ferrara, Florence, Milan, Naples, Padua, Pavia, Rome, Turin and the Legnaro National Laboratories. This collaboration worked on the radiobiology of hadron beams, detector developments and software applications for hadron therapy.

In September 1992, in order to collect funds from donors and charities and have a large and stable staff for the design of the new ion facility, the TERA Foundation (TErapia con Radiazioni Adroniche, Italian acronym that stands for 'therapy with hadronic radiations') was created in Novara, the town where the Secretary-General Gaudenzio Vanolo had his residence. At its peak activity, the foundation employed a technical staff of 25 people.

TERA's R&D activities were centred on an ion synchrotron optimised for medical purposes and capable of accelerating particles up to 400 MeV/u[3]. The centre (Figure 18.2a) – to be built close to the Novara Hospital – was called CNAO, which stands for 'Centro Nazionale di Adroterapia Oncologica', and was based on a synchrotron accelerating both carbon ions and protons, injected by two linacs. The writing of the 'Blue Book' (Amaldi and Silari, 1995) was done in the framework of the 'Programma Adroterapia', a large collaboration of about 110 physicists, engineers and radiation oncologists, which was led by TERA and involved scientists of many institutions, in particular CERN, the Italian National Institute for Nuclear Physics (INFN), the Italian National Research Council (CNR), the Italian National Health Institute (ISS) and GSI (Darmstadt).

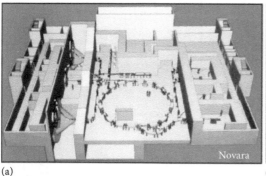
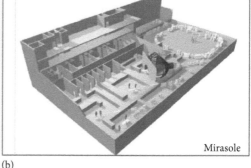

(a) (b)

FIGURE 18.2 The Novara and the Mirasole projects are described in the 'Blue Book' (Amaldi and Silari, 1995) and in the 'Red Book' (Amaldi, 1997) respectively.

---

[3] One year after the beginning of the CNAO program, TERA initiated a second and more innovative line of research by designing and prototyping three gigahertz hadron linacs, which produce beams of protons and/or carbon ions. See Amaldi et al. (2009). The long development work is now coming to fruition since the companies A.D.A.M. (Switzerland) and Advanced Oncotherapy (UK) are constructing on CERN premises the prototype of such a medical proton linac.

At the end of 1995, the Novara local government changed and its successor decided not to support the construction of the centre. Therefore, TERA set up a steering committee with five large hospitals from Milan and Pavia for the construction of the National Centre CNAO near the Mirasole Abbey, located southeast of Milan. The medical partners of TERA were two public oncological hospitals (Istituto Nazionale dei Tumori and Istituto Neurologico Besta), a private oncological centre (European Institute of Oncology) and the University Hospitals of Milano and Pavia. The Mirasole project study, completed in 1997, was designed in collaboration with Pierre Lefèvre and was based on a synchrotron similar to the LEAR built at CERN (Figure 18.2b). After initial support, at the beginning of 1997, the Italian Health Ministry rejected the by-laws of the Mirasole Foundation.

As discussed in Subsection 'The Proton and Ion Medical Machine Study – PIMMS' at the end of 1995, while working on the Mirasole project, Ugo Amaldi and Meinhard Regler proposed a study of a medical synchrotron to the CERN management. Based on the first studies, TERA prepared a third project to be built on the Mirasole site (Figure 18.3). This design also featured a carbon ion gantry equal to the one designed by GSI for the HIT Centre (Weinrich, 2006).

Since the centre depicted in Figure 18.3 was costly and had a large footprint, TERA dropped the proton and ion gantries, introduced important modifications and improvements into the original design of the synchrotron and of the beam lines developed by PIMMS, and proposed for the Mirasole site the 'PIMMS/TERA' design of Figure 18.4.

In this design,

- A single 7 MeV/u injector located inside the ring was used and the injector design made by GSI for the HIT Centre in Heidelberg was adopted

- A multi-turn injection scheme was chosen

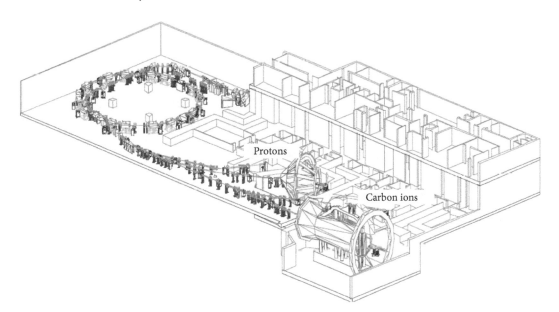

FIGURE 18.3   Perspective view of the second centre designed for the Mirasole site (Amaldi, 2004). The single injector linac is placed outside of the synchrotron ring.

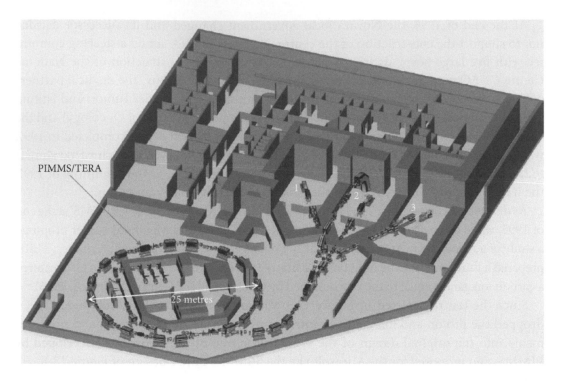

FIGURE 18.4   The layout of the Italian National Centre designed by TERA features three treatment rooms with four beams: three horizontal and one vertical. It was also foreseen that two carbon ion gantries would be added in a second construction phase.

- In addition to the betatron-core extraction, an RF Knockout system was foreseen similar to the one used by HIMAC and HIT

- A three-way switching magnet helped to make the beam lines to the treatment rooms shorter and thus less expensive than the PIMMS lines

Following a request by TERA, Italian Health Minister Umberto Veronesi created the CNAO Foundation in 2000 – for building and managing the Light Ion Centre – and attributed to it the initial construction funds. In 2001, after political elections and a new review by an ad-hoc ministerial committee, this decision was confirmed and further funds were made available to the CNAO founders: TERA and the five hospitals mentioned earlier.

One fundamental reason for the final success of this long-standing endeavour has certainly been TERA's 'political' decision not to choose the location of the centre too early. After the first two attempts in Novara and Mirasole, this decision was left to the public authorities who had to contribute the largest part of the construction funds. Thus, in 2004, the government and the Health Ministry were able to choose a site in the close vicinity of one of the five founders of the CNAO Foundation – the San Matteo University Hospital of Pavia, a town located 30 kilometres southwest of Milan.

At the end of 2003, TERA transferred 16 full-time staff members (physicists and engineers) and nine part-time consultants to the CNAO Foundation. Based on a memorandum of understanding signed with CNAO President Erminio Borloni, in 2004 TERA delivered to CNAO about 2,000 pages of specifications and technical drawings. At the same time, INFN took on significant responsibility for the construction of the synchrotron and the beam lines.

As a continuation of the long-term collaboration between CERN and TERA, the CNAO Foundation was able to sign an agreement with CERN to measure the 16 main ring bending magnets and to design and build components of the synchrotron diagnostics and injection/extraction systems. Due to these collaborations, the intellectual property of the final design was shared as follows: 55% to CNAO, 35% to INFN and 10% to CERN. In addition, GSI contributed to the design of special components, as the injection linac and the 90 degree bending magnet of the vertical beam.

The physicists and engineers who passed from TERA to CNAO constituted the core of the technical staff that followed the construction and commissioning of the facility. The group was led from the beginning by Sandro Rossi, who had been TERA Technical Director since 1995 and was nominated CNAO Technical Director, while Roberto Orecchia, TERA Vice-President, was nominated Scientific Director. The overall layout of the CNAO project is shown in Figure 18.5 (Rossi, 2015).

At the end of 2005, the construction started on a 37,000 square metres plot contiguous to the San Matteo University Hospital of Pavia. In the fall of 2011, the first patient was treated with protons, but the commissioning was very long because the Italian Health Ministry imposed experiments with cells, animals and the treatment of 130 patients with full documentation before finally issuing the CE label for a limited number of tumour indications.

FIGURE 18.5 CNAO features two separate buildings. The first contains the synchrotron and the three treatment areas, while the second one is devoted to the patient reception and the staff. The central treatment area features a horizontal and a vertical beam.

Moreover, the Lombardy Region did not introduce reimbursement for hadron therapy until 2014, and only in 2017 did the Italian National Health System introduce hadron therapy as a recognised treatment modality. This further delayed the full exploitation of the facility, which by the end of 2017 has treated 1,500 patients, 80% of whom with carbon ions.

MedAustron – The Centre for Light Ion Therapy in Wiener Neustadt

After the fall of the Iron Curtain, Austrian scientists persuaded some leading politicians, in particular Erhard Busek, Austria's Vice-Chancellor and Science Minister, to support the establishment of a major scientific facility to be called AUSTRON and to be directed at the scientists of Central Europe. This initiative was the Austrian root of PIMMS and Meinhard Regler was the main promoter of this visionary endeavour.

In April 1991, a meeting of the 'Pentagonale' in Bratislava – where representatives of Austria, Czechoslovakia, Italy, Hungary and Yugoslavia were present – came to the consensus that AUSTRON should be a neutron spallation source. The meeting was chaired by Jan Pisut (Minister of Science and Research of Czechoslovakia) and Carlo Rubbia (Nobel prize winner and then Director-General of CERN). The project was adopted and strongly promoted by the guarantee that CERN would support it and transfer the technologies needed, since Austria had no existing activity in accelerators.

From the outset, Meinhard Regler and Horst Schönauer recommended a dual use of the rapid-cycling synchrotron to include proton therapy. It turned out that a dual use of the synchrotron would not pay off so that Schönauer proposed a separate synchrotron for cancer therapy (Schönauer et al, 1994). Therefore, discussion started with prominent Austrian radiation oncologists (in particular, Horst Dieter Kogelnik and Richard Pötter) on an ion accelerator for tumour therapy as part of the much larger AUSTRON project (Baumann et al., 1995).

The history of the AUSTRON project and its transformation into MedAustron has been described in detail by Meinhard Regler (2016), who recalls the bumpy road of the project in the years 1993–1994. In particular, he writes:

> The publication of the AUSTRON feasibility study was not without problems. At the end of November 1994, Rudolf Scholten took over from Erhard Busek as Science Minister. On Friday 2 December, Meinhard Regler was instructed to withdraw the printing contract of the feasibility study. Nonetheless, many copies were printed over the weekend. This turned out to be a vital point in the history of MedAustron. During 1995 and until June 1996 the project office was reduced to winding up the current tasks before being shut down, but an association, the 'Verein Austron', was founded by Peter Skalicky and Meinhard Regler in January 1996 to ensure the long term continuity of the spallation source project.

The AUSTRON studies continued until 1998 (Bryant et al., 1997) but, unfortunately, the project was never financed and Europe will have to wait for its advanced neutron spallation source until 2019, when the European Spallation Source (ESS) facility should be completed in Lund.[4]

However, the part of Austron concerning hadron therapy survived under the name MedAustron, given the interest of the Austrian oncologists. A detailed feasibility study was presented in 1998 with the support of the city of Wiener Neustadt and the government of the Province of Lower Austria, which had become interested in the project. The provincial government entrusted the firm Forschungs- und Technologietransfer GmbH (FOTEC) in Wiener Neustadt with the further development of the project and MedAustron became the subject of a new design study under Thomas Auberger and Erich Griesmayer (Auberger et al., 2004).

In 2004, the Austrian federal government, the government of Lower Austria and the city of Wiener Neustadt prepared a plan to fund the non-clinical research part of MedAustron, and in February 2005, PEG MedAustron GmbH, led by Theodor Krendelsberger, was created to put in place a public/private partnership for the funding of the clinical part. Since this did not materialise, in April 2007, the government of Lower Austria assumed the role of primary investor and created EBG MedAustron GmbH to oversee the construction and operation of the facility, under the direction of Martin Schima and later Bernd Mösslacher. The non-clinical research remained funded by the federal government.

Following discussions with then CERN Director-General Robert Aymar and Steve Myers, at that time Accelerators & Beams Department leader, CERN agreed to host and train the MedAustron team in all aspects of the accelerator design and construction. This was the first time that CERN had agreed not only to house a project but also to train the staff by forming work packages led by senior CERN staff and appointing a CERN staff member, Michael Benedikt, as Technical Project Leader. EBG MedAustron hired 35 young engineers and physicists to work with CERN to design the Austrian medical accelerator complex. Thus, MedAustron represents the most intensive 'head-to-head' transfer of technology that CERN has ever undertaken. The project in the component-production and construction stage and the EBG staff were transferred to Wiener Neustadt in 2013, where they formed the nucleus of the personnel responsible for the installation, running-in and operation of the accelerator.

When the design of MedAustron was started in June 2008, EBG MedAustron GmbH and the CNAO Foundation (Benedikt, 2016) signed a cooperation agreement. In this framework, CNAO made the construction drawings of the Pavia facility available and was compensated according to the division of the intellectual property given at the end of Section 'CNAO – The Centre for Light Ion Therapy in Pavia'. The synchrotron is shown in Figure 18.6 and the overall layout in Figure 18.7.

---

[4] See the ESS website: https://europeanspallationsource.se/.

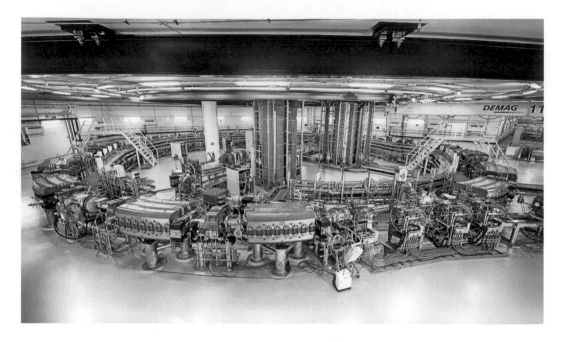

FIGURE 18.6   The synchrotron of the MedAustron facility. (Courtesy of Kaestenbauer/Ettl.)

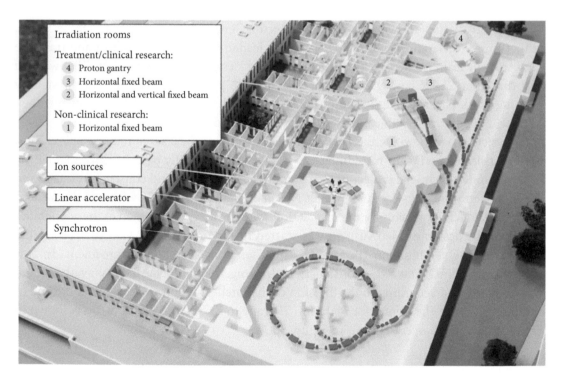

FIGURE 18.7   The MedAustron layout is similar to the one of the original PIMMS design (Figure 18.1). The standard solution of a linac located outside the ring has been adopted and the beam transport lines are long because all the modular properties of the PIMMS transport lines have been adopted. (Courtesy of EBG Medaustron Gmbh, Wiener Neustadt, Austria.)

One treatment room is served by a horizontal and a vertical beam as in CNAO. The facility also features a proton gantry, which is a copy of the PSI 'Gantry 2' (Pedroni et al., 2004), and an area with a horizontal beam used for non-clinical research, which has a maximum energy of 250 megaelectron volts. In a second phase, the maximal proton energy for research purposes will even be 800 megaelectron volts and will allow performing tests of particle detectors and irradiation of material for R&D in high-energy physics.

At the end of 2016, the first patient was treated with a proton beam and in August 2017, the second treatment room was put into operation. At the end of 2017, carbon commissioning is going on and the proton gantry is under construction. Meanwhile, children treatments were also included and compliance with the social insurance fund was able to take up the full treatment costs.

When the layouts of MedAustron and CNAO are compared (Figure 18.8), the different adopted philosophies immediately catch the eye. The two synchrotrons are practically identical, but the injection and extraction lines are very different. In the TERA design of Figure 18.4, the injector was placed inside the ring and a switching magnet was used to send the therapy beams to three treatment rooms to minimise the footprint and the cost. The MedAustron choice has instead been to build their facility on a much larger site and to follow closely the design of Badano et al. (1999) so that the beam lines have the high-performance characteristics of the PIMMS design.

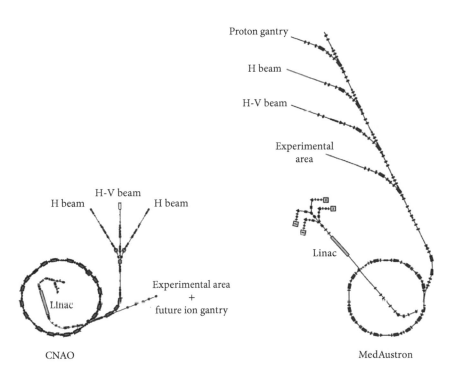

FIGURE 18.8 Comparison of the layout of the two facilities built on the basis of Proton and Ion Medical Machine Study PIMMS held at CERN between 1995 and 2000.

Conclusion

To conclude this section devoted to CERN collaborations in hadron therapy, it is worthwhile quoting a few sentences from the OECD Report (OECD, 2014), which summarises well the relation between the national efforts and the integrating role of CERN:

> *When assessing the relationship between CERN and Italy and Austria, an interesting contrast can be observed. Italy has been especially pre-eminent in high-energy physics, with many accelerators having been designed and built there on a national basis, under the aegis of INFN. Considering only the technological challenges, the PIMMS design could surely have been implemented as a purely Italian project. A CERN-based activity was able to catalyse and enable the national efforts, bringing together, within an international organisation, the elements that were missing in individual countries.*

It is obvious that the facilities described in this article owe much not only to the perseverance and tenacious efforts of a few persons but also to their willingness and ability to pool resources as done in the scientific collaborations formed to conduct physics experiments in CERN. For CERN, it will remain one of the most prominent examples of successful technology transfer to its member states.

FUTURE HADRON THERAPY ACCELERATORS

Since the time of PIMMS, many developments have occurred in the world of accelerators for hadron therapy. For carbon ion therapy, most of the important novelties have come, as was the case for PIMMS, from research centres and have focused on the optimisation of the synchrotron performance, mostly in terms of reduction in treatment delivery time. Instead, for proton therapy, the most significant development has originated from companies with the appearance of superconducting synchrocyclotrons and their drastic reduction of accelerator size and cost.

In terms of accelerator types, synchrotrons remain the state of the art for carbon ion centres and will continue to remain so in the near future. Indeed, despite their dominance on the proton therapy landscape, cyclotrons have not yet entered the carbon ion treatment realm. The design from Ion Beam Applications (IBA; Belgium) of a superconducting cyclotron for 400 MeV/u carbon ions is expected to be the first of these. The interesting newcomer in hadron therapy is the linac, with the new company ADAM (Switzerland) presently testing a prototype at CERN. This is the result of many years of research and development by the TERA Foundation and CERN. Finally, fixed-field alternating gradient (FFAG) accelerators and laser-driven accelerators (LDA) are still in the research stage but could play a relevant role in the long-term future.

The next sections will selectively detail some of the most important technical developments for synchrotrons, cyclotrons, linacs, FFAGs and LDAs and describe how these could influence the future of hadron therapy. The focus will be on carbon ion therapy machines, compared to proton therapy machines, as this is where most innovations are expected and needed. The interested reader will find more material on proton therapy accelerators in Chapter 4 on beam delivery systems of this book (Schippers, 2018).

Before discussing the various accelerator types, it is important to remark that very low currents are sufficient for tumour therapy: 1 nanoampere for protons and 0.1 nanoampere for carbon ions. The accelerator challenges are thus mostly in the machine dimensions, cost and complexities needed to obtain the energy corresponding to a Bragg peak depth of about 30 centimetres in tissue, which are about 230 megaelectron volts for protons and about 5,000 megaelectron volts (i.e. about 420 MeV/u) for carbon ions.

Synchrotrons

The accelerators for carbon ion tumour therapy are all based on the well-established technology of synchrotrons with resistive magnets. Half of all the synchrotrons in clinical operation have their origin in academic and research institutes, while the other half stem from industrial developments (PTCOG, 2017). It is interesting to note that the two facilities under construction at KIRAMS in Busan (Korea) and HITFil in Lanzhou (China) are both in-house developments. This is a testimony to the fact that the machines are still complex and costly and the field of carbon ion therapy is much less mature than that of proton therapy. Nevertheless, a new company called ProTom (USA) is presently installing at MGH a new synchrotron for proton therapy. Its elements were minimised and greatly simplified bringing the medical synchrotron footprint to its minimum and producing a 330 megaelectron volts beam for proton radiography.

Nevertheless, many developments have been made in the last few years to make the synchrotrons more compatible with clinical and economical needs. A particularly significant one has been the introduction of fast field regulation to reduce the dead time between successive flattops and, thus, beam 'spills'. Indeed, only a minor part of the machine cycle time is used to deliver the beam to the treatment rooms. The rest is used to ramp the magnets up and down and to wait for the magnetic field to be at the right level. Field regulation involves measuring the magnetic field in real time and using it to regulate the magnet power supplies. This enables a drastic reduction in cycle time, as shown in Figure 18.9.

The first implementation at the HIT Centre involves only one synchrotron dipole (out of six). It was thoroughly tested and has been used clinically since 2012 (Feldmeier et al., 2013). A simpler system using only Hall probes is also used at HIT for the transfer line dipoles and the addition of field regulation for synchrotron quadrupoles is ongoing. Despite its complexity, this innovation greatly enhances the performance of synchrotrons (with treatment delivery times reduced by 25%–30%) and will certainly become a standard in future machines. In addition, similar field regulation systems could greatly improve the performance of medical beam lines and beam delivery systems.

The National Institute for Radiological Sciences (NIRS) in Chiba (Japan) introduced another improvement to medical synchrotron performance by enabling the extraction of multiple beams of different energies within the same machine cycle (Iwata et al., 2010).

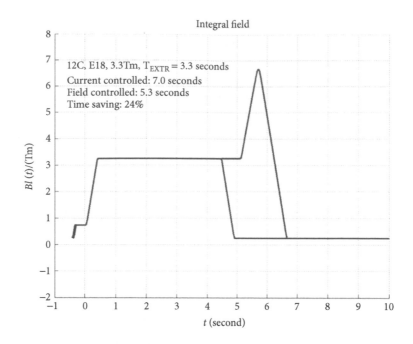

FIGURE 18.9 Carbon ion therapy synchrotron cycle in normal operating mode (blue) and reduced cycle in field-controlled mode (red). (From Feldmeier, E. et al., The first magnetic field control (B-train) to optimize the duty cycle of a synchrotron in clinical operation, *Proceedings of IPAC2012*, THPPD002, New Orleans, LA, pp. 3503–3505, 2012.)

Conventional cycles are based on sequential steps: beam injection, acceleration (to the required energy for a given irradiation layer) and extraction (until either the given irradiation layer is completed or the circulating current is reduced to zero). Even though particles are still available in the synchrotron after the extraction of a given energy, they are dumped in order to start a new cycle for a new energy layer, that is, 'one beam energy, one cycle'. By allowing the remaining beam after the first extraction to be directly brought to the next desired energy, the switching time between irradiation layers can be reduced to 100 milliseconds instead of the now typical one second, or more. The implementation, now also used in other hadron therapy synchrotrons, is limited to RF-Knockout (RF-KO) extraction schemes. An example of such a machine cycle is shown in Figure 18.10. This approach has a great impact in terms of treatment time reduction for layers where few spots or low doses are needed.

Other ongoing improvements are less mature but could play an important role in the future. Brookhaven National Laboratories (BNL) and Best Medical (USA) have designed an 'ion Rapid Cycling Medical Synchrotron' (iRCMS) based on resistive magnets with combined functions (Trbojevic, 2011). A prototype magnet is shown in Figure 18.11. The machine cycle is fixed irrespective of the extraction energy, which is only determined by the timing of the fast-pulsed extraction elements so that, during spot scanning of the tumour target, the position of the Bragg peak can be adjusted longitudinally 15–30 times per second.

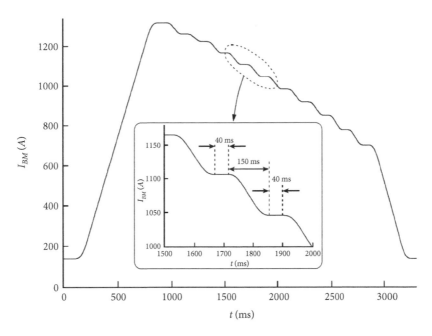

FIGURE 18.10 Current pattern for the main bending magnets in the NIRS ring as a function of time. The 11 short flattops correspond to extraction at beam energies spaced by 20–30 MeV/u. (From Iwata, Y. et al., *Nucl. Instrum. Methods Phys. Res. A*, 624, 33–38, 2010.)

FIGURE 18.11 In 2016, one of the six combined-function bending magnets was tested at BNL. (From McNulty Walsh, K., First magnet girder for prototype cancer therapy accelerator arrives for testing, 2016, https://www.bnl.gov/newsroom/news.php?a=25981.)

Finally, an ambitious research program by NIRS investigates the use of superconducting magnets for a medical synchrotron, dubbed 'Super MINIMAC'. Superconducting combined-function magnets with a five Tesla maximum field would imply a reduction of the accelerator circumference to around 20 metres, while keeping reasonable ramping speeds, which are determined by the ability to change the magnetic field without producing excessive heat (Iwata, 2016).

Superconducting Cyclotrons

The appearance of superconducting synchrocyclotrons for proton therapy has reduced the accelerator weights by a factor 10 and opened the way to the construction of *single-room proton facilities*, which many users now acquire because of the reduced investment with respect to the one needed for multi-room facilities serving five million people (or more) and consisting of one accelerator, three to four gantry rooms and patient areas. Various companies offer such single-room facilities based on cyclotrons or synchrotrons.

Despite their dominance in the world of proton therapy, there are no running cyclotron systems for carbon ion therapy. The proposals are based on superconducting isochronous cyclotrons, as high-field compact synchrocyclotrons for carbon ions would require an extremely complex external injection system. An interesting design for an isochronous cyclotron called SCENT (Maggiore et al., 2007) was produced by the group of L. Calabretta at the INFN-LNS of Catania (Italy). It has many similarities to the design of the ACCEL/VARIAN (USA) superconducting isochronous cyclotron (such as four sectors with four accelerating cavities and constant pole gaps) but uses a much higher central magnetic field of 3.2 Tesla. This results in complex geometries for the magnet and acceleration cavities (due to strong spiralling) but enables it to accelerate H_2^+ and C^{6+} up to 300 MeV/u, while keeping the magnet diameter within five metres. The extraction for H_2^+ at 250 MeV/u and C^{6+} are based on two different principles: a stripper and a system of electrostatic deflectors.

A parallel development is the C400 design (Jongen et al., 2010), which was completed by IBA in collaboration with the Joint Institute for Nuclear Research (JINR) in Dubna (Russia). This superconducting isochronous cyclotron features a central magnetic field of 2.4 Tesla, four spiralled sectors, elliptical hill pole gaps (the trademark of IBA isochronous cyclotrons) and two RF structures to accelerate and extract molecular hydrogen up to 230 MeV/u (by stripping extraction); carbon and helium ions are accelerated up to 400 MeV/u and extracted via electrostatic deflectors. The overall magnet weight is 700 tons for a total magnet diameter of about seven metres. A schematic drawing of the cyclotron elements is shown in Figure 18.12. The first of these cyclotrons is foreseen to be built at the ARCHADE Centre in France (Archade, 2017).

The concept behind these designs for carbon ion therapy is the same as that for proton therapy, that is, to accelerate the beam to the highest energy needed for treatment, to degrade as required the beam using mechanical passive absorbers and to give the resulting beam the required characteristics using dedicated beam lines (called Energy Selection Systems, ESSs).

A completely novel approach was adopted for the superconducting iron-less synchrocyclotron for proton therapy (Minervini et al., 2014). Since the superconducting coils provide all the active and passive magnetic fields without using iron, the cyclotron can be

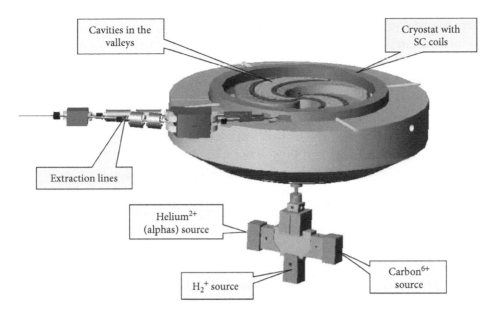

FIGURE 18.12 Main components of the C400 superconducting cyclotron design. (From Jongen, Y. et al., *Nucl. Instrum. Methods Phys. Res. A*, 624, 47–53, 2010.)

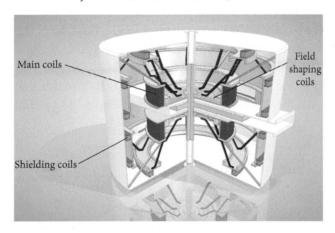

FIGURE 18.13 Artistic view of the magnet for the iron-less 250 megaelectron volts proton synchro-cyclotron by MIT. (From Bromberg, L. et al., Superconducting magnets for ultra light magnetically shielded, variable beam energy compact cyclotrons for medical applications, *Presentation at AIME-SCMED 2016*, Madrid, Spain, 2016.)

even lighter and, more importantly, could in principle extract at different energies without the need of a degrader but by changing the cyclotron magnetic field. A schematic of the 250 megaelectron volts proton accelerator is shown in Figure 18.13.

Linacs

Linacs are the most widespread medical accelerators. An operating frequency of three gigahertz has been adopted since 1947 for the construction of the more than 25,000 units, treating patients in tens of thousands of hospitals worldwide. Almost 30 years ago, the first three gigahertz medical proton linac was proposed (Lennox et al., 1989) and the many developments since then in proton and carbon ion linacs have been reviewed in Amaldi et al. (2009).

The main advantages of hadron therapy linacs with respect to (synchro)cyclotrons and synchrotrons are of two types: their very small transverse emittances, which entail smaller apertures of the magnetic elements of the beam lines and the gantries, with sizeable reductions in weights and costs, and their fast energy variation (in a few milliseconds instead of more than 100 milliseconds), which is critical for future real-time tracking in treatments of moving tumours (due to respiration and heart beat).

The TERA Foundation has long been active in this area of development and its linacs are indeed designed with short enough modules powered by independent klystrons, in order to vary the output hadron energy continuously in a couple of milliseconds by properly adjusting the power level and the phase of the last active RF power supply. With an appropriate achromatic beam transport line (e.g. with a momentum acceptance of ±1.5%), the particle range R can be varied – at a 200–400 hertz rate – by ±7%, that is, about ±15 millimetres at R = 20 centimetres. TERA has pursued two different approaches: the 'all-linac' solution (Weiss et al., 1996) and the 'cyclinac' solution (Amaldi, 2005).

At present, the company ADAM is mounting (Degiovanni et al., 2016), in collaboration with CERN, an all-linac proton therapy accelerator based on designs of a Cell-Coupled Linac structure prototyped by TERA (Amaldi et al., 2003; De Martinis et al., 2012). The injector of this structure is the novel compact RFQ from CERN (Lombardi et al., 2015) followed by the low-energy Side-Coupled Drift Tube Linac designed by the group of L. Picardi at ENEA (Ronsivalle et al., 2015). TERA and CERN have also developed a single-room facility solution for proton therapy called TULIP (Amaldi, 2006; Benedetti et al., 2017; Cuccagna et al., 2017). The layout is shown in Figure 18.14 and incorporates many

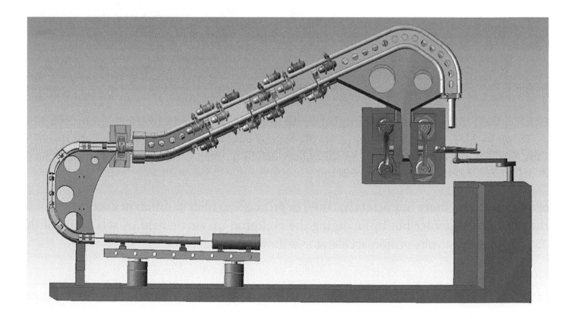

FIGURE 18.14   TULIP (Turning Linac for Proton Therapy) fits into a 22 × 9 metre bunker.

innovations, among which multi-beam klystrons and a novel linac type called Backward-Travelling Wave with 'recirculator', which consumes less power for the same acceleration gradient (around 40 MV/m).

The 'cyclinac' uses the variable-energy linacs not for the full acceleration range but only for the high-energy part where they are easier to produce and operate and, most importantly, where the advantages of the linac are needed. Thus, for proton systems the injector can be a cheap commercial proton cyclotron (Amaldi, 1994), used today for radioisotope production. The same concept has been applied to the acceleration of carbon ions (Amaldi et al., 2009). In order to produce short intense pulses of fully stripped carbon ions at high repetition rate, CERN is developing an innovative dedicated electron beam ion source (Mertzig et al., 2017). Since the energy required for carbon ion therapy is much larger than for proton therapy, higher electrical acceleration gradients, and thus shorter structures, are a natural line of development when considering carbon ion therapy. Based on the CERN CLIC group's 20 years of experience with high gradient linac structures (Wuensch, 2007; Grudiev et al., 2011), a CERN-TERA collaboration has experimentally studied the limits for safe operation of medical linacs, which at three gigahertz are found to be about 50 MV/m (Degiovanni et al., 2011). In the carbon ion cyclinac shown in Figure 18.15 (Degiovanni and Amaldi, 2014; Benedetti et al., 2016), the average gradient in the high-energy linac structure is 32 MV/m, about twice the one of previously discussed proton linacs. A similar endeavour is also being pursued at Argonne National Laboratories in collaboration with the company RadiaBeam (USA; Plastun et al., 2016).

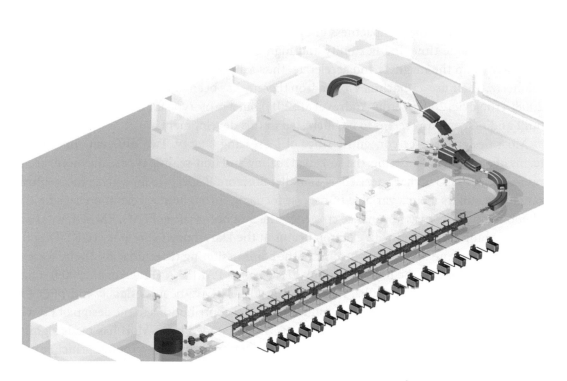

FIGURE 18.15   The CABOTO dual proton-carbon cyclinac is 35 metres long.

Finally, other examples of future linacs in hadron therapy are as boosters to existing proton therapy cyclotrons. The aim is to extend the proton energy reach from 250 mega-electron volts to at least 350 megaelectron volts, which corresponds to a water range of more than 60 centimetres. This allows the imaging of patients with proton-based radiography and opens the way to a new kind of treatment called high-energy proton therapy (Schippers and Lomax, 2011).

As far as the first possibility is concerned, despite many technical proofs of principle (Bucciantonio et al., 2013; Poludniowski et al., 2015), no commercial device is yet available to obtain computed tomography (CT) images from proton beams. The well-recognised clinical advantages are better precision for proton therapy planning – today's treatment plans are based on Hounsfield units' conversion to proton stopping power from empirical tables and the range errors are not negligible – and less dose to the patient compared to conventional X-ray CTs.

High-energy proton therapy uses the sharp lateral penumbra of high-energy proton beams to optimise the proton dose distribution at critical boundaries of the tumour volume, such as the ones near a critical healthy structure close to the tumour. Following a feasibility study for the PSI cyclotron (Degiovanni et al., 2018), a recent study is proposing a slightly different booster solution and the addition of a high-energy proton gantry (Apsimon et al., 2016).

Fixed-Field Alternating Gradient Accelerators

Fixed-Field Alternating Gradient accelerators (FFAG) are, in a way, a hybrid of a cyclotron and a synchrotron since the ring is made of separated combined-function magnets with a fixed magnetic field (as in a cyclotron) and variable extraction energy (as in a synchrotron). As such, they offer the compactness of cyclotrons without the need for degrader and ESS and, at the same time, the flexibility and high energy of synchrotrons without the need to slowly ramp the magnets and accelerate, thus achieving repetition rates in the order of kilohertz. Their use as machines for hadron therapy has been long studied (Craddock et al., 2008; Trbojevic, 2009; Verdu-Andres et al., 2011) but a prototype for medical use was never built and no company seems to be interested in this development.

Most designs foresee concentric rings of increasing energy. Two different concepts in terms of optics are competing: the so-called scaling and non-scaling types. The more recent non-scaling FFAG is characterised by a smaller horizontal aperture for a given momentum range, but suffers from a variable transverse focusing throughout acceleration causing the phenomenon known as 'resonance crossing'. The first non-scaling FFAG, EMMA, was successfully built and tested with electron beams at the Daresbury Laboratory (UK). Despite the interesting and successful outcome of the commissioning (Machida, 2013), the second stage of the project – the hadron therapy FFAG called PAMELA (Peach et al., 2009) – has still not progressed. In parallel, many scaling FFAGs have been designed and built for other purposes. The BNL team has produced original contributions to the design of FFAGs and novel rotating gantries (Trbojevic, 2009). The group led by Y. Mori has made significant contributions at KURRI in Kyoto (Japan), going from the first proof of principle one megaelectron volts proton FFAG to a larger 150 megaelectron volts proton FFAG (Mori et al., 2011), shown in Figure 18.16.

FIGURE 18.16 The 150 megaelectron volts proton FFAG at the Kyoto University Research Reactor Institute in Japan. The beam is injected from a one megaelectron volts circular accelerator. (From Osanai, A. and Tanigaki, M., Development of data-logging system for FFAG accelerator complex in KURRI, *Proceedings of PCaPAC08*, Ljubljana, Slovenia, TUP016, pp. 116–118, 2008.)

The main limitation of these systems is the complexity in manufacturing the large aperture magnets with large field gradients, the fast frequency modulation of the RF cavities and the fast extraction system. The potential gain from such machines, which can easily accelerate currents that are many orders of magnitude larger than what is needed for therapy, is still valid and FFAGs could easily have a future renaissance for medical applications, by also profiting from technological advances in cyclotrons and synchrotrons.

Laser-Driven Accelerators

A very interesting accelerator in the long-term is the LDA (Borghesi, 2014). This accelerator type makes use of very intense laser pulses impinging on thin sub-micrometre thick solid targets. The resulting stream of electrons creates electric fields in the range of TV/m, 20 times larger than maximum acceleration gradients in conventional linacs. This enables the acceleration on small distances of protons or carbon ions, depending on the nature of the solid target used. Recent developments have proven the possibility of obtaining proton beams of up to a few megaelectron volts with energy spreads of a few per cent; this spread has to reduced, using magnets and collimators, by a factor of at least five to have a sharp distal falloff of the dose. The generation of intense laser pulses requires dedicated and bulky laser systems, such as those of ELIMAIA in Prague (Czech Republic) and CALA in Munich (Germany), to generate the required large power densities, which are presently limited in repetition rate (around one hertz).

Radiobiological studies are ongoing to investigate the effect of these ultrashort (<10 nanosecond) intense pulses of radiation in *in-vitro* experiments (Zeil et al., 2013).

Until now, no important difference was observed compared to conventional irradiation fluxes. Furthermore, advanced R&D activities (Masood et al., 2014) have tackled the challenges linked to the irradiation beam lines, which for medical applications have to be adapted to the specific properties of the laser-generated beam: wide energy spectrum and large emittances. Much work remains to be done, but it is worth remarking that the special beam characteristics may also pave the way for new approaches to treatment delivery, such as multi-ion beams and ultrafast delivery for radiosurgery.

OUTLOOK

As the history of CERN collaborations in hadron therapy shows, despite the importance of industrial developments, collaborative research programs are crucial in bringing about significant pushes on technology in areas where the market pressure is not strong enough, such as that of carbon ion therapy. In that respect, CERN and GSI (and the passion of their researchers) have played a crucial role both in making PIMMS a success – paving the way for the construction of two important hadron therapy centres (CNAO and MedAustron) equipped with state-of-the-art accelerators – and in designing, constructing and continuously improving the HIT Centre, triggering a major company, Siemens (Germany), to build the Marburg and Shanghai facilities.

Recent advances in accelerators for hadron therapy have been driven by the use of superconductivity. Superconducting synchrocyclotrons have reduced the weight of proton therapy machines by a factor of 10, opening the way to a new paradigm: single-room proton therapy facilities. The trend suggests that single-room facilities attached to existing conventional radiotherapy facilities will outnumber multi-room facilities solely dedicated to proton therapy. Aside from economical and organisational considerations, this innovation could bridge the gap between proton therapy and conventional radiotherapy, facilitating better understanding and adoption of the advantages of hadron therapy. Also, in the case of carbon ion therapy, superconductivity is playing a key role with the appearance of the first superconducting gantry, which started patient treatment at NIRS in May 2017 (Iwata et al., 2012), the design for a superconducting isochronous cyclotron by IBA and the R&D program towards a superconducting synchrotron by NIRS. Among the new accelerator types, FFAGs and LDAs propose interesting characteristics although technical challenges are still to be solved. Linacs rely on a more mature technology and a first commercial company is proposing a solution for proton therapy in collaboration with CERN.

Despite the recent technological advances in proton therapy, the overall progress in technology for hadron therapy has been fairly slow. Once again, multi-disciplinary collaborative work triggered by research institutes could play a decisive role. Since 2010, proposals have been made to transform CERN's ion synchrotron LEIR into an accelerator for biomedical research, dubbed BioLEIR (Silari et al., 2010; Ghitan et al., 2017). In addition, following in the footsteps of PIMMS, a PIMMS-2 study is being discussed. It could tackle the technical challenges linked to linacs, FFAGs or fast cycling synchrotrons (Vretenar, 2017).

To be effective, these activities have to be pursued in strict connection with commercial companies, collaborative networks (Dosanjh et al., 2014) and the technology transfer units of the research centres and universities.

ACKNOWLEDGEMENTS

For the first section on PIMMS, CNAO and MedAustron, the authors very warmly thank M. Benedikt, P. J. Bryant, K. Hübner, M. Regler and S. Rossi for many useful comments and important additions. The authors are indebted to M. Dosanjh and M. Pullia for careful proofreading of the whole manuscript.

REFERENCES

Amaldi, U. and G. Magrin (Eds.), *The Path to the Italian National Centre for Ion Therapy* (*White Book*), Vercelli, Italy, Edizioni Mercurio, 2005, pp. 1–19.

Amaldi, U. and G. Tosi, For a centre of teletherapy with hadrons, TERA Internal Report, Novara, Italy, May 1991.

Amaldi, U. and M. Silari, *The TERA Project and the Centre for Oncologicl Hadrontherapy* (*Blue Book*), Frascati, Italy, INFN-LNS, 1995, Vols. 1 and 2, 2nd ed.

Amaldi, U., *CNAO*—The Italian centre for light-ion therapy, *Rad. Onc.*, 73, Supp 2 (2004) S191–S201.

Amaldi, U., *Particle Accelerators: From Big-Bang Physics to Hadron Therapy*, Springer Verlag, Berlin, Germany 2015.

Amaldi, U., Patent US 7554275, Proton accelerator complex for radioisotopes and therapy, November 2005.

Amaldi, U., The Italian hadrontherapy project, in *Hadrontherapy in Oncology*, U. Amaldi and B. Larsson (Eds.), Amsterdam, the Netherlands, Elsevier, 1994, pp. 45–58.

Amaldi, U., *The National Centre for Oncological Hadrontherapy at Mirasole* (*Red Book*), U. Amaldi (Ed.), Frascati, Italy, INFN-LNS, 1997.

Amaldi, U., S. Braccini, G. Magrin, P. Pearce and R. Zennaro, Ion acceleration system for medical and/or other applications, USPTO, US8405056 B2, 2006.

Amaldi, U., P. Berra, K. Crandall, D. Toet, M. Weiss, R. Zennaro, E. Rosso et al., LIBO—A linac-booster for proton therapy: Construction and tests of a prototype, *Nucl. Instrum. Methods Phys. Res. A* 521 (2004) 512–529.

Amaldi, U., S. Braccini and P. Puggioni, High-frequency linacs for hadrontherapy, *Rev. Acc. Sci. Tech.* 2 (2009) 111–131.

Apsimon, R., G. Burt, S. Pitman and H. Owen, ProBE proton boosting extension for imaging and therapy, in *Proceedings of International Particle Accelerator Conference,* Busan, South Korea, *IPAC16*, 2016.

Archade, 2017. ARCHADE. http:\\www.archade.fr. Accessed 22 October 2017.

Auberger, T.H. and E. Griesmayer, *Das Projekt MedAustron*, Wiener Neustadt, Austria, Wiener Neustadt: Forschungs—Und Technologietransfer GmbH, 2004.

Badano, L. and S. Rossi, *Characteristics of a Betatron Core for Extraction in a Proton-ion Medical Synchrotron*, Geneva, Switzerland, CERN/PS 97–19 (DI), 1997.

Badano, L., M. Benedikt, P.J. Bryant, M. Crescenti, P. Holy, P. Knaus, A. Maier, M. Pullia and S. Rossi, Proton-ion medical machine study (PIMMS)—*Part I*, CERN/PS 99-010 DI, Geneva, Switzerland, March 1999 and with additional authors G. Borri and S. Reimoser and contibutors F. Grammatica, M. Pavlovic and L. Weisser, *Part II*, CERN/PS 00-007 DR, Geneva, Switzerland, July 2000. A CD with drawings and other data including software is available on request, Yellow Report number CERN 2000–2006.

Baumann, F.M., B. Blind, P. J. Bryant, E. Griesmayer, J. Janic, M. Pavlovic, T. Wng et al., *The accelerator complex for the AUSTRON neutron spallation source and light-ion cancer therapy facility*, CERN-PS-95-48-DI, CERN, Geneva, Switzerland, 1995, 334 p.

Benedetti, S., A. Grudiev and A. Latina. *Design of a 750. MHz IH Structure for Medical Applications*, LINAC'16, East Lansing, MI, September 2016.

Benedetti, S., A. Grudiev and A. Latina. High gradient linac for proton therapy, *Phys. Rev. Accel. Beams* 20, 040101 (April 2017).

Benedikt, M. and C. Carli, *Optical Design of a Beam Delivery System Using a Rotator*, Geneva, Switzerland, CERN/PS 96-041 (OP), 1996.

Benedikt, M. and P.J. Bryant, Head-to-head technology transfer for hadron therapy, CERN Courier, 23 September 2011, http://cerncourier.com/cws/article/cern/47209.

Benedikt, M. MedAustron: The Austrian hadron terapy facility, in *Challenges and Goals for Accelerators in the XXI Century*, O. Brüning and S. Myers (Eds.), Singapore, World Scientific, 2016.

Benedikt, M. Optical design of a synchrotron with optimisation of the slow extraction for hadron therapy. PhD Thesis, Vienna, Austria, Vienna University of Technology, 1997.

Benedikt, M., P.J. Bryant and M. Pullia, A new concept for the control of a slow-extracted beam in a line with rotational optics: Part II, *Nucl. Instrum. Methods Phys. Res. A* 430 (1999a) 523–533.

Benedikt, M., P.J. Bryant, P. Holy and M. Pullia, Riesenrad ion gantry for hadron therapy: Part III, *Nucl. Instrum. Methods Phys. Res. A* 430 (1999b) 534–541.

Borghesi, M. Laser-driven ion acceleration: State of the art and emerging mechanisms, *Nucl. Instrum. Methods Phys. Res. A* 740 (2014) 6–9.

Bromberg, L., J.V. Minervini, P. Michael, A. Radovinsky and D. Winklehner. Superconducting magnets for ultra light magnetically shielded, variable beam energy compact cyclotrons for medical applications. *Presentation at AIME-SCMED 2016*, Madrid, Spain, 2016.

Bryant, P.J., M. Schuster and M. Regler. AUSTRON project. *Physica B: Condensed Matter.* 234–236 (1997) 1220–1223 doi: 10.1016/S0921-4526(97)00268-8.

Bucciantonio, M., U. Amaldi, R. Kieffer, F. Sauli, D. Watts, Development of a fast proton range radiography system for quality assurance in hadrontherapy, *Nucl Instr Meth.* 732 (2013) 564–567.

Craddock, M.K. and K.R. Symon, Cyclotrons and fixed-field alternating-gradient accelerators, *Rev. Acc. Science and Techn.*, I (2008) 65–97.

Crescenti, M., *RF Empty Bucket Channelling with a Betatron Core to Improve Slow Extraction in Medical Synchrotrons*, Geneva, Switzerland, CERN/PS 97–68 (DI), 1998, pp. 421–432.

Cuccagna, C., S. Benedetti, V. Bencini, D. Bergesio, P. Carrio Perez, E. Felcini, A. Garonna et al., Beam characterization for the TULIP accelerator for proton therapy through full Monte Carlo simulations, Abstract ID: 55, *Phys Med* 42/S1 (2017) 11.

De Martinis, C., D. Giove, U. Amaldi, P. Berra, K. Crandall, M. Mauri, M. Weiss et al., Acceleration tests of a 3 GHz proton linear accelerator (LIBO) for hadrontherapy, *Nucl. Instrum. Methods Phys. Res. A* 681 (2012) 10–15.

Degiovanni, A., A. Lomax, J.M. Schippers, L. Stingelin, J. Bilbao de Mendizabal and U. Amaldi, "A linac booster for high energy proton therapy and imaging", *Submitted to Phys. Rev. ST Accel. Beams in 2017*, American Physical Society (USA).

Degiovanni, A. and U. Amaldi, Proton and carbon linacs for Hadron Therapy, *Proceedings of the LINAC2014*, Geneva, Switzerland, FRIOB02, 2014, pp. 1207–1212.

Degiovanni, A. et al., A linac booster for high energy proton therapy and i`maging, 2018.

Degiovanni, A., D. Ungaro and P. Stabile, LIGHT: A linear accelerator for proton therapy,in *Proceedings of the North America Particle Accelerator Conference*, Chicago, IL, NAPAC16, 2016.

Degiovanni, A., U. Amaldi, R. Bonomi, M. Garlasché, A. Garonna, S. Verdú-Andrés, R. Wegner, TERA high gradient test program of RF cavities for medical linear accelerators, *Nucl. Instrum. Methods Phys. Res. A* 657 (2011) 55–58.

Dosanjh, M., M. Cirilli and S. Navin, ENLIGHT and LEIR biomedical facility, *Phys Med*, 30 (2014) 544–550.

Farley, F.J.M. and C. Carli, EULIMA beam delivery, *Proceedings of the Proton Radiotherapy Workshop at PSI*, 28 February—1 March 1991, PSI, Villigen, Switzerland, PSI-Report 111, 1991, pp. 21–24.

Feldmeier, E., Haberer, T., Galonska, M., Cee, R., Scheloske, S., Peters, S. The first magnetic field control (B-train) to optimize the duty cycle of a synchrotron in clinical operation, *Proceedings of IPAC2012*, New Orleans, LA, THPPD002, 2012, pp. 3503–3505.

Feldmeier, E. and T. Haberer, A. Peters, C.H. Schomers, R. Steiner, Developments of a high precision integrator for analog signals to measure magnetic fields in real-time, *Proceedings of IPAC2013*, Shanghai, China, MPWA001, 2013, pp. 661–663.

Ghithan, S., G. Roy and S. Schuh (Eds.), *Feasibility Study for BioLEIR*, Vol. 1 (2017, CERN Yellow Report, doi: 10.23731/CYRM-2017-001; CERN-2017-001-M.

Grudiev, A., S. Calatroni and W. Wuensch, New local field quantity describing the high gradient limit of accelerating structures. *Erratum Phys. Rev. ST Accel. Beams* 14 (2011) 099902.

Hardt, W., Ultraslow extraction out of LEAR, CERN, PS/DL/LEAR Note 81-6, Geneva, Switzerland, 1981.

Iwata, Y., K. Noda, T. Shirai, T. Furukawa, T. Fujita, S. Mori, K. Mizushima, et.al., Design of superconducting rotating-gantry for heavy-ion therapy, *Proceedings of the IPAC*, 012, New Orleans, LA, THPPR047, 2012.

Iwata, Y., SC Gantry at NIRS and next developments in SC magnets for gantries and synchrotrons, *Presentation at AIME-SCMED 2016*, Madrid, Spain, 2016.

Iwata, Y., T. Kadowaki, H. Uchiyama, T. Fujimoto, E. Takada, T. Shirai, T. Furukawa et al., Multiple-energy operation with extended flattops at HIMAC, *Nucl. Instrum. Methods Phys. Res. A* 624 (2010) 33–38.

Jongen, Y., M. Abs, A. Blondin, W. Kleeven, S. Zaremba, D. Vandeplassche, V. Aleksandrov et al., Compact superconducting cyclotron C400 for hadron therapy, *Nucl. Instrum. Methods Phys. Res. A* 624 (2010) 47–53.

Lennox, J., F.R. Hendrickson, D.A. Swenson, R.A. Winje, D.E. Young, *Proton Linac for Hospital-Based Fast Neutron Therapy and Radioisotope Production*, Villigen, Switzerland, Fermi National Accelerator Laboratory, TM-1622, 1989.

Lombardi, A.M., V.A. Dimov, M. Garlasche, A. Grudiev, S. Mathot, E. Montesinos, S. Myers, M. Timmins, M. Vretenar, Beam Dynamics in a high frequency RFQ, *Proceedings of the IPAC'15*, Richmond, VA, May 2015.

Machida, S. on behalf of the EMMA collaboration, What we learned from EMMA, *Proceedings of the Cyclotrons 2013*, Vancouver, BC, MO2PB01, 2013, pp. 14–16.

Maggiore, M., L. Calabretta, D. Campo, L.A.C. Piazza and D. Rifuggiato, Design studies of the 300 AMeV superconducting cyclotron for hadron therapy, in *Proceedings of the PAC07*, Albuquerque, NM, THPMN020, 2007, pp. 2748–2750.

Masood, U., M. Bussmann, T. E. Cowan, W. Enghardt, L. Karsch, F. Kroll, U. Schramm and J. Pawelke, A compact solution for ion beam therapy with laser accelerated protons, *Appl. Phys. B*, 117 (1) (2014) 41–52.

Mertzig, R., M. Breitenfeldt, S. Mathot, J. Pitters, A. Shornikov and F. Wenander, A high-compression electron gun for C6+ production: concept, simulations and mechanical design, *Nucl. Instrum. Methods Phys. Res. A* 859 (2017) 102–111.

Minervini, J., A. Radovinsky, C. Miller, L. Bromberg, P. Michael and M. Maggiore, Superconducting magnets for ultra-light and magnetically shielded,*Compact Cyclotrons for Medical, Scientific, and Security Applications*, IEEE Applied Superconductivity, 24/3 (June 2014).

Mori, Y., Y. Ishi, Y. Kuriyama, B. Qin, T. Uesugi, Present status of FFAG proton accelerators at KURRI, *Proceedings of the IPAC2011*, San Sebastian, Spain, WEPS077, 2011, pp. 2685–2687.

Osanai, A. and M. Tanigaki. Development of data-logging system for FFAG accelerator complex in KURRI, *Proceedings of PCaPAC08*, Ljubljana, Slovenia, TUP016, 2008, pp. 116–118.

OECD, Report on the impacts of large research infrastructures on economic innovation and on Society—Case studies at CERN, Organization for Economic Co-operation and Development, OECD, 2014. Available at cds.cern.ch/record/1708387?ln=en.

Particle Therapy Cooperative Group website (PTCOG) September 2017: https://www.ptcog.ch/index.php/facilities-in-operation;https://www.ptcog.ch/index.php/facilities-under-construction.

Peach, K.J., J.H. Cobb, S.L. Sheehy, H. Witte, T. Yokoi, M. Aslaninejad, M.J. Easton et al, Pamela overview: Design goals and principles, *Proceedings PAC09*, Vancouver, Canada, 2009.

Pedroni, E., R. Bearpark, T. Böhringer, A. Coray, J. Duppich, S. Forss, D. George, M. Grossmann, G. Goitein, C. Hilbes and M. Jermann, The PSI gantry 2: A second generation proton scanning gantry, *Zeit. Med. Phys.* 14 (2004) 25–34.

Plastun, A.S., B. Mustapha, A. Nassiri, P.N. Ostroumov, L. Faillace, S.V. Kutsaev and E.A. Savin, Beam dynamics studies for a compact ion linac for therapy, in *Proceedings of the LINAC2016*, East Lansing, MI, THPLR042, 2016, pp. 946–948.

Poludniowski, G., N. M. Allinson and P. M. Evans, Proton radiography and tomography with application to proton therapy, *Br. J. Radiol.* 88 (2015) 1053.

Pullia, M., Detailed dynamics of slow extraction and its influence on transfer lines design, PhD Thesis, Lyon, France, Claude Bernard University, 1999.

Regler, M., The early history of MedAustron, November 2016, unpublished: *http://info.tuwien.ac.at/austron/reports/TheHistoryofMedAUSTRON.pdf*

Reimoser, S. T., M. Pavlovic and M. Regler, Status of the riesenrad ion gantry design, *Proceedings of the EPAC*, 2000, Vienna, Austria.

Ronsivalle, C., L. Picardi, A. Ampollini, G. Bazzano, F. Marracino, P. Nenzi, C. Snels, V. Surrenti, M. Vadrucci and F. Ambrosini, First acceleration of a proton beam in a side coupled drift tube linac, *Europ. Phys. Lett.* 111 (2015) 14002, doi:10.1209/0295-5075/111/14002.

Rossi, S. The national centre for oncological hadrontherapy (CNAO)—Status and perspectives, *Phys. Med.* 31 (2015) 333–351.

Schippers, M. Advances in beam delivery techniques and accelerators in particle therapy, in *Advances in Particle Therapy: A Multidisciplinary Approach*, In Press, 2018.

Schippers, J. and A. Lomax, Emerging technologies in proton therapy, *Acta Oncol.* 50 (2011) 838–850.

Schönauer H. et al., The AUSTRON medical accelerator, in *The Austron Feasibility Study*, P.H. Bryant, M. Regler and M. Schuster (Eds.), Wien, Vienna, Im Auftrag des Bundesministeriums für Wissenschaft und Forschung, 1994.

Silari, M., U. Amaldi, M. Dosanjh, T. Eriksson and S. Maury, A proposal for an experimental facility at CERN for research in hadron-therapy, Abstract ID: 42, *Physics for Health in Europe Workshop*, February. 2010.

Trbojevic, D., J. Alessi, M. Blaskiewicz, C. Cullen, H. Hahn, D. Lowenstein, I. Marneris, et al., Lattice design of a rapid cycling medical synchrotron for carbon/proton therapy, *Proceedings of the IPAC2011*, San Sebastian, Spain, WEPS028, 2011, pp. 2541–2543.

Trbojevic, D., FFAGs as accelerators and beam delivery devices for ion cancer therapy, *Rev. Acc. Science and Techn.*, II (2009) 229–251.

Verdú-Andrés, S., U. Amaldi and Á. Faus-Golfe, Literature search on linacs and FFAGs for hadron therapy, *Int. J. Mod. Phys. A* 26 (2011) 1659–1689.

Vretenar, M., Opportunities for ion accelerators in medicine and industry, *Workshop on Ions for Cancer Therapy*, Space Research and Material Science, 26–30 August 2017, *Athens*, Greece.

Weinrich, U., Gantry design for proton and carbon ion facilities, *Proceedings of the EPAC2006*, 2006, Edinburgh, Scotland, pp. 964–968.

Weiss, M. and L. Picardi, High frequency proton linacs, in *The Rita Network and the Design of Compact Proton Accelerators* (the Green Book), U. Amaldi, M. Grandolfo and L. Picardi (Eds.), Frascati, Italy, INFN-LNS, 1996.

Wuensch, W. Progress in understanding the high-gradient limitations of accelerating structures, *Proceedings APAC 2007*, Indore, India, 2007, pp. 544–548.

Zeil, K., M. Baumann, E. Beyreuther, T. Burris-Mog, T.E. Cowan, W. Enghardt, L. Karsch, S.D. Kraft, L. Laschinsky, J. Metzkes, D. Naumburger, Dose-controlled irradiation of cancer cells with laser-accelerated proton pulses, *Appl. Phys. B* 110 (2013) 437–444.

Addressing Global Challenges for Radiation Therapy

From 'C' to Shining 'C'[1] (Cobalt to Carbon) and the 'Radiation Rotary'

C. Norman Coleman, Jeffrey Buchsbaum
and David A. Pistenmaa

CONTENTS

INTRODUCTION AND OVERARCHING CONCEPTS

Approaching key times of decision-making and unique opportunities is often referred to as being 'at a crossroads'. The preparation for this chapter comes at a time when there seem to be a number of crossroads for the future of radiation oncology, biology and physics and also

---

[1] From 'sea to shining sea' is from the song 'America the Beautiful'. See lyrics at https://en.wikipedia.org/wiki/America_the_Beautiful.

the lack of availability of radiation therapy to those in need of cancer care. As exemplified by this book on particle therapy (PT), there are outstanding opportunities for transformational thinking and effective actions. Therefore, what we now encounter is more of a wheel of choices and outcomes than a group of crossroads, hence the new concept of the 'Radiation Rotary'. All options have worldwide impact and are therefore topics of 'Addressing Global Challenges for Radiation Therapy'.

There are profound and far-reaching issues to address, including:

1. How do society, academia and industry balance investment in radiation therapy technology for those who have no cancer care at all with those who might benefit from the most sophisticated cancer treatment technology?

2. Is radiation therapy primarily, or solely, a technology involving a radiation beam or is it a biological-modifying entity that can induce important changes and therapeutic targets?

3. How can society deal with fear of the adverse effects of environmental and medical radiation that, from models of radiation injury, seems to exceed the actual risk because public concern about potential radiation health effects are real and drive energy policy, medical imaging and cancer care?

4. How can the world community effectively deal with the threat of radiological terrorism, or worse, the renewed potential of state-sponsored nuclear conflict and how can we prevent or at least mitigate health consequences should an incident occur?

So, there are a number of concepts and efforts entering this Radiation Rotary and a number of paths that exit. Figure 19.1 illustrates four pairs of choices, each with its distinct colour. All of the exits are necessary and they each reflect an important challenge and/or opportunity for the future. The Radiation Rotary is embedded in societal and global issues of enormous import as future directions in these areas will have profound impact on climate change (energy policy), nuclear proliferation (catastrophic conflagrations), healthcare shortages (hopelessness and suffering) and the potential for global collaboration versus conflict. Radiation science projects, unlike those in almost any other scientific discipline, absolutely require significant science and global partnerships such as those at the National Laboratories, CERN and space exploration agencies. So, radiation science and medicine are templates on which new knowledge and global collaboration for a common good can be built.

Society and the world are facing choices of enormous consequence, and many of the outcomes will be driven by the attitudes brought to problem-solving. An immediate challenge for civil society is how to make progress in an environment characterised by (1) anti-science and opinion-based declarations rather than knowledge-based decision-making; (2) confrontation of cultures rather than discussion of opposing ideas and understanding and

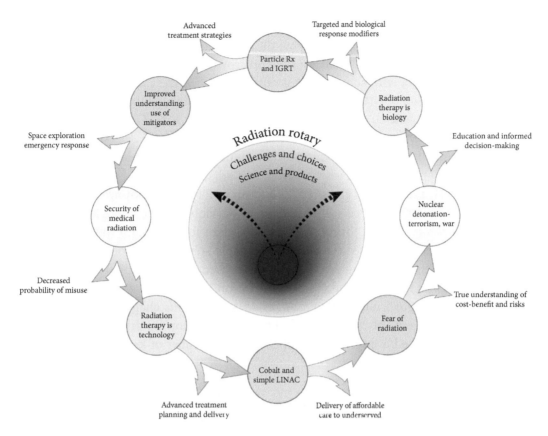

Advanced
treatment strategies

Targeted and biological
response modifiers

Particle Rx
and IGRT

Improved
understanding;
use of
mitigators

Radiation
therapy is
biology

Radiation rotary

Challenges and choices

Science and products

Space exploration
emergency response

Education and informed
decision-making

Security of
medical
radiation

Nuclear
detonation-
terrorism, war

Decreased
probability of misuse

True understanding of
cost-benefit and risks

Radiation
therapy is
technology

Fear of
radiation

Cobalt and
simple LINAC

Advanced treatment
planning and delivery

Delivery of affordable
care to underserved

FIGURE 19.1   The radiation rotary. As described in the text, there are a number of crossroads facing the field of radiation sciences best addressed as part of a rotary. Four sets of issues are illustrated with both sides of a particular issue in circles of the same colour. Talent and ideas entering the rotary face challenges and choices that will lead to science and products that, in turn, will impact all of the potential paths. None are right or wrong and all of them need to be addressed. Figure conceptualised by the faculty of the NCI, Radiation Research Program. Graphic by Alan Hoofring, NIH Medical Arts (submitted for publication as part of editorial manuscript at press).

efforts toward consensus for the common good; and (3) regression to self-interest and conflict rather than grand collaboration on global challenges. Because radiation is an integral part of our environment, the issues on the Radiation Rotary give our field a central role in solving challenges facing humanity and the planet.

Perhaps surprisingly, PT is a good vantage point from which to consider global challenges for radiation therapy and the 'Radiation Rotary'. Particles for PT are, indeed, part of the essence of our universe. The clinical application of PT is at the top of the rotary as it is complex and expensive, requiring big team science and offering unique opportunities described below. On the lower part of the rotary is cobalt and the 'simple' linear accelerator (LINAC) representing healthcare and cancer care delivery to those who now may have almost no access to care. The subtitle of this chapter comes from the song 'America the Beautiful' with the final two lines signifying the importance of working together and dealing end-to-end with broad, inclusive issues: 'And crown thy good with brotherhood/From sea to shining sea!' So, with some poetic license, we will address the incredible opportunities from 'heavy ion' PT (Carbon) and the need for the availability of effective cancer treatment for all (Cobalt-60 replacement by high-capability LINACs that work reliably in challenging environments), that is, 'C to shining C'.

This chapter will focus on ongoing efforts to build global collaboration and partnerships so that we can move ahead when exiting the Radiation Rotary. The four sections that follow address: (1) the need for global collaboration in PT; (2) how knowledge, experience and cooperation among fields involved in PT can develop new technologies for bringing cancer care to challenging environments, that is, cobalt to carbon; (3) the importance of supporting individual research to achieve sustainable solutions to global health, science and societal problems, including dealing with the consequences of radiation exposure and nuclear or radiation incidents; and (4) a consideration of how well radiation oncology fits within the world of precision medicine. A key message of this chapter is the idea of collaboration and problem-solving exemplified by the 'Radiation Rotary', so that these four issues will be addressed at high levels supported by a number of key references, not in a detailed topical review.

PARTICLE THERAPY

PT is at an inflection point. While the use of PT is no longer rare, it is not yet clear if it will or should become commonplace. Three primary factors are driving a move toward collaboration in PT at the global level:

1. Radiobiological complexity that is beyond the scope of research at even the largest medical systems

2. Decreasing cost of technology and the increased connectivity allowed by the Internet

3. Increasing numbers of world regions that recognise the potential benefit to their societies of implementing a sustainable infrastructure in radiation oncology including PT

Vastly increased complexity intrinsic to the radiobiology of PT has made it clear that collaboration in the field is absolutely required to move forward. This will allow practitioners to adapt the technology to clinical needs and adapt clinical expectations to the technology. In some ways echoes of prior radiation therapy development, such as the adoption of intensity-modulated radiation therapy (IMRT), can be seen in how progress is being made. However, in other areas, new avenues combining technology, societal good, new radiobiology and new computational capacities have made the adaption and proliferation of PT very different. The international collaboration needed to move forward due to this radiobiological complexity is supported by leaders of the radiation oncology community on multiple continents and is an area with great promise for making long needed breakthroughs. Even with current treatment planning systems that are limited to calculating only physical dose,

PT often has clear dosimetric advantages that promise to minimise normal tissue toxicity, thereby allowing safe dose escalation. PT is thus a technology whose main promise is to improve the therapeutic ratio of external beam radiation therapy. The study of PT radiobiology represents a broad space in which the next generation of radiation biologists can grow and expand the field. PT resides at the intersection of radiobiology, survivorship science, drug design and immunotherapy – making the Radiation Rotary a force for growth.

To date, collaboration in the clinical PT space has been relatively slow, with few data from prospective, randomised trials in published form despite decades of use. This is changing, but some of the impetus for clinical trial collaboration, at least in the United States, has been to generate data that justifies reimbursement from insurance carriers because centres are under enormous financial pressure due to large incurred debts from construction. In countries with nationalised health systems, science has more clearly led the way forward. Historically, the high cost of PT technology has greatly limited its implementation, has made national collaboration difficult and has required outcomes research in these environments to advance the use of PT technology.

The second aspect of the PT space that is driving collaboration and worldwide scientific expansion is the reduction of the cost of proton therapy by an order of magnitude or more in the last 10 years. This has spurred a relative explosion in the deployment of PT technology. Multiple vendors now compete in the market to provide systems with intensity-modulated proton therapy (IMPT) and cone beam computed tomography (CT) capability. Because the current generation of light ion (proton) PT systems has full commercial support, there are multiple treatment planning system options from which to choose. Gone are the days of the basic science physics lab approach with its customised hardware and software. Purchasing a proton therapy system today is similar to the process of purchasing a commercial LINAC system. The current class of PT machines is every bit as developed in appearance now as their photon machine brethren, from software branding to fancy, corporate-labelled plastic skins. Vendors truly have made remarkable progress in commercialising proton therapy and in providing high levels of support internationally.

Under the refined superficial surfaces of the PT systems there are innumerable radiobiological issues that have yet to be fully explored, let alone understood, thereby creating a paradox. PT is simultaneously our most advanced and our least well-understood radiation treatment technology. In concert with the availability of the breathtaking new PT systems, renewed interest and emphasis are being expressed at international meetings by the current generation of research leaders in the science related to normal tissue toxicity, the biologic differences between particles and photons and the unique spatial uncertainties intrinsic to PT. The radiation oncology community now understands that PT is not fully mature in terms of physical dose, biological effects and real benefit as well as how PT should be combined with other agents. Simultaneously the field has become more willing to discuss the problems in PT. A recent published review in hadron therapy contained a plot showing linear energy transfer (LET) and ranges of relative biological effectiveness (RBE) for each LET value that varied up to threefold [1].

Despite the need for a deeper understanding of the radiobiological effects of PT, its current affordability and subsequent availability have moved it more into the mainstream of radiation therapy such that PT is now incorporated into collaborative clinical research programs. The Swedish healthcare system has set up a system of feeder centres and one central PT centre to optimise the system costs and patient benefits. This new affordability has allowed PT systems to be rationalised within the nationalised medical systems around the world that are relatively less dependent on short-term revenue generation and seek to optimise global expenditures and outcomes. Many of these national health systems plan to use modelling of normal tissue complications to determine which patients are referred to the central PT centre. The British initially envisioned a similar system with two large PT centres, one in Manchester and the other in London, to optimise the cost/benefit ratio to society. However, due to the decreased cost of PT centres, several small centres have been constructed in addition to the two large English centres [2]. These centralised systems are important because they plan to institute national, prospective trials to fully test PT [3,4]. PT has not yet been tested robustly against the best treatment with brachytherapy, IMRT and surgery. In the United States, the marketing and promised benefits of PT have been presented to patients so effectively and without strong, supportive prospective data that doing true randomised protocols will now be very challenging. Despite this challenge, NRG Oncology is conducting randomised trials for low- and intermediate-risk prostate cancer [5] and inoperable stage II-IIB lung cancer [6], and a novel 'pragmatic' randomised trial (RADCOMP) is comparing photons with proton PT, not in classical survival terms but in terms of adverse effects and quality of life [7].

While the issues of hidden biologic complexity were known to some investigators early in the development of PT, the limited collaboration among investigators allowed those data to remain cloistered from others in the field and from the public at large. With increasing availability of PT, there is a new drive to better evaluate the role of PT despite the historical suggestions of its clear superiority. Initially when PT was so costly, studies that might have failed to demonstrate benefit were not considered options, often due to understandable physician preference based on the computer image of the treatment plan that had dose distributions which we now know are far more complex when subtleties of physics and biology are incorporated. The decreasing cost of PT has, ironically, encouraged some scientists in the field to ask when it is appropriate to treat patients with PT. The combination of increasing collaboration and decreasing PT machine costs is allowing the user community to expand. The PT community is inherently motivated to do good and the use of PT in this era of affordability is being evaluated more critically. Meanwhile, technological developments continue and PT will have to prove itself against new technologies. Ultimately, the world will benefit.

The discussion to this point applies mostly to light ion or proton PT. Collaboration in heavy ion PT, the current frontier of PT with its increasing complexity at every level compared to proton therapy, has been robust in the physics and particle engineering space but costs have not fallen to the same degree as for light ion PT. Much of the technological success in PT can be attributed to the leadership teams at CERN (Geneva) who pioneered the design and implementation of open, collaborative centres for heavy ion PT. The first CERN-designed heavy ion PT centre was constructed in Pavia, Italy, and the second in Vienna, Austria. More will likely follow. Each centre represents the joint efforts of scores of scientists to place global PT technology into the hands of users in an open-source fashion, much like the operating system and set of applications we commonly call LINUX. Not unlike LINUX, a huge number of person-years went into designing, implementing and testing these clinical systems, including their control software, both to make them available and safe [8].

The third aspect of the current environment driving global collaboration is the increasing awareness of the extreme inequity in the distribution of radiation therapy technology across the planet. Data from recent studies show that increasing the access to radiation therapy saves money and that to truly create a long-term solution in low- and middle-income countries (LMICs), the most just solution is placement of high technology as evenly as possible [9]. To accomplish this end, there is a great need for trained experts, the commodity in shortest supply. Deployment of PT in a rational fashion might allow expansion of the scientific frontier into all areas. In the PT space, having regional PT centres would promote the retention of talent needed to teach and to mentor. Additionally, placing PT centres to optimally provide state-of-the-art care for specific clinical indications, such as paediatric central nervous system tumours, would save lives and prevent needless suffering. The sharing and propagation of healthcare technology is a just and good thing and would make the world a better and safer place. PT centres could serve as hubs for scientific efforts in a region and as anchors to retain expertise.

Both light and heavy ion PTs share the issue of complexity. Unlike traditional radiation therapy, with more than a century of empirical clinical experience and its ability to address dosimetry mathematically, PT has biologic uncertainties that allow dose effects to vary tremendously based on subtleties in beam geometry and clinical context. This complexity of PT demands that collaborations expand. For the future of worldwide radiation oncology, the stakes could not be higher: the field is moving on from the relative simplicity of photon radiobiology. We can best serve the world by sharing science across borders openly and without regard to individual gain. PT science is simply too complex to solve by working alone and the benefits it promises of fewer adverse effects and new biological strategies require working together to accelerate acquiring the needed scientific information.

NEW LINACS AND CANCER CARE FOR CHALLENGING ENVIRONMENTS: LINKING COBALT TO CARBON

New Linear Accelerators and Cancer Care for Challenging Environments

The annual global incidence of cancer is estimated to rise from 15 million cases in 2015 to ~25 million cases in 2035 with over two-thirds occurring in LMICs [10] where there is a severe shortfall in radiation treatment capacity. A comprehensive report on global access to radiotherapy, the Lancet Oncology Commissioned Global Task Force on Radiotherapy for Cancer Control (GTFRCC) of the Union for International Cancer Control (UICC) [9], documented the global demand for radiotherapy, the inadequacy of current equipment coverage, the resources required and the economic and societal benefits that could be achieved by additional investment to provide that coverage. It was estimated that as many as 3,500–5,000 megavoltage treatment machines will be needed by 2035 in LMICs to meet the radiotherapy demands. Yap and colleagues estimated that if this demand for radiotherapy is met in LMICs by 2035, an additional 1.3 million people would obtain local disease control and over 615,000 patients would derive a survival benefit [11]. In addition to highlighting the shortfall in radiation treatment capacity in LMICs, the GTFRCC report presented a health technology assessment (HTA) that predicted a sustained investment in radiotherapy over 20 years will lead to dramatic societal and economic benefits that far outweigh the cost of the investment [9,12].

Unique Partnerships to Improve Radiation Oncology in LMICs

A goal of the Office of Radiological Security (ORS) of the U.S. National Nuclear Security Administration (NNSA) [13,14] is to reduce global reliance on medical radioactive sources, as well as to protect existing sources such as cobalt-60 from unauthorised access. A report from the Centre for Nonproliferation Studies (CNS) [15] emphasises the benefit of replacing cobalt-60 radiation treatment units with linear accelerators (LINACs) to decrease the risk of malicious use by non-state (terrorist) actors, as well as the benefits of the sophisticated treatment capability of LINACs compared to cobalt-60 radiotherapy machines. This new paradigm is embodied in the concept of 'treatment, not terror' [16,17].

Challenges in Implementing High Technology in LMICs

It has long been recognised that the often harsh environmental conditions encountered in many LMICs and the shortage of adequately trained technical personnel affect the introduction of high-technology equipment as well as the continued operation of such equipment in these circumstances. This has been experienced time and again in low-resourced countries where barriers to the implementation of accessible radiotherapy include the initial cost of equipment, the sustainability of equipment and infrastructure and the shortage of trained personnel needed to deliver safe, effective and high-quality treatment.

In recognition of these impediments to expending access to radiotherapy for cancer patients in LMICs, the 'Design Characteristics of a Novel Linear Accelerator for Challenging Environments' workshop was convened 7–8 November 2016 by the International Cancer Expert Corps (ICEC) [18] and hosted by CERN [19]. Participants addressed (1) the potential difficulties associated with treating cancer patients with radiotherapy in the challenging environments encountered in many LMICs (unstable electrical supply, lack of air-conditioning, unclean water sources etc.); (2) the security concerns related to medical radiological materials noted above; (3) the design characteristics of LINACs and related technologies that are robust enough to function in challenging environments; (4) the education, training and mentoring required to train and sustain the workforce needed to utilise novel radiation treatment systems; and (5) the costs and financing to implement the recommendations from the workshop.

Among the general design considerations of a LINAC for challenging environments proposed were well-recognised factors such as ease of operation, reliability, robustness, easy reparability, self-diagnosing of malfunctions, insensitivity to power interruptions, low power requirement, reduced heat production and a basic unit capable of modular upgrading. Achieving most of these design considerations in a relatively short period of time would require a systems solution based on current hardware technology and evolving software that exploits automation to the fullest, including auto-planning and operator monitoring and training. This could be a treatment system that requires limited on-site human involvement with remote telemedicine-based expertise for the complex aspects of treatment and quality assurance (QA). In this manner, high-quality treatment could be delivered by an on-site team with less technical expertise. The goal is to develop a treatment machine that delivers state-of-the-art radiotherapy rather than provides substandard treatment with a stripped-down LINAC. This approach would not only provide higher-quality treatment but also be an incentive for recruitment and retention of high-quality faculty and staff. Improved hardware such as a power generator in conjunction with energy management for control of electrical network fluctuations should be provided.

The theme of 'cobalt to carbon' is that the implementation of a systems solution to expand radiation therapy in LMICs can start now, building up the required expertise, mentorship and education networks, as well as increasing innovation in LINAC design.

Future Directions

Three ICEC task forces established as a result of the November 2016 workshop hosted by CERN are continuing to work towards their goals and objectives. The ICEC Technology Task Force (TF#1) is evaluating the feasibility of mounting CT and/or MR scanners as well as radiation therapy machines and treatment planning systems in shipping containers (the 'BoxCare' approach[2]) for ease in shipping, installation and commissioning. TF#1 is also identifying and analysing the shortfalls – weaknesses – in current LINACs with the goal of developing more durable, robust and reliable sub-system components for challenging environments. New software is envisioned that will not only simplify the work of the radiation therapists but also provide opportunities for teaching therapists while performing up-to-the-minute QA. To follow up on the development of new or better performing LINACs for challenging environments, a workshop sponsored by the UK Science and Technology Finance Council (STFC) and ICEC to be hosted by CERN is being planned for October 2017. Invitees will include academics, industry, government laboratories, professional societies and NGOs.

Vendors have recently introduced more user-friendly radiation treatment systems that may be attractive to users in high-income countries as well as to those in less developed countries. Time will tell whether the cost of the new machines will encourage or discourage the purchase of the new machines by LMICs.

The ICEC Education, Training and Mentoring Task Force (TF#2) is evaluating the current capabilities and deficiencies in training and education for radiation oncologists and medical physicists as well as for other essential radiotherapy and nursing staff. Examples of the use of telemedicine, including interactive systems such as National Institutes of Health's (NIH) TELESYNERGY®, for education, training and mentoring were presented at the International Conference on Advances in Radiation Oncology (ICARO2) in June 2017. Among the systems currently in use are 'Chart Rounds' [20] and the ASTRO contouring 'EduCase' system [21].

---

[2] Concept presented by David Jaffray, University of Toronto, at the ICEC–CERN meeting November 2016.

At ICARO2 over 20 international organisations and professional societies discussed their global health activities that demonstrated extensive interest, energy and commitment. Each of these excellent activities is limited in scope considering the extent of the global need for radiation oncology services. It appears that these and innumerable other small programs around the world would benefit from greater connectivity that could amplify their efforts by sharing resources, minimizing duplication of effort and reducing expenses. Improving connectivity is the goal of the Global Connectivity and Development Task Force (TF#3).

GLOBAL HEALTH AND RADIATION 'INCIDENTS' AND THE NEED FOR *SUSTAINABLE EFFORTS* TO ADDRESS 'HARD PROBLEMS'

The opportunities at both ends of the radiation therapy technology spectrum – PT and advanced image-guided technology on the one end and safe, sophisticated, sturdy technology for providing cancer care globally in challenging environments on the other – illustrate the need for a broad range of talent. As much as there is opportunity for breadth, there is a need for sustainable careers in that programs which involve patient care, including the prevention and mitigation of adverse impacts of radiation to the general population, astronauts and cancer patients, are poorly served by interruptions or incomplete projects. Lack of sustainability wastes investment and leaves potential beneficiaries as bad off as when the projects started.

The global shortage of available healthcare focused primarily on infectious diseases throughout the twentieth century. Naturally occurring epidemics and pandemics are readily spread through both global trade and travel and also by climate change that alters the life cycle and location of the vectors, such as birds and mosquitoes. The threat from global terrorism escalated markedly in the first year of the twenty-first century. The United States established 15 National Planning Scenarios [22], including two with radiation, an improvised nuclear device and a radiological dispersal device. This planning demonstrated the need to prepare for a nuclear threat, something that had been dormant since the end of the Cold War. The Fukushima nuclear power plant disaster [23] further raised the concern about the adverse impact of radiation on humans and the environment, thereby thrusting radiation science and the mitigation of radiation injury to the top of the list of topics of interest. The changing face of illness in the developing world has begun to look like the developed world with the striking rise of non-communicable diseases (NCDs), including respiratory, metabolic, cardiovascular and oncologic diseases. The section on new linacs and cancer care for challenging environments outlined the shortages in global radiation therapy and the potential systems solution. Interestingly, the pattern of diseases in LMICs [10] is similar to that in geographically isolated indigenous populations in high-income countries, providing a clear avenue for mutual problem-solving among rich and poor nations [24]. The United Nations agencies, particularly the International Atomic Energy Agency including its Program for Action for Cancer Treatment [25], have helped define the gaps in global treatment of NCDs. The Union for International Cancer Control's GTFRCC [9] contributions are described earlier. That this is an area of great need and rapidly increasing interest is seen in recent issues of *Clinical Oncology* and of *Seminars in Radiation Oncology* that summarise the potential role of radiation therapy in global health [26,27]. An update of the status of radiation oncology programs in the developing world was the topic of ICARO2 in June 2017 [28].

As the result of our assessing what is needed to fill the gap in global healthcare, the consistent weak link is the availability of expertise, particularly in the underserved sites, but also of mentors to implement change and sustain care. This shortage of expertise is not limited to cancer care but the benefit of focusing on cancer is that providing the range of services and programs needed for the care of cancer patients – from epidemiology to prevention to treatment to supportive care to long-term health – also addresses the other NCDs and many infectious diseases associated with cancer. This includes the role of HPV in cervical and head and neck cancers, of hepatitis in liver cancer and EBV in the lymphomas and nasopharynx cancer. As demonstrated for the infectious diseases, non-governmental organisations can be central to problem-solving by complementing governmental efforts and also by being more flexible. To address the role of radiation therapy in protocol/guideline-based cancer care as a key component of the continuum of cancer control, a diverse group of international collaborators and altruists have created the International Cancer Expert Corps [18], which has as its two major aims: (1) a sustainable mentorship model and (2) development of novel LINACs for challenging environments, discussed earlier.

A key reason for the gap in care for underserved populations, certainly so in the United States, is that service to health disparity populations, particularly those outside a government healthcare system, does not bring in sufficient revenue to be considered part of most healthcare jobs. With the exception of a few visionary institutions, there is almost a complete lack of support for trainees and faculty or staff working on global health projects as part of their career path. There are a few programs that aim to address this issue, such as the NIH Fogarty International Center [29] and National Cancer Institute (NCI) Center for Global Health (CGH) [30]. However, while the support they provide is welcome and sets an example, it is a tiny fraction of the overall need. An encouraging surge of interest in global health is demonstrated by the rapid growth of the Consortium of Universities for Global Health [31]. Means by which investment in issues of global health and nuclear threats can be viewed as a positive rather than a waste of funds or people's time have been proposed [32]. These broaden the issue of what should be supported in modern academia to adjust the balance between the generation of revenue on the one hand and science as well as service to society on the other [33].

RADIATION AS PART OF PRECISION MEDICINE

Two issues on the Radiation Rotary are (1) fear of radiation and mitigation of radiation injury and (2) radiation as technology or biology. The first pairing follows from the discussion earlier with the need to understand the mechanisms of radiation injury that could better define the risk for radiation-induced cancers in accidentally or intentionally exposed populations and the stepwise lesions in normal tissue injury following radiation therapy. These issues are addressed by the Radiation and Nuclear Countermeasures Program in the National Institutes of Allergy and Infectious Diseases [34] and in the NCI Radiation Research Program [35].

The second issue, radiation oncology, as a key player in the era of precision medicine, is a new frontier that will require understanding the biological perturbations induced by radiation that have a function in cell killing, immune modulation and cancer cell adaptation that may provide new means to explore susceptibility to cell death with radiation as a key component. This has been dubbed 'focused biology' [36]. In this approach, the radiation effect differs by (1) the type of radiation, (2) dose, (3) dose-rate, (4) fractionation and (5) targeting. Targets of interest range from a 'system', that is, the immune system, to specific tissues which respond to radiation differently, to tumour types based on anatomy, to molecular abnormalities and to radiation-adaptive responses. A few examples relevant to PT are its role in altering the immune system [37] and its impact on immunotherapy [38]. Radiation given by a large single dose can produce very different changes from those seen after multi-fraction radiation [39], including susceptibility to drug targeting [39] and immune modulation [40]. There are many examples of combining radiation therapy with molecular targeted drugs, two recent reviews being examples [41,42]. Addressing the new paradigm of possibly defining radiation dose in both terms of energy imparted, the gray, and also in terms of molecular changes or radiobiological effectiveness as used in proton therapy with the term 'cobalt gray equivalent' (CGE) [43,44] was the subject of an NCI Radiation Research Program Workshop in September 2017, addressing 'Shades of Gy' (manuscripts in preparation).

SUMMARY

Addressing global challenges in radiation therapy opens up remarkably diverse and broadly impactful opportunities, as illustrated in the Radiation Rotary (Figure 19.1). The ubiquitous nature of radiation stems from the fabric of our universe being immersed in radiation, including the ions relevant to cancer care (the 'so-called' heavy ions that are heavier than protons) and those critical to the role of humans in space exploration. The breadth of the underlying science of radiation interactions with living tissue requires not only big but also team science that is the 'brotherhood' in 'C to shining C'. The broad field of the radiation sciences, particularly those involved in health-related issues, presents opportunities, has societal responsibilities and requires a unique set of tools and talent to transform problems in society well beyond cancer care.

REFERENCES

1. Durante, M, Brenner, DJ, and Formenti, SC, Does heavy ion therapy work through the immune system? *Int J Radiat Oncol Biol Phys.* 2016;96(5): 934–936.
2. Crellin, AM, and Burnet, NG, Proton beam therapy: The context, future direction and challenges become clearer. *Clin Oncol (R Coll Radiol).* 2014;26(12): 736–738.
3. Widder, J et al., The quest for evidence for proton therapy: Model-based approach and precision medicine. *Int J Radiat Oncol Biol Phys.* 2016;95(1): 30–36.
4. Langendijk, JA et al., Selection of patients for radiotherapy with protons aiming at reduction of side effects: The model-based approach. *Radiother Oncol.* 2013;107(3): 267–273.
5. NRG prostate cancer trial. Available at: https://www.rtog.org/ClinicalTrials/ProtocolTable/StudyDetails.aspx?study=1326. Accessed 1 July 2017.
6. NRG lung cancer proton trial. Available at: https://www.rtog.org/ClinicalTrials/ProtocolTable/StudyDetails.aspx?study=1308. Accessed 1 July 2017.

7. Bekelman, J, Pragmatic randomized trial of proton versus photon therapy for patients with non-metastatic breast cancer receiving comprehensive nodal radiation: A radiotherapy comparative effectiveness (RADCOMP) trial. 2017. Available at: http://www.pcori.org/research-results/2015/pragmatic-randomized-trial-proton-vs-photon-therapy-patients-non-metastatic. Accessed 22 February 2018.

8. Evans, P, and Wolf, B, Collaboration rules. *Harv Bus Rev.* 2005;83(7): 96–104, 192.

9. Atun, R, Jaffray, DA, Barton, MB, Bray, F, Baumann, M, Vikram, B, Hanna, TP et al., Expanding global access to radiotherapy. *Lancet Oncol.* 2015;16(10): 1153–1186. doi: 10.1016/S1470-2045(15)00222-3.

10. Review. World Health Organization. Cancer fact sheet. Available at: http://www.who.int/mediacentre/factsheets/fs297/en/. Accessed 1 July 2017.

11. Yap, ML, Hanna, TP, Shafiq, J, Ferlay, J, Bray, F, Delaney, GP, and Barton, M, The benefits of providing external beam radiotherapy in low- and middle-income countries. *Clin Oncol (R Coll Radiol).* 2017;29(2): 72–83. doi:10.1016/j.clon.2016.11.003.

12. Rodin, D, Aggarwal, A, Lievens, Y, and Sullivan, R, Balancing equity and advancement: The role of health technology assessment in radiotherapy resource allocation. *Clin Oncol (R Coll Radiol).* 2017;29(2): 93–98. doi:10.1016/j.clon.2016.11.001.

13. Radiological Security, National Nuclear Security Agency. Available at: https://nnsa.energy.gov/aboutus/ourprograms/dnn/gms/rs. Accessed 23 December 2016.

14. Transitioning from high-activity radioactive sources to non-radioisotopic (alternative) technologies. Available at: https://www.hsdl.org/?abstract&did=797521. Accessed 2 February 2018.

15. James Martin Center for Nonproliferation Studies. Available at: http://www.nonproliferation.org/. Accessed 1 July 2017.

16. Pomper, M, Dalnoki-Veress, F, and Moore, G, Treatment, not terror: Strategies to enhance external beam therapy in developing countries while permanently reducing the risk of radiological terrorism. 2016. Available at: http://www.stanleyfoundation.org/publications/report/TreatmentNotTerror212.pdf. Accessed 23 December 2016.

17. Coleman CN, Pomper MA, Chao NL, Dalnoki-Veress F, and Pistenmaa DA, Treatment, not terror: Time for unique problem-solving partnerships for cancer care in resource-challenged environments. *J. Glob. Oncol.* doi:10.1200/JGO.2016.007591.

18. International Cancer Expert Corps (ICEC). Available at: http://www.iceccancer.org/about-icec/. Accessed 1 July 2017.

19. A new approach for global access to radiotherapy. *CERN Courier.* Available at: https://home.cern/scientists/updates/2016/11/new-approach-global-access-radiation-therapy. Accessed 1 July 2017.

20. Chart rounds. Available at: https://chartrounds.com/default.aspx. Accessed 1 July 2017.

21. EduCase. Available at: http://www.astro.educase.com/. Accessed 1 July 2017.

22. National planning scenarios. Available at: https://emilms.fema.gov/IS800B/lesson5/NRF0105060t.htm. Accessed 11 June 2017.

23. The Fukushima Daiichi Accident. Report by the Director General and Technical Volumes, International Atomic Energy Agency. Available at: https://www.iaea.org/newscenter/news/iaea-releases-director-generals-report-on-fukushima-daiichi-accident. Accessed 11 June 2017.

24. Guadagnolo, BA, Petereit, DG, and Coleman, CN, Cancer care access and outcomes for American Indian populations in the United States: Challenges and models for progress. *Semin Radiat Oncol.* 2017;27(2): 143–149. doi:10.1016/j.semradonc.2016.11.006.

25. International Atomic Energy Agency, Program of Action for Cancer Treatment. Available at: https://www.iaea.org/services/key-programmes/programme-of-action-for-cancer-therapy-pact. Accessed 11 June 2017.

26. Yap, ML, Hanna, TP, Shafiq, J, Ferlay, J, Bray, F, Delaney, GP, Barton, M, Radiotherapy in low and middle income countries. *Clin Oncol.* 2017;29(2): 72–83.

27. Tim, RW, and Coleman, CN, Global health disparities. *Semin Radiat Oncol.* 2017;27(2): 95–188.

28. International Conference on Advances in Radiation Oncology (ICARO2). Available at: http://www-pub.iaea.org/iaeameetings/50815/International-Conference-on-Advances-in-Radiation-Oncology-ICARO2. Accessed 11 June 2017.

29. Fogarty International Center, National Institutes of Health. Available at: https://www.fic.nih.gov/Pages/Default.aspx. Accessed 11 June 2017.

30. NCI Center for Global Health, National Cancer Institute. Available at: https://www.cancer.gov/about-nci/organization/cgh. Accessed 11 June 2017.

31. Consortium of Universities for Global Health. Available at: http://www.cugh.org/. Accessed 11 June 2017.

32. Coleman CN. Masters of our destiny: From jazz quartet to symphony orchestra. *Int J Radiat Oncol Biol Phys.* 2016;96(3): 511–513. doi:10.1016/j.ijrobp.2016.07.006.

33. Coleman, CN, and Love, RR, Transforming science, service, and society. *Sci Transl Med.* 2014;6(259): 259fs42. doi:10.1126/scitranslmed.3009640.

34. Radiation and Nuclear Countermeasures Program, National Institutes of Allergy and Infectious Diseases: Available at: https://www.niaid.nih.gov/research/radiation-nuclear-countermeasures-program. Accessed 11 June 2017.

35. Radiation Research Program, National Cancer Institute. Available at: https://rrp.cancer.gov/. Accessed 11 June 2017.

36. Coleman, CN, Linking radiation oncology and imaging through molecular biology (or now that therapy and diagnosis have separated, it's time to get together again!). *Radiology* 2003;228(1): 29–35.

37. Wage, J, Ma, L, Peluso, M, Lamont, C, Evens, AM, Hahnfeldt, P, Hlatky, L, Beheshti, A, Proton irradiation impacts age-driven modulations of cancer progression influenced by immune system transcriptome modifications from splenic tissue. *J Radiat Res.* 2015;56(5): 792–803. doi:10.1093/jrr/rrv043.

38. Ebner, DK, Tinganelli, W, Helm, A, Bisio, A, Yamada, S, Kamada, T, Shimokawa, T, and Durante, M, The immunoregulatory potential of particle radiation in cancer therapy. *Front Immunol.* 2017;8: 99. doi:10.3389/fimmu.2017.00099.

39. Makinde, AY, Eke, I, Aryankalayil, MJ, Ahmed, MM, and Coleman, CN, Exploiting gene expression kinetics in conventional radiotherapy, hyperfractionation, and hypofractionation for targeted therapy. *Semin Radiat Oncol.* 2016;26(4): 254–260. doi:10.1016/j.semradonc.2016.07.001.

40. Vanpouille-Box, C, Alard, A, Aryankalayil, MJ, Sarfraz, Y, Diamond, JM, Schneider, RJ, Inghirami, G, Coleman, CN, Formenti, SC, Demaria, S. DNA exonuclease Trex1 regulates radiotherapy-induced tumour immunogenicity. *Nat Commun.* 2017;8: 15618. doi:10.1038/ncomms15618.

41. Wahl, DR, and Lawrence, TS, Integrating chemoradiation and molecularly targeted therapy. *Adv Drug Deliv Rev.* 2017;109: 74–83. doi:10.1016/j.addr.2015.11.007.

42. Deng, L, Liang, H, Fu, S, Weichselbaum, RR, and Fu, YX, From DNA damage to nucleic acid sensing: A strategy to enhance radiation therapy. *Clin Cancer Res.* 2016;22(1): 20–25. doi:10.1158/1078-0432.CCR-14-3110.

43. Paganetti, H, Niemierko, A, Ancukiewicz, M, Gerweck, LE, Goitein, M, Loeffler, JS, and Suit, HD, Relative biological effectiveness (RBE) values for proton beam therapy. *Int J Radiat Oncol Biol Phys.* 2002;53(2): 407–421.

44. Paganetti, H. Relative biological effectiveness (RBE) values for proton beam therapy. Variations as a function of biological endpoint, dose, and linear energy transfer. *Phys Med Biol.* 2014;59(22): R419–R472. doi: 10.1088/0031-9155/59/22/R419.

Index

Note: Page numbers followed by f and t refer to figures and tables respectively.

T - #0182 - 111024 - C298 - 254/178/14 - PB - 9780367571559 - Gloss Lamination